Mesothelioma Heterogeneity

Mesothelioma Heterogeneity

Potential Mechanisms

Special Issue Editor

Emanuela Felley-Bosco

MDPI • Basel • Beijing • Wuhan • Barcelona • Belgrade

MDPI

Special Issue Editor
Emanuela Felley-Bosco
Zurich University Hospital
Switzerland

Editorial Office
MDPI
St. Alban-Anlage 66
4052 Basel, Switzerland

This is a reprint of articles from the Special Issue published online in the open access journal *International Journal of Molecular Sciences* (ISSN 1422-0067) from 2017 to 2018 (available at: https://www.mdpi.com/journal/ijms/special_issues/mesothelioma)

For citation purposes, cite each article independently as indicated on the article page online and as indicated below:

LastName, A.A.; LastName, B.B.; LastName, C.C. Article Title. *Journal Name* **Year**, *Article Number*, Page Range.

ISBN 978-3-03897-473-4 (Pbk)
ISBN 978-3-03897-474-1 (PDF)

Contents

About the Special Issue Editor

Emanuela Felley-Bosco, PhD PD, received a PhD degree in Pharmacology and Toxicology from the University of Lausanne in Switzerland (1986). She was then post-doc for one year at Occupational Health Institute in Lausanne; then for two and a half years at the Swiss Institute for Experimental Cancer Research, Switzerland; and, finally, for three years at National Cancer Institute, Bethesda, USA. Emanuela Felley-Bosco was a group leader (1994–2006) at the Department of Pharmacology and Toxicology at the University of Lausanne thanks partly to a women-academic promotion award. She has been a lecturer at Lausanne University since 1998. Since 2007, she has been group leader in the Laboratory of Molecular Oncology at Zürich University Hospital, Zurich, Switzerland. Her major interest is inflammation/injury-related cancer with a more recent focus on mesothelioma. Her group is involved in translational research ongoing in parallel with clinical trials for the treatment of patients with mesothelioma and in pre-clinical studies aimed at a better understanding of mesothelioma biology. In this context, targeted therapies are also tested.

International Journal of
Molecular Sciences

MDPI

Editorial

Special Issue on Mechanisms of Mesothelioma Heterogeneity: Highlights and Open Questions

Emanuela Felley-Bosco

Laboratory of Molecular Oncology, University Hospital Zurich, Sternwartstrasse 14, 8091 Zürich, Switzerland; emanuela.felley-bosco@usz.ch

Received: 24 October 2018; Accepted: 11 November 2018; Published: 12 November 2018

Abstract: This editorial aims to synthesize the eleven papers that have contributed to this special issue, where the mechanisms of mesothelioma heterogeneity have been tackled from different angles.

Keywords: mesothelioma heterogeneity; NF2/Hippo pathway; BAP1; non-coding RNA; tumor microenvironment; experimental models

---

A general feature of a tumor is that it comprises tumor cells and stroma containing immune cells, fibroblasts, matrix and blood vessels. Therefore, it is not surprising that in this special issue, the mechanisms of mesothelioma heterogeneity have been addressed extensively at the level of tumoral cells, highlighting differences in genetic alterations [1–5] or temporal differences during tumor progression [2].

In this context, it is worth noting that besides the two pathways widely mutated in cancer, namely, cell cycle control (cyclin-dependent kinase Inhibitor 2A, *CDKN2A*) and genome integrity (*TP53*), there are also two specific pathways frequently mutated in MPM, namely, the neurofibromatosis type 2 (*NF2*)/Hippo and the Breast-Repair-associated-Cancer 1(BRCA)-associated protein 1 (*BAP1*) pathways.

With regard to NF2/Hippo, as pointed out by Sato and Sekido [5], it is intriguing that if their downstream targets are activated yes-associated protein 1 (YAP1) and transcriptional co-activator with PDZ domain-binding motif (TAZ), no mutations that result in their activation have been observed in mesothelioma. Mutations that result in their constitutive activation would involve mutations of individual or multiple phosphorylation sites, allowing YAP and TAZ retention in the cytosol preventing activation of YAP/TAZ-dependent transcription. However, there are well-known examples, like Phosphatase and tensin homolog (PTEN), where loss of control of phosphorylation targets are tumorigenic. In addition, both YAP and TAZ have multiple phosphorylation sites so it is likely that deregulation of the upstream kinase would be more efficient. As reviewed by Sato and Sekido [5], YAP has been largely investigated in mesothelioma, however, Hagenbeeck et al. [6] recently noted that YAP and TAZ have slightly different transcriptional profiles, whereby TAZ increases, for example, the expression of wound-healing-associated, pro-tumorigenic genes such as *Arginase 1*. This gene was one of the genes with the highest expression in tissues from asbestos exposed mice and remained high in tumors [7]. Therefore, there remains an open question about a possibly synergistic mode of action where TAZ modifies the tumor microenvironment while YAP promotes tumor cell proliferation.

While the understanding of the mechanisms behind the contribution of the NF2/Hippo pathway to mesothelioma has progressed greatly since the seminal observation of the high frequency of *NF2* mutations in mesothelioma [8,9], understanding of the mechanisms underlying BAP1 are less advanced. This is to be expected as this mutational event was discovered more recently [10,11]. Interestingly, in the analysis of TCGA samples, *BAP1* status was associated with differential gene expression [12] as originally described in Drosophila (fruit fly). Here the BAP1 homolog was responsible for repression of *HOX* genes in the fly embryo while also increasing *HOX* expression in particular tissues in central nervous system [13].

Because of the known role of long non-coding RNA (lncRNA) in assembling and controlling transcriptional complexes (reviewed in [14]), it would be of interest to explore if lncRNA associated with BAP1 show differential transcriptional profiles that are associated with better clinical outcome [12,15]. In fact, their expression may, for example, point to a given cell of origin and commitment to epithelial differentiation phenotype. This was observed in patients' samples by Felley-Bosco and Rehrauer [16] for *FENDRR*, a lncRNA found to be overexpressed in tumors developing in mice after exposure to asbestos fibers, and which also clusters with better outcomes in human mesothelioma patients [12]. Similarly, *Meg3*, another lncRNA found to be overexpressed in tumors developing in mice after exposure to asbestos fibers [16] is overexpressed in TCGA cluster 1, which was characterized by better overall survival [12] compared to the other 3 clusters of patients with different transcription profiles.

Other non-coding RNA of interest that have been extensively reviewed [17] include microRNA (miR), which have been deeply investigated for diagnostic and prognostic purposes and reviewed by Martinez-Rivera et al. [17]. They highlight the challenges to come with the investigation of circulating miR in total plasma/serum vs exosomal vesicles. In this context, additional complexity has been recently added by the investigation of expression obtained through RNA-seq data. This has revealed how classical analysis approaches may miss isomiRs [18].

Even though peritoneal mesothelioma is less frequent compared to pleural mesothelioma, the mutational landscape is similar, with *BAP1* frequently being mutated [19]. The reported case of long-survivor peritoneal mesothelioma by Serio et al. [4] did not display any of the mutations in the frequently mutated genes *BAP1*, *CDKN2A*, or *NF2* and was treated with oxaliplatin, a known inducer of immunogenic cell death [20]. Therefore, if more tissue were available from mesothelioma patients treated with oxaliplatin, it would be interesting to establish a cohort where potential neoantigens generation and immune response could be explored.

Heterogeneity in the tumor environment has been widely reviewed [1,21] with more emphasis on heterogeneity in immune cell content in the tumor microenvironment, which is also in line with the intensive exploration of immunotherapy in mesothelioma treatment [22]. Minnema-Luiting and colleagues [21] emphasize how several studies point to the important role of M2-polarized macrophages in mesothelioma. Interestingly, according to the interactive web-based platform https://www.cri-iatlas.org/ [23], which was established as an analytic tool for studying the interactions between tumors analyzed in TCGA and the immune microenvironment, the best relationship with leukocytes tumoral infiltration is observed for the signature known as the "macrophage regulation" (Figure 1a) This is better when compared to the relationship with the signature called the "IFN-gamma response" (Figure 1b). Altogether, these observations point to the macrophage population as a major regulator of the immune system in mesothelioma.

Besides immune cells, the mesothelioma tumor environment also contains cancer-associated fibroblasts and a matrix, likely produced by the tumor, immune cells and fibroblasts themselves. However, both cancer-associated fibroblasts and the matrix, which are likely to be major contributors of stiffness-dependent effects such as modulation of YAP/TAZ transcriptional regulators [24], remain to be explored.

As highlighted by Tolani et al. [1], a stem cell signaling pathway that should be further explored is the Notch signaling pathway, especially since it is expressed in patients with predominantly non-epithelioid histologies with poorer outcomes [12] compared to patients in cluster 1, who are characterized by better overall survival.

MESO

MESO

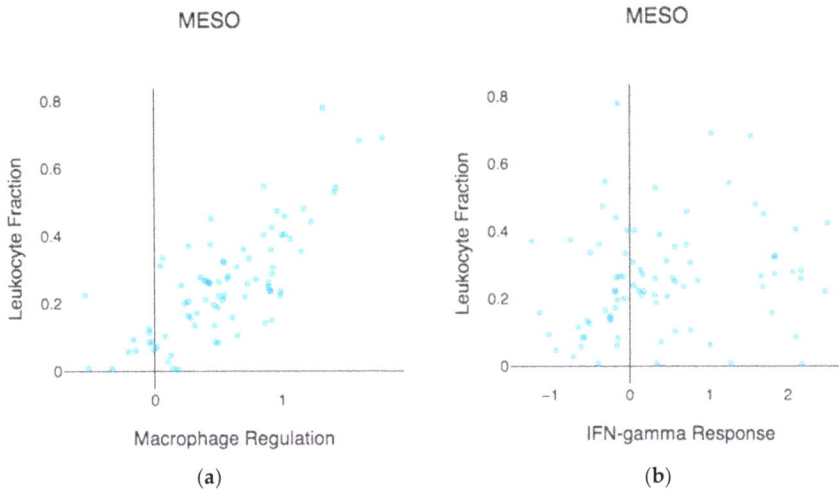

Figure 1. Mesothelioma leukocyte fraction is highly correlated with the signature "Macrophage regulation" (**a**) compared with the correlation with IFN-gamma response (**b**). These graphics were obtained using the interactive web-based platform https://www.cri-iatlas.org/ [23].

Jean and Jaurand wrote a timely, comprehensive review on how experimental murine mesothelioma models [25] have helped in understanding the mechanism of mesothelioma development using tissue specific targeted gene disruption using injections of AdenoCre or exposure to asbestos fibers. Genetic alteration signatures observed in mice exposed to asbestos resemble what is observed in human clinical samples and is mostly associated with copy number variations. This is in line with the lack of detection of a specific point mutation signature (https://cancer.sanger.ac.uk/cosmic/signatures), besides aging, in the two-human high-through-put studies [12,26]. These models are useful for the investigation of other relevant changes, such as epigenetic modifications.

Finally, yet importantly, Colin et al. [27] developed a human orthotopic (intrapleural) xenograft model in athymic mice, where it is possible to investigate the role of macrophage migration inhibiting factor (MIF) because this particular model expresses both MIF and its functional receptor CD74. The authors show the presence of M2-polarized macrophages in this model. Therefore, the model allows not only investigating the role of MIF but also testing drugs acting on macrophage polarization, thus allowing testing of the effect of macrophage polarization on tumor growth.

Funding: E. Felley-Bosco research is supported by the Stiftung für Angewandte Krebsforschung, the Krebsliga Zürich and the Swiss National Science Foundation 320030_182690.

Conflicts of Interest: The author declares no conflict of interest.

References

1. Tolani, B.; Acevedo, L.A.; Hoang, N.T.; He, B. Heterogeneous contributing factors in mpm disease development and progression: Biological advances and clinical implications. *Int. J. Mol. Sci.* **2018**, *19*. [CrossRef]
2. Oehl, K.; Vrugt, B.; Opitz, I.; Meerang, M. Heterogeneity in malignant pleural mesothelioma. *Int. J. Mol. Sci.* **2018**, *19*, 1603. [CrossRef] [PubMed]
3. Sarun, K.H.; Lee, K.; Williams, M.; Wright, C.M.; Clarke, C.J.; Cheng, N.C.; Takahashi, K.; Cheng, Y.Y. Genomic deletion of BAP1 and CDKN2A are useful markers for quality control of malignant pleural mesothelioma (MPM) primary cultures. *Int. J. Mol. Sci.* **2018**, *19*. [CrossRef] [PubMed]

4. Serio, G.; Pezzuto, F.; Marzullo, A.; Scattone, A.; Cavone, D.; Punzi, A.; Fortarezza, F.; Gentile, M.; Buonadonna, A.L.; Barbareschi, M.; et al. Peritoneal mesothelioma with residential asbestos exposure. Report of a case with long survival (seventeen years) analyzed by Cgh-array. *Int. J. Mol. Sci.* **2017**, *18*. [CrossRef] [PubMed]
5. Sato, T.; Sekido, Y. Nf2/merlin inactivation and potential therapeutic targets in mesothelioma. *Int. J. Mol. Sci.* **2018**, *19*. [CrossRef] [PubMed]
6. Hagenbeek, T.J.; Webster, J.D.; Kljavin, N.M.; Chang, M.T.; Pham, T.; Lee, H.J.; Klijn, C.; Cai, A.G.; Totpal, K.; Ravishankar, B.; et al. The hippo pathway effector TAZ induces TEAD-dependent liver inflammation and tumors. *Sci. Signal* **2018**, *11*. [CrossRef] [PubMed]
7. Rehrauer, H.; Wu, L.; Blum, W.; Pecze, L.; Henzi, T.; Serre-Beinier, V.; Aquino, C.; Vrugt, B.; de Perrot, M.; Schwaller, B. How asbestos drives the tissue towards tumors: Yap activation, macrophage and mesothelial precursor recruitment, RNA editing, and somatic mutations. *Oncogene* **2018**, *37*, 2645–2659. [CrossRef] [PubMed]
8. Sekido, Y.; Pass, H.I.; Bader, S.; Mew, D.J.; Christman, M.F.; Gazdar, A.F.; Minna, J.D. Neurofibromatosis type 2 (NF2) gene is somatically mutated in mesothelioma but not in lung cancer. *Cancer Res.* **1995**, *55*, 1227–1231. [PubMed]
9. Bianchi, A.B.; Hara, T.; Ramesh, V.; Gao, J.; Klein-Szanto, A.J.; Morin, F.; Menon, A.G.; Trofatter, J.A.; Gusella, J.F.; Seizinger, B.R.; et al. Mutations in transcript isoforms of the neurofibromatosis 2 gene in multiple human tumour types. *Nat. Genet.* **1994**, *6*, 185–192. [CrossRef] [PubMed]
10. Bott, M.; Brevet, M.; Taylor, B.S.; Shimizu, S.; Ito, T.; Wang, L.; Creaney, J.; Lake, R.A.; Zakowski, M.F.; Reva, B.; et al. The nuclear deubiquitinase BAP1 is commonly inactivated by somatic mutations and 3p21.1 losses in malignant pleural mesothelioma. *Nat. Genet.* **2011**, *43*, 668–672. [CrossRef] [PubMed]
11. Testa, J.R.; Cheung, M.; Pei, J.; Below, J.E.; Tan, Y.; Sementino, E.; Cox, N.J.; Dogan, A.U.; Pass, H.I.; Trusa, S.; et al. Germline bap1 mutations predispose to malignant mesothelioma. *Nat. Genet.* **2011**, *43*, 1022–1025. [CrossRef] [PubMed]
12. Hmeljak, J.; Sanchez-Vega, F.; Hoadley, K.A.; Shih, J.; Stewart, C.; Heiman, D.I.; Tarpey, P.; Danilova, L.; Drill, E.; Gibb, E.A.; et al. Integrative molecular characterization of malignant pleural mesothelioma. *Cancer Discov.* **2018**. [CrossRef] [PubMed]
13. Scheuermann, J.C.; de Ayala Alonso, A.G.; Oktaba, K.; Ly-Hartig, N.; McGinty, R.K.; Fraterman, S.; Wilm, M.; Muir, T.W.; Muller, J. Histone H2A deubiquitinase activity of the polycomb repressive complex PR-DUB. *Nature* **2010**, *465*, 243–247. [CrossRef] [PubMed]
14. Renganathan, A.; Felley-Bosco, E. Long noncoding RNAS in cancer and therapeutic potential. *Adv. Exp. Med. Biol.* **2017**, *1008*, 199–222. [PubMed]
15. Singh, A.S.; Heery, R.; Gray, S.G. In silico and in vitro analyses of lncRNAs as potential regulators in the transition from the epithelioid to sarcomatoid histotype of malignant pleural mesothelioma (MPM). *Int. J. Mol. Sci.* **2018**, *19*. [CrossRef] [PubMed]
16. Felley-Bosco, E.; Rehrauer, H. Non-coding transcript heterogeneity in mesothelioma: Insights from asbestos-exposed mice. *Int. J. Mol. Sci.* **2018**, *19*. [CrossRef] [PubMed]
17. Martinez-Rivera, V.; Negrete-Garcia, M.C.; Avila-Moreno, F.; Ortiz-Quintero, B. Secreted and tissue mirnas as diagnosis biomarkers of malignant pleural mesothelioma. *Int. J. Mol. Sci.* **2018**, *19*. [CrossRef] [PubMed]
18. Telonis, A.G.; Magee, R.; Loher, P.; Chervoneva, I.; Londin, E.; Rigoutsos, I. Knowledge about the presence or absence of miRNA isoforms (isomirs) can successfully discriminate amongst 32 tcga cancer types. *Nucleic Acids Res.* **2017**, *45*, 2973–2985. [CrossRef] [PubMed]
19. Leblay, N.; Lepretre, F.; Le Stang, N.; Gautier-Stein, A.; Villeneuve, L.; Isaac, S.; Maillet, D.; Galateau-Salle, F.; Villenet, C.; Sebda, S.; et al. Bap1 is altered by copy number loss, mutation, and/or loss of protein expression in more than 70% of malignant peritoneal mesotheliomas. *J. Thorac. Oncol.* **2017**, *12*, 724–733. [CrossRef] [PubMed]
20. Garg, A.D.; More, S.; Rufo, N.; Mece, O.; Sassano, M.L.; Agostinis, P.; Zitvogel, L.; Kroemer, G.; Galluzzi, L. Trial watch: Immunogenic cell death induction by anticancer chemotherapeutics. *Oncoimmunology* **2017**, *6*. [CrossRef] [PubMed]
21. Minnema-Luiting, J.; Vroman, H.; Aerts, J.; Cornelissen, R. Heterogeneity in immune cell content in malignant pleural mesothelioma. *Int. J. Mol. Sci.* **2018**, *19*. [CrossRef] [PubMed]

22. Alley, E.W.; Lopez, J.; Santoro, A.; Morosky, A.; Saraf, S.; Piperdi, B.; van Brummelen, E. Clinical safety and activity of pembrolizumab in patients with malignant pleural mesothelioma (keynote-028): Preliminary results from a non-randomised, open-label, phase 1b trial. *Lancet Oncol.* **2017**, *18*, 623–630. [CrossRef]
23. Thorsson, V.; Gibbs, D.L.; Brown, S.D.; Wolf, D.; Bortone, D.S.; Ou Yang, T.H.; Porta-Pardo, E.; Gao, G.F.; Plaisier, C.L.; Eddy, J.A.; et al. The immune landscape of cancer. *Immunity* **2018**, *48*, 812–830. [CrossRef] [PubMed]
24. Dupont, S.; Morsut, L.; Aragona, M.; Enzo, E.; Giulitti, S.; Cordenonsi, M.; Zanconato, F.; Le Digabel, J.; Forcato, M.; Bicciato, S.; et al. Role of yap/taz in mechanotransduction. *Nature* **2011**, *474*, 179–183. [CrossRef] [PubMed]
25. Jean, D.; Jaurand, M.C. Mesotheliomas in genetically engineered mice unravel mechanism of mesothelial carcinogenesis. *Int. J. Mol. Sci.* **2018**, *19*. [CrossRef] [PubMed]
26. Bueno, R.; Stawiski, E.W.; Goldstein, L.D.; Durinck, S.; De Rienzo, A.; Modrusan, Z.; Gnad, F.; Nguyen, T.T.; Jaiswal, B.S.; Chirieac, L.R.; et al. Comprehensive genomic analysis of malignant pleural mesothelioma identifies recurrent mutations, gene fusions and splicing alterations. *Nat. Genet.* **2016**, *48*, 407–416. [CrossRef] [PubMed]
27. Colin, D.J.; Cottet-Dumoulin, D.; Faivre, A.; Germain, S.; Triponez, F.; Serre-Beinier, V. Experimental model of human malignant mesothelioma in athymic mice. *Int. J. Mol. Sci.* **2018**, *19*. [CrossRef] [PubMed]

International Journal of
Molecular Sciences

MDPI

Review
Heterogeneity in Malignant Pleural Mesothelioma

Kathrin Oehl [1,2,*], Bart Vrugt [2], Isabelle Opitz [1] and Mayura Meerang [1,*]

[1] Department of Thoracic Surgery, University Hospital Zurich, 8091 Zürich, Switzerland;
Isabelle.Schmitt-Opitz@usz.ch
[2] Institute of Pathology and Molecular Pathology, University Hospital Zürich, 8091 Zürich, Switzerland;
Bart.Vrugt@usz.ch
* Correspondence: Kathrin.Oehl@usz.ch (K.O.); mayura.meerang@usz.ch (M.M.)

Received: 9 May 2018; Accepted: 26 May 2018; Published: 30 May 2018

Abstract: Despite advances in malignant pleural mesothelioma therapy, life expectancy of affected patients remains short. The limited efficiency of treatment options is mainly caused by inter- and intra-tumor heterogeneity of mesotheliomas. This diversity can be observed at the morphological and molecular levels. Molecular analyses reveal a high heterogeneity (i) between patients; (ii) within different areas of a given tumor in terms of different clonal compositions; and (iii) during treatment over time. The aim of the present review is to highlight this diversity and its therapeutic implications.

Keywords: mesothelioma; inter-tumor heterogeneity; temporal intra-tumor heterogeneity; spatial intra-tumor heterogeneity; chemoresistance; cancer stem cells; targeted therapy

1. Introduction

Malignant Pleural Mesothelioma (MPM) is a rare and aggressive neoplasm arising from a layer of mesothelial cells lining the pleura. The main cause of MPM is exposure to asbestos fibers that provoke constant inflammation and malignant transformation of mesothelial cells by direct mitotic spindle interference, reactive oxygen species release, and macrophage attraction [1]. The latency of the cancer is about 40 years, but once diagnosed, the life expectancy without treatment is less than 12 months [2]. The treatment usually includes chemotherapy followed by surgery, which can prolong the median survival to 22 months [3]. However, the chemotherapy is only effective in approximately 30–40% of the patients [4]. In addition, an effective alternative treatment or second line treatment has not yet been established [5]. With the exception of a recent phase three trial combining Bevacizumab with the standard cisplatin and pemetrexed chemotherapy in newly diagnosed MPM [6], clinical trials aiming for a targeted therapy approach in common cancer signaling pathways have not resulted in a better overall survival (OS) [7]. These studies stress the need for new biomarkers to predict the clinical response to chemotherapy as well as to find new possible targets for alternative therapy approaches. The search for new treatment options is complicated by the genetic composition of the tumor. Mutations are mainly found in tumor suppressors (COSMIC [8]), but common oncogenes such as *PI3K*, *EGFR*, and *VEGFR* are, if any, rarely found to be mutated in MPM, which limits the choice of targeted inhibitors. Although studies have shown that loss of tumor suppressors, such as *NF2* and *CDKN2A/p16*, lead to upregulation of associated oncogenic pathways, the translation of this knowledge into effective treatments has not yet occurred.

The mechanisms underlying the poor response of patients with MPM to a wide range of therapeutic interventions is still unknown. One reason for the inefficacy of the treatment regimens is the molecular inter-tumor heterogeneity, describing the diverse mutational (referred to as "genetic" in this review), epigenetic, expressional, and macroscopic (summarized as "phenotypical") changes between patients. Many mutations, such as in *EGFR* or *TP53*, are undetectable in the majority of MPM cases (COSMIC [8]). In contrast to non-small cell lung cancer [9], the relatively low number of MPM cases combined with the low prevalence of drug-targetable EGFR mutations in MPM compromises the investigation and use of selective EGFR inhibitors in the treatment of mesothelioma.

Adding to the complexity that arises due to inter-tumor heterogeneity, patient tumors also display intra-tumor heterogeneity. The existence of several tumor clones and subclones within one tumor sample of the same patient significantly limits the ability to devise logical treatment strategies. Intra-tumor heterogeneity appears during the course of the disease (temporal intra-tumor heterogeneity) as well as in different locations within the tumor at one time point (spatial intra-tumor heterogeneity).

Histologically, temporal and spatial intra-tumor heterogeneity in MPM manifests with a morphological spectrum, ranging from epithelioid to sarcomatoid tumors with the biphasic subtype containing a combination of both epithelioid and sarcomatoid components, each constituting at least 10% of the tumor. Adding to the complexity of histological subtyping, morphological biomarkers in epithelioid MPM, including nuclear atypia and number of mitoses, have been used to determine a total score which independently correlates with overall survival [10]. This further supports the existence of tumor heterogeneity, even within morphological well-defined subgroups of MPM. Furthermore some MPMs show a change of histology during the course of the disease, which represents temporal heterogeneity [11]. Besides this microscopic diversity, an increasing number of publications highlight the importance of genetic intra-tumor heterogeneity for therapeutic resistances in several cancer types [12]. Until now, this phenomenon has attracted little attention in MPM.

The aim of the present review is to highlight the different forms of heterogeneity in MPM with emphasis on the genetic and phenotypic intra-tumor heterogeneity. We summarize evidence of the spatial and temporal evolution of MPM, during the treatment with standard of care chemotherapy, and discuss the implications of heterogeneity on treatment decisions.

2. Inter-Tumor Heterogeneity

MPMs are known to have a high degree of molecular inter-tumor heterogeneity. In terms of genetic alterations, MPM generally displays a low number of mutations and recurrent mutations compared to other cancers [13]. The genes that were reported to be most often mutated are *BAP1* and *NF2*. Other commonly detected SNVs are found in *LATS1/2*, *TP53*, and *TERT* [14,15]. More prominent than SNVs are large chromosomal aberrations, which are thought to arise from direct interference with asbestos fibers or general chromosomal instability due to dysfunctional DNA damage response [1]. Chromosomal losses are the most frequent alterations in MPM, mostly affecting the chromosomal arms 3p, 9p, and 22q, where, amongst others genes, *BAP1*, *CDKN2A*, and *NF2* are located, respectively [8,16,17]. A high number of patients even harbor homozygous deletions of the *CDKN2A* region [18].

Despite these common alterations, the composition and gene locations of the mutations vary considerably between patients. A large sequencing study by Lo Iacono and colleagues, using 123 FFPE samples, sequenced 50 genes using the AmpliSeq Cancer Hotspot Panel plus another custom-designed amplicon panel covering the exons of the *NF2* and *BAP1* genes [19]. Although the authors reported a higher number of mutations clustering in exon 13 and 17 of the *BAP1* gene, which are the two largest exons, it did not seem that those were common hotspots for *BAP1* mutations (COSMIC [8]); there was more of an enrichment found in the N-terminal Ubiquitin Hydrolase domain (COSMIC [8]). Another study by Guo et al. compared 22 MPM tumor samples with matched blood samples using exome sequencing [13]. In total, they detected 490 somatic protein-altering mutations of which 477 were private alterations. Another working group led by Mäki-Nevala also performed exome-sequencing on 21 patients (two of them with peritoneal mesothelioma) and only found two non-private mutations in TTLL6 and MRPL1 occurring in two asbestos-exposed MM patients [20]. Ugurluer and colleagues as well as Kato and colleagues [21,22] both used a large gene panel covering 236 genes. Both groups, analyzing 11 [21] and 42 [22] mesothelioma patients, also failed to find any non-private alterations. Other groups working with smaller gene panels [14,23] also showed only private mutations. The results from these publications clearly illustrate that, in contrast to e.g., the L858R mutation in EGFR in lung cancer [24], there are no commonly mutated amino acid positions or "hotspot" regions in any of the genes tested.

In summary, these molecular analyses highlight the high inter-patient variability of locations and compositions of mutational patterns. This heterogeneity compromises the use of targeted therapy for mesothelioma patients and necessitates a personalized approach (Table 1). Clinical trials inhibiting for example the EGFR receptor in MPM patients using Erlotinib (NCT01592383, NCT00137826, NCT00039182), Gefitinib (NCT00787410, NCT00025207), Vandetanib (NCT00597116), or Cetuximab (NCT00996567) did not reveal any beneficial effects of the treatment. Although the mutational rate of EGFR is below 1% in MPM (COSMIC [8]), the rationale of those studies were the overexpression of EGFR which is found in over 50% of cases [25,26]. Destro and colleagues stained tumor tissue of 61 patients, whereby positive staining in 0–10% of tumor cells was regarded as negative expression, in 10–50% as low, and in >50% as high [25]. Only 9/61 (14.8%) showed a high EGFR expression, whereas 41.0% (21/61) only showed a staining in less than 50% of tumor cells, indicating that only a subpopulation of tumor cells overexpress EGFR. Enomoto and colleagues also stained 22 MPM cases, setting the thresholds for score 1+ for <5% positive tumor cells, score 2+ for 5–50% and score 3+ for >50% [26]. They scored 50% (11/22) of tumors as 3+ expression. Based on the assumption that high EGFR expression predicts the success of EGFR inhibiting drugs such as Erlotinib, detection of strong positive staining should be used as inclusion criteria in future studies. However, it was already shown that in many cancers, EGFR expression levels are not associated with a positive response to targeted therapy [27]. This was also documented in MPM by Garland and colleagues assessing EGFR expression in 57 patients with MPM [28]. A score of 0 was given for negative staining, score 1 for weak and focal staining, score 2 for positive and homogenous staining and score 3 for intense staining. In their cohort, 75% of the tumors stained score 2 or 3 for EGFR. However, no objective clinical responses to Erlotinib treatment was noted. Similar results were shown using Gefitinib [29] and Cetuximab [30], which strongly indicates that high EGFR expression cannot be used to predict response to EGFR inhibitors in patients with mesothelioma.

In our sequencing studies (Oehl et al., manuscript in preparation), we could see *EGFR* mutations at low allele frequency in the tissues, indicating a subclonal origin. Further, EGFR staining, as described above, often shows a focal pattern. Both findings suggest that there could be, additionally to the inter-tumor variability, a high intra-tumor heterogeneity additively influencing the outcome of anti-EGFR treatments in a negative way.

Table 1. Selection of finished studies using targeted therapy approaches in malignant pleural mesothelioma (MPM). The mutational rate in MPM was taken from the cosmic database. Data on expression were taken from indicated references.

Target	Drug	Study	Year of Completion	Status	# Patients	Results	Marker	Mutational Rate in MPM	Expression in MPM
mTor	Everolimus	NCT00770120	2014	completed	61	primary endpoint not reached	NF2 (Merlin)	17% (105/629)	4% [31]–8% [32] negative
FAK	Defactinib	NCT01024946	2012	completed	11	none published			
		NCT01870609	2016	terminated	344	lack of efficiency			
ALK1	PF-03446962	NCT01486368	2015	completed	17	primary endpoint not reached	ALK	0% (1/343)	0% [33]–20% [34] positive
EGFR	Erlotinib	NCT00039182	2007	completed	55	primary endpoints not reached	EGFR	1% (8/652)	15% [25], 50% [26], 75% [28] high
	Cetuximab	NCT00996567	2015	completed	22	primary endpoint not reached			
c-Met	Tivantinib	NCT01861301	2015	terminated	18	lack of efficiency	MET	1% (3/448)	17% [35]–40% [36] high

Background color highlights groups of studies that employed the same marker for patient stratification.

3. Spatial Intra-Tumor Heterogeneity

3.1. Spatial Genetic Heterogeneity

MPM is known to show intrinsic therapy resistances and is so far non-curable. The high number of non-responders to chemotherapy [4] as well as the frequent recurrences of the disease [37,38] suggest a substantial degree of resistant clones within an MPM patient. In silico modeling of spatial tumor growth suggests that the number of driver gene mutations, as well as the speed of cell turnover, greatly influences the degree of heterogeneity within a tumor [39]. Interestingly, the model proposed by Waclaw et al. shows that fewer driver mutations and a slow cell turnover lead to an increased level of heterogeneity [39]. Given that mesothelioma is supposed to develop over many years, the replication rate is in most cases quite low, indicating that there should be a very high degree of molecular diversity within the tumor.

Indeed, Comertpay and colleagues assessed the clonality of malignant mesothelioma in 14 female patients using a HUMARA assay [40]. This assay is based on X-chromosome inactivation by methylation and the *HUMARA* gene which is located on the X-chromosome. This gene encodes for the Human Androgen Receptor and harbors a varying number of CAG repeats, which usually differs between the maternal and paternal allele. One allele gets deactivated in healthy females; therefore, if a cancer was of monoclonal origin, only one allele would be detected in the tumor. However, when using the HUMARA assay on MPM tissue, Comertpay et al. detected paternal and maternal *HUMARA* alleles within most of the tumors, indicative of a polyclonal origin of MPM.

As described above, a common molecular alteration is the homozygous deletion of *CDKN2A* (p16) on chromosome 9. However, when measured by fluorescent in-situ hybridization (FISH) on tumor tissue, it is well known that the homozygous deletion cannot be detected in all cells of the tumor. Indeed, the status of the *CDKN2A* gene is highly variable, with no detectable loss, hemizygous losses and homozygous losses of *CDKN2A* within the same tumor. Defining a tumor as "homozygously deleted for *CDKN2A*" therefore requires defined cut-offs, such as 14.4% in a study by Wu et al. that compared the homozygous deletion patterns of *CDKN2A* between sarcomatoid mesothelioma and fibrous pleuritis [41]. These detections of non-homogenous deletions of *CDKN2A* suggest that besides the polyclonal origin, several genetic subclones might also exist within one tumor.

However, the only study so far describing genetic spatial heterogeneity was recently conducted by Kiyotani and colleagues [42]. From the surgical specimens of six MPM patients, they extracted DNA and RNA from fresh frozen tissue from three different locations within the tumor, namely from anterior, posterior, and diaphragm positions. They then conducted whole-exome sequencing, resulting in 19–47 non-synonymous mutations per sample. When looking at the SNVs that were detected at the three different locations within one patient, they found clearly distinct mutational patterns. Comparing the allele frequencies of these mutations, they detected some high variant allele frequency mutations in every examined location of the respective tumor, indicative of mutations of early clonal origin. Moreover, they saw a high degree of intra-tumoral spatial heterogeneity represented by varying amounts of subclonal fractions. The addition of TCRβ sequencing data and immune-related gene expression analysis revealed that this heterogeneity also extends to the immune microenvironment.

3.2. Spatial Phenotypic and Tumor Microenvironment Heterogeneity

As mentioned above, tumor heterogeneity is not only described by a heterogeneous genetic makeup of tumor cells within the same patient. The heterogeneity can also arise from selective environmental pressure such as nutrient, oxygen, tumor stroma, and immune microenvironment that can induce tumor heterogeneity by altering their phenotypes. This selective pressure of the microenvironment can govern the tumor phenotype by altering signaling pathways, regulating gene and protein expression. These intra-tumoral differences in the environment could result in therapy resistances [43]. To support this idea, it has been clearly demonstrated that hypoxic tumors are more resistant to chemotherapy and radiotherapy [44]. A recent study visualized tumor hypoxia by non-invasive imaging, [F-18] fluoromisonidazole (FMISO) PET-CT, and demonstrated that MPM has a visible area of hypoxia, predominantly in bulky tumor masses [45]. Thus, tumor cells in different regions of the tumor nodule may respond differentially to treatment. In view of immunotherapeutic approaches, for example using PD-1 or PD-L1 blocking antibodies, the heterogeneity of PD-L1 protein levels between primary and metastatic sites was recently studied in 64 MPM patients [46]. It was shown that PD-L1 expression, measured by immunohistochemical staining, was discordant in up to 31% of the cases (depending on the reviewer), which pronounces the limits of successful immunotherapy using anti-PD-L1 antibodies in mesothelioma. However, as seen in the example of the focal EGFR staining described above, the heterogeneity of protein expression is not only found in primary tumors and metastases but it also occurs within different regions within the same primary tumor. For example, loss of BAP1 expression in biphasic MPM can only be found in the epithelioid part of the tumor (Figure 1), whereas BAP1 is retained in the sarcomatoid component. The different expression profiles in different parts of the tumor could be related to the genetic or epigenetic background and require further investigation.

Although some studies indicate that genetic, phenotypic, and microenvironmental intra-tumor heterogeneity in MPM exist, our knowledge is still limited. More studies assessing spatial MPM heterogeneity are needed to improve our understanding of the pathophysiological mechanisms underlying MPM and to develop new treatment approaches that circumvent the impact of intra-tumor heterogeneity.

However, the question remains, which mechanisms lead to the development of spatial heterogeneity in MPM? A widely accepted theory (besides the cancer stem cell (CSC) theory which we will discuss later in this review) is that of clonal tumor evolution [47]. Hereby, a tumor accumulates somatic mutations and chromosomal aberrations in a stepwise manner. Some of these alterations are so-called "driver mutations", conferring a fitness advantage to the respective clones and leading to the development of several subclones, which are ultimately detected as spatial heterogeneity (Figure 2). Therefore, the spatial heterogeneity within a tumor can be seen as the result of a temporal heterogeneity, which we will discuss in the next chapters.

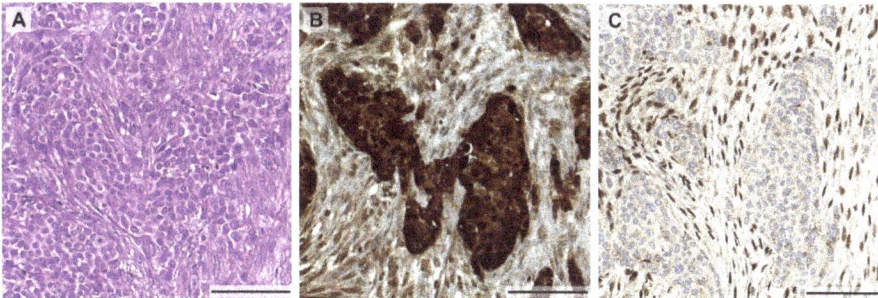

Figure 1. Morphological and immunohistochemical heterogeneity in mesothelioma. (**A**) Biphasic mesothelioma consisting of an epithelioid and sarcomatoid component (H&E stain), highlighted by a calretinin staining (**B**) showing a weaker expression in the sarcomatoid proliferation. (**C**) Heterogeneous expression of BAP-1 with positive nuclear staining in the sarcomatoid component and loss of BAP-1 expression in the epithelioid areas of the tumor. All pictures were taken at 10× magnification (scale bar: 100 µm).

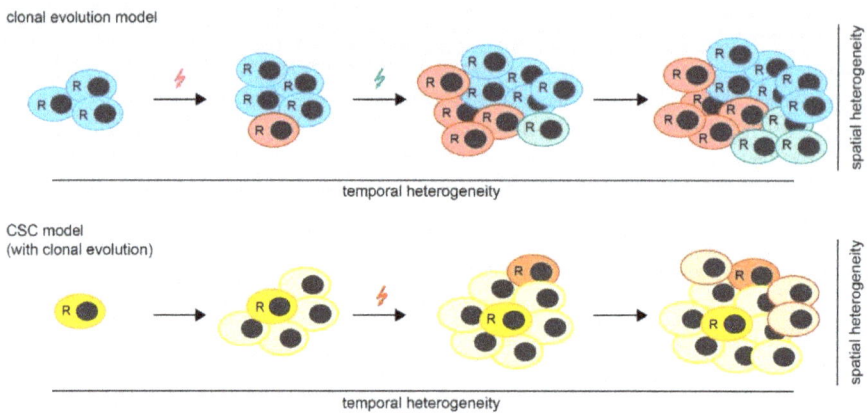

Figure 2. Models of tumor heterogeneity. In the clonal evolution model (upper panel) all cells are able to replicate (indicated by "R"). Mutations (colored arrows) are gained over time, leading to the formation of subclones which results in a heterogeneous tumor. In the cancer stem cell (CSC) model with clonal evolution, only CSCs are able to replicate. However, mutations occurring over time lead to the formation of additional CSCs.

4. Temporal Intra-Tumor Heterogeneity

Mesothelioma is often diagnosed at an advanced stage due to late onset and non-specific symptoms. Surgery and cytotoxic chemotherapy, platinum (*cis-* or carboplatin) plus pemetrexed, are standard first-line treatment for patients with MPM. Nevertheless, the prognosis of MPM remains poor because of tumor recurrence within a median time of 10–18 months after initial treatment [2].

As described above, tumors can evolve over time during multiple rounds of cell division. The presence of selective pressure such as treatment with an anti-cancer drug is an additional factor driving tumor clonal evolution (Figure 2). This temporal heterogeneity has severe implications for treatment decisions, as seen in various cancer entities. For example, a study in medulloblastoma, revealed that genetic aberrations of recurrent tumor tissues following the treatment with chemotherapy and radiotherapy diverged from that of diagnostic (treatment naive) tissues. This was due to the selection of the preexisting subclones that were already present before the treatment [48]. Another study in breast cancer observed an enrichment of slow proliferating cell populations with different molecular and biological characteristics following the treatment with chemotherapy, depending on the subtype of cancer. Employing single cell FISH analysis, they further demonstrated that breast cancer patients with low genetic heterogeneity responded better to the treatment [49]. To date, there is no study directly assessing temporal MPM heterogeneity and its implication on treatment outcomes.

4.1. Chemotherapy and Tumor Heterogeneity

Cisplatin and pemetrexed are the common cytotoxic agents given to MPM patients as a first-line treatment. Cisplatin is a genotoxic drug that induces intrastrand DNA cross-linking, inducing DNA damage, growth arrest, and cell death [50]. Pemetrexed is a folate antimetabolite that inhibits three enzymes involved in purine and pyrimidine synthesis. The lack of purine and pyrimidine results in the inhibition of deoxyribonucleic acid (DNA) and ribonucleic acid (RNA) synthesis, essential elements for cell proliferation and survival. Treatment with cisplatin-pemetrexed may eradicate sensitive and highly proliferative cells, but the resistant cells remain and can still grow or regrow following the treatment. Thus, this selective pressure can alter the composition of the tumor cell population and the extent of tumor heterogeneity following the treatment. Moreover, these cytotoxic agents can change the biological characteristics of tumor cells. For example, changes in the expression of genes associated with both cellular senescence (*PAI-1* and *IL-6*) and a gene identified as a cancer stem cell marker (*Thy-1*) were detected in primary MPM cells following exposure to cisplatin-pemetrexed. Moreover, increased Thy-1 expression was observed in acquired cisplatin pemetrexed resistant cells in vitro [51]. In addition to inducing phenotypical changes, DNA damage caused by cisplatin, if incorrectly repaired, can generate novel mutations or chromosome alterations, thereby increasing genetic diversity of tumors. Nonetheless, chemotherapy may generate a novel targetable mutation or cause an enrichment of targetable mutations that are not detectable prior to the treatment.

Thus, knowing the genetic and phenotypical changes of relapse tumors can be useful for the selection of effective second line treatment. For MPM, there has been no study assessing the effect of chemotherapy on the temporal heterogeneity. To study whether the genetic makeup of MPM changes over the course of treatment and disease progression, we compare the mutation profile of MPM in tumor tissues collected at three different time points during treatment (pre-chemo, post-chemo, and recurrence). Preliminary data (Oehl et al., manuscript in preparation), suggests that the genetic basis of some of the MPM tumors change over the course of treatment with chemotherapy.

Phenotypic changes of MPM tumor cells following the treatment with chemotherapy have also been observed in some studies. Indeed, the histologic subtype of the tumor changes over the course of therapy [11]. We also observed protein expression changes of MPM tumors following the treatment with chemotherapy. In our previous study [52], we compared protein expression of markers of the PI3K-mTOR pathway namely PTEN, p-mTOR, and p-S6 in matched MPM tissues pre- and post- cisplatin-based induction chemotherapy. Staining of the tissue microarray (TMA) revealed a reduction of protein expression of PTEN, p-mTOR, and p-S6 in the tumor tissues following chemotherapy. A decrease in tumor proliferative activity (Ki-67 expression) and slightly increased numbers of apoptotic cells (cleaved Caspase-3 staining) were also detected following the treatment. In another study [53] using the same patient cohort, we observed increased NF2 (Merlin) expression and decreased Survivin labelling in post chemotherapy treatment tissues (unpublished data). Employing a cohort of 34 patients with pre- and post- chemotherapeutic tissues available, Sidi et al. demonstrated that expression of senescence marker genes such as *PAI-1* and *p21* was significantly increased after chemotherapy [54].

These studies show that sampling at different time points during MPM treatment might reveal new potential treatment targets, which were not detectable at the time of diagnosis. Thus, longitudinal analysis of tumor tissues may be useful for the selection of subsequent effective therapies for MPM patients.

4.2. MPM Cancer Stem Cell

Cancer stem cells also play a role in tumor heterogeneity. Cancer stem cells represent a small population of tumor cells that are able to self-renew (Figure 2). Upon cell division, cancer stem cells give rise to progeny cells that maintain self-renewal properties or differentiate into various cell entities. It is widely demonstrated that cancer stem cells are commonly resistant to various anti-cancer drugs. Although the stem cell model remains controversial, cells with stem-cell like properties have been shown to contribute to tumor heterogeneity and therapy resistance in many solid tumors [55].

Relying on the basis that cancer stem cells express high levels of membrane drug transporter ABCG2, a study by Kai et al., employed Hoechst 33342 dye efflux assay to identify a MPM stem cell population [56]. Using this assay, they detected a small subset of cells (side population; SP) that can exclude Hoechst dye in three MPM cell lines and a transformed mesothelial cell line (MeT-5A) (number ranging from 0.05–1.32%). Treatment with cisplatin substantially increased the SP fraction of MPM cells. Interestingly, this SP population expressed higher levels of stem cell related genes namely *BMI1*, *OCT4*, and *NOTCH1* compared to non-SP (NSP) cells. However, despite exhibiting enhanced proliferation in vitro, there was no difference in in vivo tumorigenicity of both SP and NSP when implanted subcutaneously in NOD/SCID mice. Another study by Frei et al., also employed the same functional assay (using DyeCycleViolet) to identify SP on MPM cell lines and primary cells [57].

This study could also detect a small population of SP in all cell lines tested (ranging from 0.2–1%). Similar to the previous study, there was no difference in tumorigenicity between SP and NSP when implanted under the renal capsular of NOD/SCID mice. However, when SP cells were sorted from the in vivo tumor tissues, these tended to be more tumorigenic compared to NSP (although this was not statistically significant). MPM SP were however more resistant to cisplatin and expressed increased level of *PTCH1*, a gene of the sonic hedgehog signaling pathway.

Kim Chul et al. further characterized MPM SP using genome wide DNA methylation profiling coupled with mRNA expression [58]. This study described increased DNA methylation in CpG islands, gene flanking, and intragenic regions in SP cells compared to NSP. They identified 1130 genes differentially expressed in SP compared to NSP, among which, 122 genes are known to be regulated by aberrant DNA methylation. Importantly, these candidate genes, such as *YAP1* and *NOTCH2*, are known to play an important role in the maintenance of stem cell and the regulation of differentiation and development.

High level of ALDH1A has been used as a marker of CSCs. Thus, a study by Shapiro, et al. identified a CSC population of Merlin (NF2) negative MPM cell lines using the Aldefluor assay [59]. Here, they detected MPM CSCs (Aldefluor + cells) with increased tumor initiating potential compared to non-CSCs when implanted into immunodeficient mice. Treatment with cisplatin-pemetrexed increased this CSC population of MPM cell lines while treatment with the FAK inhibitor VS-4718 reduced the number of CSCs in vitro. Using a preclinical patient derived xenograft (PDX) model, they further demonstrated that treatment with the FAK inhibitor targeting CSC populations was effective in the control of tumor growth following cisplatin-pemetrexed treatment, that had enriched the CSC population.

In conclusion, SP displaying CSC characteristics have been identified in MPM cell lines and primary cells. In a preclinical model, CSCs survive during the treatment with cisplatin and pemetrexed and thus can give rise to a new generation of tumor cell population and create the diversity of recurrent tumor clones that differ from tumor at diagnosis. Thus far, all studies isolated potential MPM CSCs from cell lines, thus more evidence on the existence of CSCs in MPM clinical specimens and their role in treatment resistant is needed. An important factor when investigating CSCs in MPM is to take into account the intrinsic self-renewing capacity of mesothelial tissues which is conferred by mesothelial progenitor cells [60]. Given the heterogeneous and polyclonal nature of MPM, one can speculate that there exist several MPM CSC clones with different genetic alterations. These can also stem from clonal evolution of CSCs that acquire mutations over time during disease progression (Figure 2). This scenario would further increase the complexity of MPM heterogeneity.

5. Implications for Therapy

Results from clinical trials testing targeted treatments in MPM have so far been discouraging. As illustrated above, one factor responsible for poor treatment outcome is inter-patient variability in the expression or mutational status of the target molecules. Thus, appropriate predictive markers for targeted treatments are needed for the design and implementation of clinical studies. Although stratification of patients regarding to predictive markers is difficult to realize given the low incidence of MPM, this personalized treatment is probably the only way to overcome the high inter-patient heterogeneity of the disease.

As discussed above, another level of heterogeneity, namely the intra-tumor heterogeneity observed on both the genetic and phenotypic level in MPM, is another hurdle for the success of MPM treatment. MPM tumor may comprise of heterogeneous variants of tumor cells possessing different levels of chemosensitivity that can affect macroscopic response outcomes. The remaining chemoresistant cells can progress and reestablish tumor heterogeneity in the relapse tumor. Cancer stem cells may also play a role in chemoresistance of MPM. Their role in actual tumor and disease progression, however, remains to be elucidated. If CSCs can be defined in the patient tumor tissue, they might represent a potent target for therapy approaches to overcome chemotherapy resistance of MPM. Intra-tumor heterogeneity is also a major factor that complicates the development of new targeted agents. Treatment targeting only a small subpopulation of tumor will not be effective, but targeting the core gene or pathway alteration that are shared across all tumor cells will provide the most efficient way to eradicate tumors. To address genetic heterogeneity, an obvious approach would be to model the tumor evolution by either single cell or tumor bulk sequencing to define driving mutations of early clonal origin. Understanding the pathways behind those mutations could then be exploited to develop new therapeutic approaches targeting most, if not all, of the tumor cells. However, using this approach it will be inevitable to monitor the development of the tumor over the course of treatment, since the probability of selecting resistant clones is quite high, as seen previously for example in lung adenocarcinoma [61]. This monitoring will further provide important information for the selection of subsequent effective treatment.

In general, the tracking and multi-level analysis of the tumor tissue implies the need for sufficient material. This is usually available after surgery (if conducted), but in order to reveal alternative treatment options prior to or instead of operative interventions and chemotherapy, it will be necessary to analyze the tissue derived from the diagnostic biopsy. Therefore, it is critical to remove an adequate amount of tissue.

Furthermore, knowing that MPM is a heterogeneous disease, single-region sampling is unlikely to reflect the complete genetic and phenotypic landscape of the tumor. Therefore, multi-region sampling is required in order to determine dominant clones and potential therapy resistant subclones.

Since this approach might be difficult to be carried out in clinics, an alternative option could be the detection of circulating cell free tumor DNA (ctDNA) in the blood or pleural fluid of MPM patients [62]. In this scenario, ctDNA is a pool of DNA released from tumor cells at different locations thus could serve as a good representation of genomic heterogeneity. A recent study in colorectal cancer for example collected plasma samples at different time points during multimodality treatment [63]. They could show that patients that tested mutation-positive in their ctDNA after chemoradiotherapy or surgery had an increased risk of recurrence. A previous study by the same group, also in colorectal cancer, could even show that the median lead time between ctDNA detection and radiological recurrence was higher than five months [64]. However, both studies relied on the detection of mutations that were previously found in the primary cancers of the respective patients, and thus could not identify novel mutations or mutations that were enriched during the treatment course. In order to address this drawback, other groups like Shu et al. used a panel of 382 cancer-associated genes for sequencing of ctDNA from various cancer entities and reported several mutations that were not found in the corresponding primary tumor, probably representing temporal or spatial heterogeneity [65]. A similar approach using MPM associated genes could be envisioned for the future monitoring of mesothelioma treatment. New next generation sequencing techniques and improved bioinformatical pipelines hereby enable the reliable detection of low frequency alleles and possible subclones also in clinical settings [66].

Taken together, intra-tumor heterogeneity and its importance in the treatment of cancer has been clearly demonstrated in recent years. It is predictable that MPM is a heterogeneous tumor as it takes 30–40 years from asbestos exposure to disease development with the tumor developing on a large surface of the mesothelial layer lining the thoracic cavity. Nevertheless, there has only been a limited number of publications addressing the intra-tumor heterogeneity of MPM, an aggressive malignancy with limited therapeutic options. Thus, more studies are required to dissect different levels of MPM intra-tumor heterogeneity at different time points during treatment. Clearly, a better understanding of MPM evolution is essential for designing more effective treatment regimens.

Author Contributions: K.O. and M.M. performed literature search, designed and wrote the manuscript. K.O. designed and prepared the table and figures. B.V. provided and interpreted Figure 1 and proof-read the article. I.O. proof-read the article.

Funding: This research was supported by the Swiss National Science Foundation (grant number: PP00P3_159269) and Polianthes foundation.

Acknowledgments: We thank Ailsa J. Christiansen for critically reading the manuscript.

Conflicts of Interest: The authors declare no conflicts of interest.

References

1. Jaurand, M.C.; Fleury-Feith, J. Pathogenesis of malignant pleural mesothelioma. *Respirology* **2005**, *10*, 2–8. [CrossRef] [PubMed]

2. Opitz, I. Management of malignant pleural mesothelioma-the european experience. *J. Thorac. Dis.* **2014**, *6* (Suppl. 2), S238–S252. [PubMed]

3. Opitz, I.; Friess, M.; Kestenholz, P.; Schneiter, D.; Frauenfelder, T.; Nguyen-Kim, T.D.; Seifert, B.; Hoda, M.A.; Klepetko, W.; Stahel, R.A.; et al. A new prognostic score supporting treatment allocation for multimodality therapy for malignant pleural mesothelioma: A review of 12 years' experience. *J. Thorac. Oncol.* **2015**, *10*, 1634–1641. [CrossRef] [PubMed]

4. Vogelzang, N.J.; Rusthoven, J.J.; Symanowski, J.; Denham, C.; Kaukel, E.; Ruffie, P.; Gatzemeier, U.; Boyer, M.; Emri, S.; Manegold, C.; et al. Phase III study of pemetrexed in combination with cisplatin versus cisplatin alone in patients with malignant pleural mesothelioma. *J. Clin. Oncol.* **2003**, *21*, 2636–2644. [CrossRef] [PubMed]

5. Buikhuisen, W.A.; Hiddinga, B.I.; Baas, P.; van Meerbeeck, J.P. Second line therapy in malignant pleural mesothelioma: A systematic review. *Lung Cancer* **2015**, *89*, 223–231. [CrossRef] [PubMed]

6. Zalcman, G.; Mazieres, J.; Margery, J.; Greillier, L.; Audigier-Valette, C.; Moro-Sibilot, D.; Molinier, O.; Corre, R.; Monnet, I.; Gounant, V.; et al. Bevacizumab for newly diagnosed pleural mesothelioma in the mesothelioma avastin cisplatin pemetrexed study (maps): A randomised, controlled, open-label, phase 3 trial. *Lancet* **2016**, *387*, 1405–1414. [CrossRef]

7. Hiddinga, B.I.; Rolfo, C.; van Meerbeeck, J.P. Mesothelioma treatment: Are we on target? A review. *J. Adv. Res.* **2015**, *6*, 319–330. [CrossRef] [PubMed]

8. COSMIC database, w.s.i. Available online: Https://cancer.Sanger.Ac.Uk/cosmic (accessed on 30 April 2018).

9. Dahabreh, I.J.; Linardou, H.; Siannis, F.; Kosmidis, P.; Bafaloukos, D.; Murray, S. Somatic egfr mutation and gene copy gain as predictive biomarkers for response to tyrosine kinase inhibitors in non-small cell lung cancer. *Clin. Cancer Res.* **2010**, *16*, 291–303. [CrossRef] [PubMed]

10. Rosen, L.E.; Karrison, T.; Ananthanarayanan, V.; Gallan, A.J.; Adusumilli, P.S.; Alchami, F.S.; Attanoos, R.; Brcic, L.; Butnor, K.J.; Galateau-Salle, F.; et al. Nuclear grade and necrosis predict prognosis in malignant epithelioid pleural mesothelioma: A multi-institutional study. *Mod. Pathol.* **2018**, *31*, 598–606. [CrossRef] [PubMed]

11. Vrugt, B.; Felley-Bosco, E.; Simmler, S.; Storz, M.; Friess, M.; Meerang, M.; Soltermann, A.; Moch, H.; Stahel, R.; Weder, W.; et al. Sarcomatoid differentiation during progression of malignant pleural mesothelioma. *Zent. Chir. Z. Allg. Vis. Thorac. Gefäßch.* **2015**, *140*, FV21. [CrossRef]

12. Marusyk, A.; Almendro, V.; Polyak, K. Intra-tumour heterogeneity: A looking glass for cancer? *Nat. Rev. Cancer* **2012**, *12*, 323–334. [CrossRef] [PubMed]

13. Guo, G.; Chmielecki, J.; Goparaju, C.; Heguy, A.; Dolgalev, I.; Carbone, M.; Seepo, S.; Meyerson, M.; Pass, H.I. Whole-exome sequencing reveals frequent genetic alterations in bap1, nf2, cdkn2a, and cul1 in malignant pleural mesothelioma. *Cancer Res.* **2015**, *75*, 264–269. [CrossRef] [PubMed]

14. Bott, M.; Brevet, M.; Taylor, B.S.; Shimizu, S.; Ito, T.; Wang, L.; Creaney, J.; Lake, R.A.; Zakowski, M.F.; Reva, B.; et al. The nuclear deubiquitinase bap1 is commonly inactivated by somatic mutations and 3p21.1 losses in malignant pleural mesothelioma. *Nat. Genet.* **2011**, *43*, 668–672. [CrossRef] [PubMed]

15. Bueno, R.; Stawiski, E.W.; Goldstein, L.D.; Durinck, S.; De Rienzo, A.; Modrusan, Z.; Gnad, F.; Nguyen, T.T.; Jaiswal, B.S.; Chirieac, L.R.; et al. Comprehensive genomic analysis of malignant pleural mesothelioma identifies recurrent mutations, gene fusions and splicing alterations. *Nat. Genet.* **2016**, *48*, 407–416. [CrossRef] [PubMed]

16. Borczuk, A.C.; Pei, J.; Taub, R.N.; Levy, B.; Nahum, O.; Chen, J.; Chen, K.; Testa, J.R. Genome-wide analysis of abdominal and pleural malignant mesothelioma with DNA arrays reveals both common and distinct regions of copy number alteration. *Cancer Biol. Ther.* **2016**, *17*, 328–335. [CrossRef] [PubMed]

17. Ivanov, S.V.; Miller, J.; Lucito, R.; Tang, C.; Ivanova, A.V.; Pei, J.; Carbone, M.; Cruz, C.; Beck, A.; Webb, C.; et al. Genomic events associated with progression of pleural malignant mesothelioma. *Int. J. Cancer* **2009**, *124*, 589–599. [CrossRef] [PubMed]

18. Illei, P.B.; Rusch, V.W.; Zakowski, M.F.; Ladanyi, M. Homozygous deletion of cdkn2a and codeletion of the methylthioadenosine phosphorylase gene in the majority of pleural mesotheliomas. *Clin. Cancer Res.* 2003, *9*, 2108–2113. [PubMed]

19. Lo Iacono, M.; Monica, V.; Righi, L.; Grosso, F.; Libener, R.; Vatrano, S.; Bironzo, P.; Novello, S.; Musmeci, L.; Volante, M.; et al. Targeted next-generation sequencing of cancer genes in advanced stage malignant pleural mesothelioma: A retrospective study. *J. Thorac. Oncol.* **2015**, *10*, 492–499. [CrossRef] [PubMed]

20. Maki-Nevala, S.; Sarhadi, V.K.; Knuuttila, A.; Scheinin, I.; Ellonen, P.; Lagstrom, S.; Ronty, M.; Kettunen, E.; Husgafvel-Pursiainen, K.; Wolff, H.; et al. Driver gene and novel mutations in asbestos-exposed lung adenocarcinoma and malignant mesothelioma detected by exome sequencing. *Lung* **2016**, *194*, 125–135. [CrossRef] [PubMed]

21. Ugurluer, G.; Chang, K.; Gamez, M.E.; Arnett, A.L.; Jayakrishnan, R.; Miller, R.C.; Sio, T.T. Genome-based mutational analysis by next generation sequencing in patients with malignant pleural and peritoneal mesothelioma. *Anticancer Res.* **2016**, *36*, 2331–2338. [CrossRef] [PubMed]

22. Kato, S.; Tomson, B.N.; Buys, T.P.; Elkin, S.K.; Carter, J.L.; Kurzrock, R. Genomic landscape of malignant mesotheliomas. *Mol. Cancer Ther.* **2016**, *15*, 2498–2507. [CrossRef] [PubMed]

23. Shukuya, T.; Serizawa, M.; Watanabe, M.; Akamatsu, H.; Abe, M.; Imai, H.; Tokito, T.; Ono, A.; Taira, T.; Kenmotsu, H.; et al. Identification of actionable mutations in malignant pleural mesothelioma. *Lung Cancer* **2014**, *86*, 35–40. [CrossRef] [PubMed]

24. Sharma, S.V.; Bell, D.W.; Settleman, J.; Haber, D.A. Epidermal growth factor receptor mutations in lung cancer. *Nat. Rev. Cancer* **2007**, *7*, 169–181. [CrossRef] [PubMed]

25. Destro, A.; Ceresoli, G.L.; Falleni, M.; Zucali, P.A.; Morenghi, E.; Bianchi, P.; Pellegrini, C.; Cordani, N.; Vaira, V.; Alloisio, M.; et al. Egfr overexpression in malignant pleural mesothelioma. An immunohistochemical and molecular study with clinico-pathological correlations. *Lung Cancer* **2006**, *51*, 207–215. [CrossRef] [PubMed]

26. Enomoto, Y.; Kasai, T.; Takeda, M.; Takano, M.; Morita, K.; Kadota, E.; Iizuka, N.; Maruyama, H.; Haratake, J.; Kojima, Y.; et al. A comparison of epidermal growth factor receptor expression in malignant peritoneal and pleural mesothelioma. *Pathol. Int.* **2012**, *62*, 226–231. [CrossRef] [PubMed]

27. Dancey, J.E. Predictive factors for epidermal growth factor receptor inhibitors—The bull's-eye hits the arrow. *Cancer Cell* **2004**, *5*, 411–415. [CrossRef]

28. Garland, L.L.; Rankin, C.; Gandara, D.R.; Rivkin, S.E.; Scott, K.M.; Nagle, R.B.; Klein-Szanto, A.J.; Testa, J.R.; Altomare, D.A.; Borden, E.C. Phase ii study of erlotinib in patients with malignant pleural mesothelioma: A southwest oncology group study. *J. Clin. Oncol.* **2007**, *25*, 2406–2413. [CrossRef] [PubMed]

29. Govindan, R.; Kratzke, R.A.; Herndon, J.E., 2nd; Niehans, G.A.; Vollmer, R.; Watson, D.; Green, M.R.; Kindler, H.L. Gefitinib in patients with malignant mesothelioma: A phase II study by the cancer and leukemia group B. *Clin. Cancer Res.* **2005**, *11*, 2300–2304. [CrossRef] [PubMed]

30. Paepe, A.D.; Vermaelen, K.Y.; Cornelissen, R.; Germonpre, P.R.; Janssens, A.; Lambrechts, M.; Bootsma, G.; Meerbeeck, J.P.V.; Surmont, V. Cetuximab plus platinum-based chemotherapy in patients with malignant pleural mesothelioma: A single arm phase ii trial. *J. Clin. Oncol.* **2017**, *35*, e20030.

31. Sheffield, B.S.; Lorette, J.; Shen, Y.; Marra, M.A.; Churg, A. Immunohistochemistry for nf2, lats1/2, and yap/taz fails to separate benign from malignant mesothelial proliferations. *Arch. Pathol. Lab. Med.* **2016**, *140*, 391. [CrossRef] [PubMed]

32. Hylebos, M.; Van Camp, G.; van Meerbeeck, J.P.; Op de Beeck, K. The genetic landscape of malignant pleural mesothelioma: Results from massively parallel sequencing. *J. Thorac. Oncol.* **2016**, *11*, 1615–1626. [CrossRef] [PubMed]

33. Varesano, S.; Leo, C.; Boccardo, S.; Salvi, S.; Truini, M.; Ferro, P.; Fedeli, F.; Canessa, P.A.; Dessanti, P.; Pistillo, M.P.; et al. Status of anaplastic lymphoma kinase (alk) in malignant mesothelioma. *Anticancer Res.* **2014**, *34*, 2589–2592. [PubMed]

34. Mönch, D.; Bode-Erdmann, S.; Kalla, J.; Sträter, J.; Schwänen, C.; Falkenstern-Ge, R.; Klumpp, S.; Friedel, G.; Ott, G.; Kalla, C. A subgroup of pleural mesothelioma expresses alk protein and may be targetable by combined rapamycin and crizotinib therapy. *Oncotarget* **2018**, *9*, 20781. [CrossRef] [PubMed]

35. Jagadeeswaran, R.; Ma, P.C.; Seiwert, T.Y.; Jagadeeswaran, S.; Zumba, O.; Nallasura, V.; Ahmed, S.; Filiberti, R.; Paganuzzi, M.; Puntoni, R.; et al. Functional analysis of c-met/hepatocyte growth factor pathway in malignant pleural mesothelioma. *Cancer Res.* **2006**, *66*, 352–361. [CrossRef] [PubMed]

36. Bois, M.C.; Mansfield, A.S.; Sukov, W.R.; Jenkins, S.M.; Moser, J.C.; Sattler, C.A.; Smith, C.Y.; Molina, J.R.; Peikert, T.; Roden, A.C. C-met expression and met amplification in malignant pleural mesothelioma. *Ann. Diagn. Pathol.* **2016**, *23*, 1–7. [CrossRef] [PubMed]

37. Politi, L.; Borzellino, G. Second surgery for recurrence of malignant pleural mesothelioma after extrapleural pneumonectomy. *Ann. Thorac. Surg.* **2010**, *89*, 207–210. [CrossRef] [PubMed]

38. Baldini, E.H.; Richards, W.G.; Gill, R.R.; Goodman, B.M.; Winfrey, O.K.; Eisen, H.M.; Mak, R.H.; Chen, A.B.; Kozono, D.E.; Bueno, R.; et al. Updated patterns of failure after multimodality therapy for malignant pleural mesothelioma. *J. Thorac. Cardiovasc. Surg.* **2015**, *149*, 1374–1381. [CrossRef] [PubMed]

39. Waclaw, B.; Bozic, I.; Pittman, M.E.; Hruban, R.H.; Vogelstein, B.; Nowak, M.A. A spatial model predicts that dispersal and cell turnover limit intratumour heterogeneity. *Nature* **2015**, *525*, 261–264. [CrossRef] [PubMed]

40. Comertpay, S.; Pastorino, S.; Tanji, M.; Mezzapelle, R.; Strianese, O.; Napolitano, A.; Baumann, F.; Weigel, T.; Friedberg, J.; Sugarbaker, P.; et al. Evaluation of clonal origin of malignant mesothelioma. *J. Transl. Med.* **2014**, *12*, 301. [CrossRef] [PubMed]

41. Wu, D.; Hiroshima, K.; Matsumoto, S.; Nabeshima, K.; Yusa, T.; Ozaki, D.; Fujino, M.; Yamakawa, H.; Nakatani, Y.; Tada, Y.; et al. Diagnostic usefulness of p16/cdkn2a fish in distinguishing between sarcomatoid mesothelioma and fibrous pleuritis. *Am. J. Clin. Pathol.* **2013**, *139*, 39–46. [CrossRef] [PubMed]

42. Kiyotani, K.; Park, J.H.; Inoue, H.; Husain, A.; Olugbile, S.; Zewde, M.; Nakamura, Y.; Vigneswaran, W.T. Integrated analysis of somatic mutations and immune microenvironment in malignant pleural mesothelioma. *Oncoimmunology* **2017**, *6*, e1278330. [CrossRef] [PubMed]

43. Junttila, M.R.; de Sauvage, F.J. Influence of tumour micro-environment heterogeneity on therapeutic response. *Nature* **2013**, *501*, 346–354. [CrossRef] [PubMed]

44. Horsman, M.R.; Overgaard, J. The impact of hypoxia and its modification of the outcome of radiotherapy. *J. Radiat. Res.* **2016**, *57* (Suppl. 1), i90–i98. [CrossRef] [PubMed]

45. Francis, R.J.; Segard, T.; Morandeau, L.; Lee, Y.C.; Millward, M.J.; Segal, A.; Nowak, A.K. Characterization of hypoxia in malignant pleural mesothelioma with fmiso pet-ct. *Lung Cancer* **2015**, *90*, 55–60. [CrossRef] [PubMed]

46. Terra, S.B.S.P.; Mansfield, A.S.; Dong, H.; Peikert, T.; Roden, A.C. Temporal and spatial heterogeneity of programmed cell death 1-ligand 1 expression in malignant mesothelioma. *Oncoimmunology* **2017**, *6*, e1356146. [CrossRef] [PubMed]

47. McGranahan, N.; Swanton, C. Biological and therapeutic impact of intratumor heterogeneity in cancer evolution. *Cancer Cell* **2015**, *27*, 15–26. [CrossRef] [PubMed]

48. Morrissy, A.S.; Garzia, L.; Shih, D.J.; Zuyderduyn, S.; Huang, X.; Skowron, P.; Remke, M.; Cavalli, F.M.; Ramaswamy, V.; Lindsay, P.E.; et al. Divergent clonal selection dominates medulloblastoma at recurrence. *Nature* **2016**, *529*, 351–357. [CrossRef] [PubMed]

49. Almendro, V.; Cheng, Y.K.; Randles, A.; Itzkovitz, S.; Marusyk, A.; Ametller, E.; Gonzalez-Farre, X.; Munoz, M.; Russnes, H.G.; Helland, A.; et al. Inference of tumor evolution during chemotherapy by computational modeling and in situ analysis of genetic and phenotypic cellular diversity. *Cell Rep.* **2014**, *6*, 514–527. [CrossRef] [PubMed]

50. Dasari, S.; Tchounwou, P.B. Cisplatin in cancer therapy: Molecular mechanisms of action. *Eur. J. Pharmacol.* **2014**, *740*, 364–378. [CrossRef] [PubMed]

51. Oehl, K.; Kresoja-Rakic, J.; Opitz, I.; Vrugt, B.; Weder, W.; Stahel, R.; Wild, P.; Felley-Bosco, E. Live-cell mesothelioma biobank to explore mechanisms of tumor progression. *Front. Oncol.* **2018**, *8*, 40. [CrossRef] [PubMed]

52. Bitanihirwe, B.K.; Meerang, M.; Friess, M.; Soltermann, A.; Frischknecht, L.; Thies, S.; Felley-Bosco, E.; Tsao, M.S.; Allo, G.; de Perrot, M.; et al. Pi3k/mtor signaling in mesothelioma patients treated with induction chemotherapy followed by extrapleural pneumonectomy. *J. Thorac. Oncol.* **2014**, *9*, 239–247. [CrossRef] [PubMed]

53. Meerang, M.; Berard, K.; Friess, M.; Bitanihirwe, B.K.; Soltermann, A.; Vrugt, B.; Felley-Bosco, E.; Bueno, R.; Richards, W.G.; Seifert, B.; et al. Low merlin expression and high survivin labeling index are indicators for poor prognosis in patients with malignant pleural mesothelioma. *Mol. Oncol.* **2016**, *10*, 1255–1265. [CrossRef] [PubMed]

54. Sidi, R.; Pasello, G.; Opitz, I.; Soltermann, A.; Tutic, M.; Rehrauer, H.; Weder, W.; Stahel, R.A.; Felley-Bosco, E. Induction of senescence markers after neo-adjuvant chemotherapy of malignant pleural mesothelioma and association with clinical outcome: An exploratory analysis. *Eur. J. Cancer* **2011**, *47*, 326–332. [CrossRef] [PubMed]

55. Kreso, A.; Dick, J.E. Evolution of the cancer stem cell model. *Cell Stem Cell* **2014**, *14*, 275–291. [CrossRef] [PubMed]

56. Kai, K.; D'Costa, S.; Yoon, B.I.; Brody, A.R.; Sills, R.C.; Kim, Y. Characterization of side population cells in human malignant mesothelioma cell lines. *Lung Cancer* **2010**, *70*, 146–151. [CrossRef] [PubMed]

57. Frei, C.; Opitz, I.; Soltermann, A.; Fischer, B.; Moura, U.; Rehrauer, H.; Weder, W.; Stahel, R.; Felley-Bosco, E. Pleural mesothelioma side populations have a precursor phenotype. *Carcinogenesis* **2011**, *32*, 1324–1332. [CrossRef] [PubMed]

58. Kim, M.C.; Kim, N.Y.; Seo, Y.R.; Kim, Y. An integrated analysis of the genome-wide profiles of DNA methylation and mrna expression defining the side population of a human malignant mesothelioma cell line. *J. Cancer* **2016**, *7*, 1668–1679. [CrossRef] [PubMed]

59. Shapiro, I.M.; Kolev, V.N.; Vidal, C.M.; Kadariya, Y.; Ring, J.E.; Wright, Q.; Weaver, D.T.; Menges, C.; Padval, M.; McClatchey, A.I.; et al. Merlin deficiency predicts fak inhibitor sensitivity: A synthetic lethal relationship. *Sci. Transl. Med.* **2014**, *6*, 237ra268. [CrossRef] [PubMed]

60. Lachaud, C.C.; Lopez-Beas, J.; Soria, B.; Hmadcha, A. Egf-induced adipose tissue mesothelial cells undergo functional vascular smooth muscle differentiation. *Cell Death Dis.* **2014**, *5*, e1304. [CrossRef] [PubMed]

61. Lin, L.; Asthana, S.; Chan, E.; Bandyopadhyay, S.; Martins, M.M.; Olivas, V.; Yan, J.J.; Pham, L.; Wang, M.M.; Bollag, G.; et al. Mapping the molecular determinants of braf oncogene dependence in human lung cancer. *Proc. Natl. Acad. Sci. USA* **2014**, *111*, E748–E757. [CrossRef] [PubMed]

62. Peng, M.; Chen, C.; Hulbert, A.; Brock, M.V.; Yu, F. Non-blood circulating tumor DNA detection in cancer. *Oncotarget* **2017**, *8*, 69162–69173. [CrossRef] [PubMed]

63. Tie, J.; Cohen, J.D.; Wang, Y.; Li, L.; Christie, M.; Simons, K.; Elsaleh, H.; Kosmider, S.; Wong, R.; Yip, D.; et al. Serial circulating tumour DNA analysis during multimodality treatment of locally advanced rectal cancer: A prospective biomarker study. *Gut* **2018**. [CrossRef] [PubMed]

64. Tie, J.; Wang, Y.; Tomasetti, C.; Li, L.; Springer, S.; Kinde, I.; Silliman, N.; Tacey, M.; Wong, H.-L.; Christie, M.; et al. Circulating tumor DNA analysis detects minimal residual disease and predicts recurrence in patients with stage ii colon cancer. *Sci. Transl. Med.* **2016**, *8*, 346ra392. [CrossRef] [PubMed]

65. Shu, Y.; Wu, X.; Tong, X.; Wang, X.; Chang, Z.; Mao, Y.; Chen, X.; Sun, J.; Wang, Z.; Hong, Z.; et al. Circulating tumor DNA mutation profiling by targeted next generation sequencing provides guidance for personalized treatments in multiple cancer types. *Sci. Rep.* **2017**, *7*, 583. [CrossRef] [PubMed]

66. Salk, J.J.; Schmitt, M.; Loeb, L. Enhancing the accuracy of next-generation sequencing for detecting rare and subclonal mutations. *Nat. Rev. Genet.* **2018**, *19*, 269–285. [CrossRef] [PubMed]

International Journal of
Molecular Sciences

MDPI

Review

NF2/Merlin Inactivation and Potential Therapeutic Targets in Mesothelioma

Tatsuhiro Sato [1] and Yoshitaka Sekido [1,2,*]

[1] Division of Molecular Oncology, Aichi Cancer Center Research Institute, 1-1 Kanokoden, Chikusa-ku, Nagoya 464-8681, Japan; satot@aichi-cc.jp
[2] Department of Cancer Genetics, Nagoya University Graduate School of Medicine, 65, Tsurumai-cho, Showa-ku, Nagoya 464-8603, Japan
*   Correspondence: ysekido@aichi-cc.jp; Tel.: +81-52-762-6111

Received: 28 February 2018; Accepted: 19 March 2018; Published: 26 March 2018

Abstract: The neurofibromatosis type 2 (*NF2*) gene encodes merlin, a tumor suppressor protein frequently inactivated in schwannoma, meningioma, and malignant mesothelioma (MM). The sequence of merlin is similar to that of ezrin/radixin/moesin (ERM) proteins which crosslink actin with the plasma membrane, suggesting that merlin plays a role in transducing extracellular signals to the actin cytoskeleton. Merlin adopts a distinct closed conformation defined by specific intramolecular interactions and regulates diverse cellular events such as transcription, translation, ubiquitination, and miRNA biosynthesis, many of which are mediated through Hippo and mTOR signaling, which are known to be closely involved in cancer development. MM is a very aggressive tumor associated with asbestos exposure, and genetic alterations in *NF2* that abrogate merlin's functional activity are found in about 40% of MMs, indicating the importance of *NF2* inactivation in MM development and progression. In this review, we summarize the current knowledge of molecular events triggered by *NF2*/merlin inactivation, which lead to the development of mesothelioma and other cancers, and discuss potential therapeutic targets in merlin-deficient mesotheliomas.

Keywords: malignant mesothelioma; neurofibromatosis type 2 (*NF2*); merlin; Hippo signaling pathway; PI3K/AKT/mTOR signaling pathway

1. Introduction

Mutations in the neurofibromatosis type 2 (*NF2*) gene are responsible for neurofibromatosis 2, a dominantly inherited familial cancer syndrome characterized by the formation of bilateral vestibular schwannomas and meningiomas [1,2]. Besides sporadic schwannomas [3] and meningiomas [4], frequent biallelic inactivation of *NF2* was also found in malignant mesothelioma (MM), a very aggressive tumor which is not associated with the *NF2* cancer syndrome [5,6]. Tumors carrying *NF2* mutations are also observed, albeit infrequently, in multiple organs such as the breast, the prostate, the liver, and the kidney [7,8], indicating a significant role of *NF2* in the development of various human malignancies.

Findings in mouse models support the biological function of *NF2* as a tumor suppressor gene. Since it was shown that a homozygous mutation in the *NF2* gene of mice causes embryonic death by day 6.5 of their development [9], the role of *NF2* as a tumor suppressor gene has been studied in mice that are heterozygous for *NF2* mutations. It was found to develop a variety of malignant tumors, including lymphoma, sarcoma, and carcinoma [10,11]. Furthermore, some studies revealed the involvement of *NF2* in the development of malignant plural mesothelioma after asbestos exposure. Thus, heterozygous $NF2^{+/-}$ mice had a higher sensitivity to asbestos, which resulted in an increased risk of malignant mesothelioma formation compared to wild-type $NF2^{+/+}$ mice [2,12]. A direct injection of the Adeno-Cre virus into the pleural cavity of adult mice resulted in a conditional knockout of oncosuppressor genes, which further demonstrated that the loss of *NF2*, together with *Tp53* or *Ink4a/Arf*, frequently causes the development of mesothelioma which closely mimicked human MM [13]. It was also shown that the restoration of *NF2* expression in *NF2*-deficient mesothelioma cells significantly inhibited their growth [14–16]. These in vitro and in vivo data strongly support the role of *NF2* inactivation in mesothelioma development.

2. Domain Organization and Functions of Merlin

2.1. NF2 Transcript Variants

The *NF2* gene is located in the chromosomal region 22q12 [1,17]; the gene contains 17 exons and spans approximately 95 kb of DNA. *NF2* transcripts undergo alternative splicing, thereby generating multiple isoforms [18], and variable *NF2* transcripts are observed in human mesotheliomas [5,12]. Two transcripts, one lacking exon 16 and the other containing all 17 exons, are the predominant variants encoding isoforms I and II; the first contains 595 amino acids, while the second, which is generated by the insertion of exon 16 into mRNA which creates a new stop codon, contains 590 amino acids and is identical to isoform I in the first 579 residues (Figure 1A). Initially, it was thought that isoform II lacked anticancer activity [19,20]; however, later studies showed that both isoforms exhibited the function of tumor suppression [21–23].

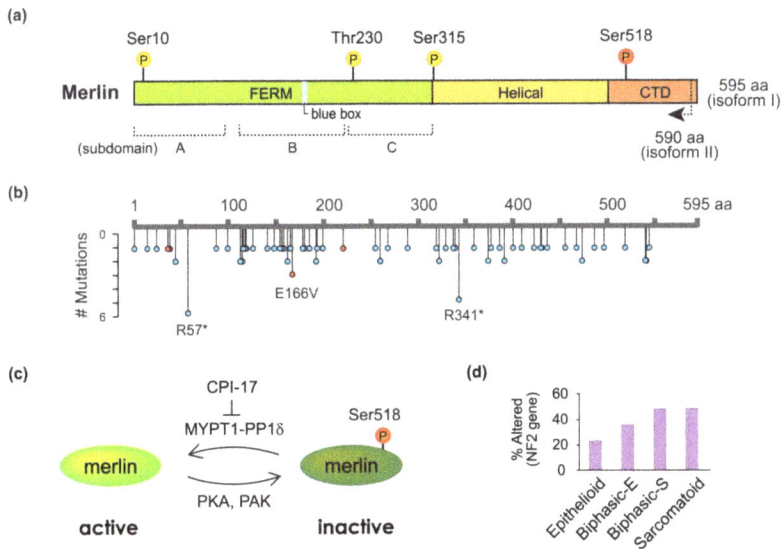

Figure 1. Mechanisms underlying the activation/inactivation of merlin. (**a**) Domain organization of merlin. The protein consists of the N-terminal FERM (band 4.1/ezrin/radixin/moesin) domain (green) comprising three subdomains (A, B, and C), a central helical domain (yellow), and a C-terminal domain (CTD, orange). Major phosphorylation sites are indicated; (**b**) *NF2* mutations and their frequency in pleural and peritoneal cancers. Nonsense/frameshift (blue) and missense (red) mutations registered in COSMIC (Catalogue of Somatic Mutations in Cancer; http://cancer.sanger.ac.uk/cosmic/) as of 27 February 2018, are mapped; (**c**) Phosphorylation-dependent inactivation of merlin. Phosphorylation at Ser518 inactivates merlin and inhibits its growth suppression activity; (**d**) Frequency of genetic alterations in the *NF2* gene, including mutations, fusions, and copy number variations in different subtypes of malignant pleural mesothelioma based on an analysis of 211 malignant plural mesothelioma samples. The data were adapted from Bueno et al. [24].

2.2. Domain Organization

The *NF2* gene product, named merlin, is widely expressed in various human tissues and is most closely related to the ezrin/radixin/moesin (ERM) family proteins, which are localized at cell-surface structures such as ruffling membranes and cell–cell adhesion sites, and connect actin filaments to the plasma membrane. The significant similarity in amino acid sequences between merlin and ERM proteins suggests that merlin can be associated with the actin cytoskeleton and the organization of membrane domains [25].

A structural analysis shows that merlin consists of three domains: the N-terminal FERM (band 4.1, ezrin, radixin, moesin) domain containing three subdomains (A, B, and C), the central helical domain, and the C-terminal domain (CTD) (Figure 1A). Merlin shares 45–47% sequence similarity with the ERM family members, especially in the conserved FERM domain (60–70%). The FERM of merlin binds to membrane proteins such as hyaluronate receptor CD44 [26,27], adaptor molecule Na^+/H^+ exchanger three, regulating factor one (NHERF/EBP50) [28,29], and E-cadherin [30]. Furthermore, the FERM mediates protein binding to phospholipids such as phosphatidylinositol 4,5-bisphosphate (PIP2) [31,32]. Despite the similarity in the binding properties between merlin and ERM proteins, their CTDs show distinct binding preferences. The CTDs of ERM proteins have actin-binding sites [33] linking the plasma membrane to the actin cytoskeleton, whereas merlin lacks the region corresponding to the C-terminal F-actin-binding site [34] and interacts with actin fibers through residues 1–27 and 280–323, which seem to be sufficient for the binding [35]. Moreover, merlin has a unique seven-amino-acid stretch (residues 177–183) in the FERM domain, named the 'blue box', which is conserved from fly to mammalian proteins but is lacking in ERM family members [34,36]. Alanine substitution in, or deletion of, this region produces unique merlin mutants, which have dominant-negative activity and result in an excessive proliferation of wing epithelial cells in flies [36] and a loss of contact inhibition in mammalian cells [26,37]. The unique characteristics of merlin domains suggest that the regulation of merlin is distinct from that of ERM proteins.

2.3. Molecular Conformation and Phosphorylation

ERM proteins have a 'closed' inactive conformation formed by the binding of CTD to the N-terminal FERM, whereas the phosphorylation of C-terminal residues disrupts the interaction, resulting in the 'open' active state, where the released FERM and CTD can bind to cell adhesion molecules and actin filaments, respectively [31]. Although a C-terminal phosphorylation site, threonine 576, critical for the conformational change in ERM proteins, is also conserved in merlin; the Thr576Ala substitution does not affect merlin's ability to suppress cell growth and motility [38]. Aside from this, the phosphorylation of merlin at serine 518 abrogates its growth inhibition activity [38,39]. These findings indicate that in merlin, phosphorylation causes inactivation, which is in contrast to its effect in ERM proteins (Figure 1C).

Merlin phosphorylation at Ser518 was frequently observed in mesothelioma cells expressing full-length merlin [12]. Moreover, CPI-17, a cellular inhibitor of myosin phosphatase targeting subunit 1 (MYPT1-PP1δ), was increased in mesothelioma cells with full-length *NF2* compared to normal pleura or mesothelioma with truncated *NF2* [38]. As MYPT1-PP1δ dephosphorylates merlin at Ser518 [40], CPI-17 upregulation would result in an increased phosphorylation and inactivation of merlin (Figure 1C). These findings suggest that merlin can be inactivated not only by mutations but also through posttranslational modifications occurring in mesothelioma cells.

The Ser518 phosphorylation in merlin is independently catalyzed by distinct protein kinases such as p21-activated kinase (PAK) [41–43] and protein kinase A (PKA) [44]. PAK causes a phosphorylation-dependent inactivation of merlin and promotes the loss of contact inhibition of proliferation [45], whereas PKA, in addition to Ser518, also phosphorylates Ser10 that is not conserved in ERM proteins, which results in increased cell migration [46]. Another protein kinase, AKT, phosphorylates merlin at Thr230 and Ser315, which appears to stimulate ubiquitin-dependent protein degradation [47].

Given the data on ERM proteins, Ser518 phosphorylation in merlin has been suggested to change its conformation from a 'closed' to an 'open' state [48]. Although the FERM and the CTD of merlin bind each other, their mutual affinity is low compared to that in ERM proteins [49], suggesting that merlin may not form a fully closed form. Instead, phosphorylation was shown to rather strengthen the head-to-tail folding in merlin [23]. Analysis by fluorescence resonance energy transfer (FRET) suggests that phosphorylation causes a subtle conformation change in merlin [50]. Furthermore, although merlin isoform II does not form the 'closed' state since it lacks five C-terminal residues [51], both isoforms I and II exhibit antitumor activity [21–23]. Cumulatively, these findings suggest that the phosphorylation at Ser518 would inactivate merlin without the accompanying dynamic conformational change observed in ERM proteins.

2.4. NF2 Inactivation in Mesothelioma

In addition to a frequent loss of the 22q12 region, which is the locus of the *NF2* gene, mutations within the entire *NF2* coding region are common for mesothelioma (Figure 1B). Nonsense mutations either totally abolish merlin expression or lead to the production of truncated forms. The functional activity of the truncated merlin variants, especially those with a short deletion at the C-terminus, has not been fully characterized. However, it was shown that the mutant with a C-terminal deletion of 40 residues was incapable of restoring proper growth inhibition in *NF2*-null mesothelioma cells [15], and that merlin truncated by 63 residues at the C-terminus did not cause growth arrest of primary Schwann cells [13], indicating the importance of the CTD for the antitumor activity of merlin. Therefore, nonsense mutations in *NF2*, even those occurring close to the C-terminus, are suggested to produce functional defects and are responsible for mesothelioma development. In contrast, the impact of missense mutations that cause amino acid substitutions is less understood, and it is unclear as to how and to what extent individual mutations affect merlin tumor-suppressive function. Although the pathogenic activity of several missense mutants identified in tumors have been studied [14], further investigation is required for a complete understanding of the effect produced by merlin mutations on tumor progression. In addition, *NF2* gene rearrangements are also frequently detected in MM, and each *NF2* gene fusion variant was thought to cause functional inactivation [24].

Regarding gene mutation frequency in MM, gene alterations in *NF2* are considered to be the second most common after those in BAP1. Developed mesothelioma tumors have different histological subtypes: epithelioid, sarcomatoid, and biphasic MMs. An expression analysis of 211 malignant plural mesothelioma samples suggested that among the subtypes, sarcomatoid tumors had the highest *NF2* mutation rate, while epithelioid tumors had the lowest *NF2* mutation rate [24] (Figure 1D). Furthermore, hemizygous *NF2* loss has been shown to decrease both the overall survival and the progression-free survival in a cohort of 86 peritoneal mesothelioma patients [52]. These data suggest that *NF2* inactivation might be involved in the epithelial–mesenchymal transition during metastasis, and that the development of sarcomatoid mesotheliomas is characterized by a poorer overall survival compared to the epithelioid subtype.

2.5. Loss of Contact Inhibition in NF2-Deficient Cells

Contact inhibition, a regulatory mechanism providing cell growth arrest at confluence in tissue culture, is frequently disrupted in cancer cells [53], and *NF2*-null cells grow to a significantly higher density compared to wild-type cells, suggesting that *NF2* controls tumor progression. The mechanism underlying the merlin regulation of growth arrest in response to cell confluence has been addressed in several studies. For example, it has been shown that merlin forms a complex with CD44, which is activated by the stimulation of extracellular hyaluronate, resulting in growth inhibition of rat schwannoma cells in vitro. Other studies have suggested that merlin regulates contact inhibition through small GTPase Rac1 [45,54], α-catenin, cell-polarity protein Par3 [55], and a tight-junction-associated complex composed of angiomotin (AMOT), Patj, and Pals1 [48] (Figure 2, shown in pink). These findings suggest that merlin could sense its environmental conditions and control cell growth via complex interactions with signaling proteins involved in cell–cell adhesion.

2.6. Subcellular Localization

FERM domain-containing proteins link plasma membrane receptors to cytoskeleton components [12]. Consistent with this notion, immunostaining with merlin-specific antibodies detects merlin at the cell membrane or the ruffling edges in human fibroblasts, meningioma cells, and Schwann cells [56,57]. Although the localization of the wild-type or the mutant merlin in mesothelial and mesothelioma cells is not defined, we have observed exogenously expressed full-length V5-tagged merlin both at the plasma membrane and in the cytoplasm of merlin-negative mesothelioma cells [58]. However, as merlin localization is dynamically regulated in response to various signals (described below), a further detailed investigation is necessary.

3. Proteins and Signaling Related to Merlin's Functions

3.1. Hippo Signaling Pathway

Merlin exerts its tumor-suppressive effects by controlling the expression of oncogenic genes through the activation of Hippo signaling (Figure 2, shown in orange). The Hippo pathway is composed of core proteins including MST1/2 (Mammalian STE20-Like Protein Kinases), SAV1 (Salvador Family WW Domain Containing Protein 1), MOB (MOB Kinase Activators), and LATS1/2 (Large Tumor Suppressor Kinase 1/2) [59]. At the plasma membrane, merlin recruits LATS1/2 kinases which directly phosphorylate the downstream effectors of the Hippo pathway, YAP (Yes-Associated Protein) and its paralogue TAZ (WW Domain-Containing Transcription Regulator 1, alternatively WWTR1), thus preventing their translocation to the nucleus and inhibiting their function as transcription co-activators. Alternatively, Hippo pathway inactivation induces an accumulation of underphosphorylated YAP and TAZ in the nucleus and their association with DNA-binding TEAD (TEA Domain Transcription Factor) family proteins, which upregulates the transcription of multiple oncogenic genes [60]. Along with *NF2* mutations, gene alterations are also frequently observed in Hippo pathway components, including LATS1/2, SAV1, and LIM-domain containing protein AJUBA, a *Drosophila* djub homolog and LATS1/2 binding partner [61,62]. High-level amplification of the 11q22 locus encompassing the *YAP* gene was also observed in a small subset of MMs [58]. These results indicate that the disruption of Hippo signaling plays a central role in the transformation of mesothelial cells.

YAP activation in mesothelial cells drastically changes their behavior. Kakiuchi et al. [63] have shown that the expression of constitutively active YAP Ser127Ala mutants in immortalized mesothelial cells promotes their growth in vitro, as well as tumor formation after their transplantation in mice. Conversely, YAP knockdown inhibits cell growth, motility, and invasion in mesothelioma cells with activated YAP, but did not show any effects in cells without YAP activation [64]. Furthermore, these studies showed that the YAP-dependent transcriptional activations of cyclin D2 (CCND2), forkhead box M1 (FOXM1), and phospholipase C beta 4 (PLCB4) are involved in mesothelioma cell growth [63,64], suggesting that activated YAP influences diverse cellular processes, thereby resulting in mesothelial cell transformation. The role of TAZ in mesothelioma has not been defined yet, but considering its functional redundancy with YAP, the oncogenic function of TAZ could be predicted.

3.2. DCAF1

It has been reported that merlin can translocate into the nucleus, where it binds to DCAF1 (also known as VprBP) through the N-terminal FERM domain [65]. DCAF1 is a substrate adaptor of E3 ubiquitin ligase CRL4DCAF1 containing CUL4 and DDB1. The interaction between merlin and DCAF1 depends on merlin activation, since neither the Ser518Asp phosphomimetic mutant, nor the Ser64Ala mutant, which lacks tumor-suppressor activity, bind to CRL4DCAF1. Merlin inhibits the activity of CRL4DCAF1, which regulates ubiquitination of target proteins. It was shown that LATS1/2 are functional targets of CRL4^{DCAF1} and that in tumors with mutated *NF2*, such as mesothelioma, activated CRL4 induces LATS1/2 ubiquitination to promote their degradation and YAP/TAZ activation, thus stimulating oncogenesis [66] (Figure 2, shown in purple). These results suggest that DCAF1 and CRL4^{DCAF1} are potential therapeutic targets for merlin-deficient mesothelioma. Cooper et al. [67] tested whether CRL4^{DCAF1} inhibition with NEDD8-activating enzyme (NAE) inhibitor MLN4924 could suppress the growth of tumor cells carrying *NF2* mutations. MLN4924 alone caused only a moderate inhibition of mesothelioma cell growth, but the combination of MLN4924 and GDC-0980, an mechanistic target of rapamycin/phosphatidylinositol 3-kinase (mTOR/PI3K) inhibitor, strongly suppressed cell proliferation. Despite blocking a broad spectrum of Cullin–RING E3 ligases including CRL4^{DCAF1}, NF2–NAE inhibitors could be a promising target for therapeutic intervention in patients with merlin-negative mesothelioma.

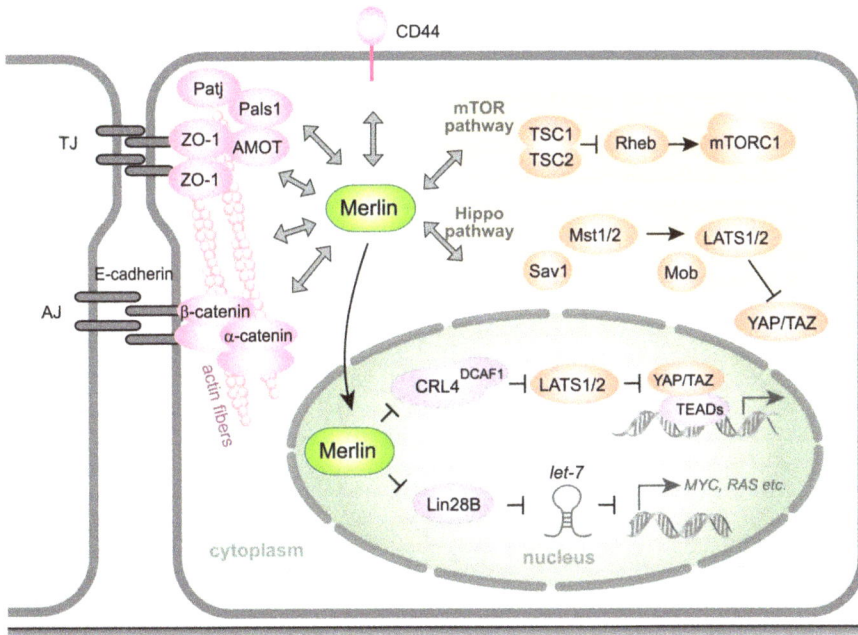

Figure 2. A model of the *NF2*/merlin signaling pathway. Merlin is involved in contact inhibition by interacting with many membrane-associated proteins such as CD44 [26,27], the angiomotin (AMOT) –Patj–Pals1 complex [48], E-cadherin–α-catenin [30,55], and actin fibers. A loss of merlin expression disrupts cancer-related signaling through the Hippo and mTOR pathways. Merlin is also localized in the nucleus where it binds to and inhibits E3 ubiquitin ligase CRL4^{DCAF1}, which promotes LATS1/2 degradation [66,67], and RNA-binding protein Lin28B, which suppresses let-7 miRNAs that are involved in the silencing of oncogenes such as *MYC* and *RAS* [68]. TJ: tight junction; AJ: adherens junction; ZO-1: Zonula occludens-1; AMOT: angiomotin; mTOR: mechanistic target of rapamycin; TSC1/2: tuberous sclerosis complex 1/2; Rheb: Ras homolog enriched in brain; Sav1: Salvador Family WW Domain Containing Protein 1; Mst1/2: mammalian Ste20-like kinase 1/2; Mob: Mps one binder kinase activator-like protein; YAP: yes-associated protein 1; TAZ: WW domain-containing transcription regulator 1; CRL4: Cullin-RING ubiquitin ligase 4; DCAF1: DDB1- and CUL4-associated factor 1; LATS1/2: large tumor suppressor kinase 1/2; TEAD: TEA domain transcription factor; Lin28B: lin-28 homolog B.

3.3. PI3K/AKT/mTOR Signaling Pathway

mTOR is a serine/threonine kinase that plays a key role in cell growth and proliferation. The mTOR signaling pathway has been reported to be frequently activated in a variety of human malignancies, indicating its close involvement in carcinogenesis.

mTOR is composed of two distinct complexes, mTOR complex 1 (mTORC1) and mTOR complex 2 (mTORC2) [69,70]; both of them contain mTOR kinase and a mTORC subunit mLST8, which is suggested to stabilize the structure of the mTOR catalytic domain [71]. mTORC1 binds to raptor, whereas mTORC2 binds to rictor and Sin1, forming functional kinase complexes. mTORC1 and its activator Rheb have been shown to enhance protein translation and pyrimidine nucleotide biosynthesis, thereby promoting cell growth and proliferation [72–74].

The involvement of the mTOR pathway in mesothelioma formation has been suggested in several studies. Thus, López-Lago et al. [75] showed, using a panel of malignant mesothelioma cell lines, that the loss of merlin correlated with the activation of mTORC1 signaling and the sensitivity to rapamycin. Similarly, James et al. [76] reported that merlin-deficient meningioma cells also exhibited constitutive mTORC1 activation and increased growth. Furthermore, it was demonstrated that the concurrent loss of Tp53 and tuberous sclerosis 1 (TSC1), a negative regulator of Rheb–mTORC1 signaling, induces the development of peritoneal mesothelioma in mice [77]. Immunohistochemical analysis of human mesotheliomas revealed the hyperactivation of mTORC1 and the reduced expression of TSC2, which binds to TSC1 and negatively regulates the activation of mTORC1 by Rheb. These findings suggest that mTOR activation caused by merlin inactivation plays a significant role in mesothelioma development (Figure 2, shown in orange).

The deregulation of mTORC1 signaling in mesothelioma cells can be attributed to changes in the state of various upstream effectors. AKT, an mTORC1 activator and mTORC2 substrate, is stimulated in more than 60% of malignant mesothelioma cell lines and tumors [78,79]; furthermore, the homozygous deletion of PTEN, a negative regulator of AKT signaling, has also been reported in mesothelioma cells [78,79]. PTEN loss leads to an increase in phosphatidylinositol (3,4,5)-triphosphate (PIP3), resulting in the activation of both mTORC1 and mTORC2 signaling. These data suggest that the activation of mTORC1, as well as mTORC2, may be involved in mesothelioma development. On the other hand, no activating mutations in the *MTOR*, nor the *RHEB* genes, have been identified in mesothelioma cells to date, although such mutations were shown to cause the hyper-activation of mTORC1 [80,81] observed in mesothelioma. The biological role of mTORC1 in mesothelioma formation is now beginning to be examined.

3.4. Lin28B and let-7 miRNAs

An RNA-binding protein, Lin28B, has been recently reported to be an alternative binding partner of merlin. Lin28B is involved in cell growth and reprogramming [82,83] and suppresses the biogenesis of the let-7 microRNAs (miRNAs) that function as tumor suppressors by silencing the expression of several oncogenes such as *MYC* and *RAS* [84,85]. Hikasa et al. [68] found that merlin bound to Lin28B through the FERM domain and translocated Lin28B from the nucleus to cytoplasm, leading to let-7 miRNA maturation (Figure 2, shown in purple). The association between merlin and Lin28B is induced when merlin is dephosphorylated, which occurs at high cell density, suggesting a novel mechanism in which merlin exerts cell-density-dependent tumor suppression through let-7 miRNA maturation.

3.5. TRAF7

Recurrent mutations in the *TRAF7* gene are observed in mesothelioma cells. *TRAF7* belongs to tumor necrosis factor (TNF) receptor-associated factors (TRAFs) possessing E3 ubiquitin ligase activity [86], and it was shown to promote ubiquitination of an apoptosis inhibitor, FLIP [87], which is increased in mesothelioma cells [88]. FLIP inhibition by small interfering RNA (siRNA) sensitizes mesothelioma cells to Fas- and TRAIL-induced apoptosis, suggesting a role of FLIP in protecting cells from death signals. Interestingly, *TRAF7* and *NF2* mutations are mutually exclusive in malignant pleural mesothelioma [24] as well as in meningioma [89], suggesting that merlin and *TRAF7* may use a common signal transduction pathway.

4. Potential Molecular Targets in Merlin-Negative Mesothelioma

4.1. FAK Inhibitors

Focal adhesion kinase (FAK) is a serine/threonine kinase that mediates signals from focal adhesion complexes to the cell growth and migration machinery. FAK is elevated in most human cancers, and its inhibition has been recognized as a novel approach to targeted anticancer therapy against various types of solid tumors. In 2014, Shapiro and colleagues [90] reported that a beneficial effect of an FAK inhibitor, VS-4718 (alternatively PND-1186), on MM cells lacking merlin expression was the increased sensitivity of MM cells to VS-4718 in vitro and in tumor xenograft models. Therefore, FAK inhibitors were considered as potential candidates for mesothelioma therapy. However, a phase II clinical trial investigating the effects of an FAK inhibitor, defactinib (VS-6063), on merlin-deficient mesotheliomas was terminated early due to its futility, and the reason for the poor clinical performance is currently unclear. A recent study on the pharmacological effects of FAK inhibitors has demonstrated a significant correlation between *E*-cadherin mRNA levels and VS-4718 in merlin-negative mesothelioma [15], suggesting that E-cadherin may serve as a promising biomarker for predicting the response to FAK inhibitors in mesothelioma, which should be tested in clinical settings.

4.2. YAP Inhibitors

The screening of more than 3300 Food and Drug Administration (FDA)-approved small molecules resulted in the identification of verteporfin as a novel compound that disrupts the YAP–TEAD interaction and inhibits YAP oncogenic activity [91]. Verteporfin, a benzoporphyrin derivative, is currently used in clinics as a photosensitizer in photodynamic therapy for macular degeneration. The compound is activated by 690 nm far-red light, generating reactive oxygen species (ROS) which eliminate abnormal blood vessels; however, its inhibition of YAP–TEAD interactions does not require light activation. Several in vitro studies revealed that verteporfin can suppress the growth, the migration, and the tumorsphere formation of cultured MM cells [92,93]. Recently, CIL56 (also named CA3), a small molecule that induces cellular ferroptosis through ROS production [94], has been identified as a novel YAP inhibitor. By preventing the interaction between YAP and TEAD, CIL56 strongly inhibited esophageal adenocarcinoma cell growth both in vitro and in vivo [95]. Therefore, YAP inhibitors may be effective anticancer drugs for mesothelioma and other tumors in which YAP/TAZ are activated through the disruption of the Hippo pathway.

In addition to the nucleocytoplasmic shuttling of YAP/TAZ, a recent study revealed a mechanism for the intracellular translocation of TEAD proteins. The findings of Lin et al. [96] suggest that environmental stresses such as osmotic stress, high cell density, and cell detachment promote TEAD translocation from the nucleus to the cytoplasm via p38 mitogen-activated protein kinase (MAPK). Interestingly, TEAD nucleocytoplasmic transfer occurred in a Hippo-independent manner and suppressed YAP and YAP-dependent cancer cell growth, suggesting that the regulation of TEAD translocation might serve as another therapeutic strategy in merlin-deficient tumors.

4.3. mTOR Inhibitors

Given the emerging role of mTOR in mesothelioma development and proliferation, mTOR inhibitors are thought to be promising drugs against merlin-negative mesothelioma. Unfortunately, however, a phase II clinical trial of everolimus, a first-generation mTOR inhibitor rapamycin analog (so-called rapalog), demonstrated that there was insufficient activity in patients with advanced mesothelioma [97]. The reason for this limited success is not fully understood; it is possible that certain rapamycin-resistant functions of mTORC1 [98] may account for the low efficacy of rapalogs.

To date, various types of improved mTOR inhibitors have been developed. Second-generation mTOR inhibitors (also called ATP-competitive mTOR inhibitors) directly compete with ATP for the binding to the mTOR kinase domain, thus completely inhibiting both mTORC1 and mTORC2. PI3K/mTOR dual inhibitors also target the ATP-binding pocket; the advantage of these compounds is that they recognize the ATP-binding site not only in mTOR but also in PI3K. Considering the reports of mTORC2 activation in mesothelioma cells due to the loss of PTEN, which is an mTORC2 negative regulator, or the increased phosphorylation of mTORC2 substrate AKT [79,99], PI3K/mTOR dual inhibitors are predicted to be more effective in suppressing mesothelioma growth than rapalogs. Moreover, a combination treatment with mTOR or PI3K/mTOR inhibitors together with other antitumor drugs appears to be a reasonable approach, because mTOR signaling is involved in a compensatory pathway that renders cancer cells drug resistant; thus, increased mTORC1 activity in breast and pancreatic cancer cells confers resistance to cyclin-dependent kinase 4/6 (CDK4/6) inhibitors [100,101]. Although in vivo experiments are lacking, in vitro data indicate that a combination treatment with the CDK4/6 inhibitor, palbociclib, and a PI3K/mTOR dual inhibitor exerts a synergistic effect on mesothelioma cell growth [102].

Recently, a third-generation mTOR inhibitor which overcomes the resistance to first- and second-generation mTOR inhibitors has been developed [103] and already showed promise by exhibiting a higher efficacy in glioblastomas compared to previous mTOR inhibitors [104]. Although the antitumor activity of the new mTOR inhibitors against mesothelioma has yet to be demonstrated, enhanced clinical benefits can be expected.

4.4. Statins

Statins are inhibitors of 3-hydroxyl-3-methyl coenzyme A (HMG–CoA) reductase, the rate-limiting enzyme of the mevalonate pathway for the biosynthesis of mevalonate and downstream isoprenoids, which are suggested to have beneficial effects on several cancers, including colorectal cancer, breast cancer, and melanoma [105]. The therapeutic potential of statins for suppressing mesothelioma cell growth has been reported in vitro and in mouse xenografts [106,107]. Furthermore, statins are suggested to have synergistic or additive antitumor effects when used with other drugs [108–110]. Recently, it was reported that mesothelioma cells with *NF2* and/or *LATS2* mutations were more sensitive to fluvastatin compared to those with *BAP1* mutations [111], whereas merlin-negative breast cancer cells showed sensitivity to simvastatin [112]. The regulation of YAP and TAZ through the mevalonate pathway [113] suggests that statins may show a more significant effect on cell growth in *NF2*-deficient mesothelioma and other types of tumors.

4.5. COX2 Inhibitors

It has been demonstrated that YAP activation in *NF2*-null Schwann cells promotes the transcription of the *PTGS2* gene encoding cyclooxygenase 2 (COX-2), the key enzyme in prostaglandin biosynthesis. Interestingly, the treatment of NF2-null Schwann or schwannoma cells with a COX-2 inhibitor, celecoxib, dramatically inhibited cell growth in vitro and in vivo [114], which suggests that COX-2 is a potential therapeutic target in *NF2*-null tumors. However, a recent study showed that celecoxib failed to prevent the generation of schwannomas in a genetically engineered mouse model of *NF2* inactivation, although COX-2 expression was increased in tumors that developed in these mice [115]. Considering the controversial results on COX-2 as a target in *NF2*-inactive tumors, further investigations are required in this direction.

5. Conclusions

Malignant mesothelioma is highly refractory to conventional therapies, and the current chemotherapeutic approach approved in clinics is still based on a combination of platinum and

Int. J. Mol. Sci. **2018**, *19*, 988

an antifolate, pemetrexed [116]. *NF2* is one of the most frequently mutated genes in mesothelioma; therefore, the restoration of *NF2* functions is expected to cure a large population of mesothelioma patients. However, the introduction of tumor suppressor genes in every tumor cell and the subsequent expression of the encoded proteins at levels that are comparable to those of normal cells remain highly challenging [117]. Growing evidence demonstrates that merlin is distributed in multiple subcellular compartments and suppresses a number of proteins and signaling pathways that are related to tumor progression [118]. Once *NF2* is inactivated, these oncogenic mechanisms are constitutively induced, conferring malignant phenotypes to the cells; therefore, targeting merlin-dependent molecular pathways is a promising strategy for the treatment of *NF2*-deficient cancers. The restoration of the Hippo signaling and the inhibition of the PI3K/AKT/mTOR pathway are predicted to exert potent anticancer effects, but the clinical performance of the perspective drugs has not yet been evaluated, and the mechanisms underlying *NF2* control of these signaling pathways in mesothelial and other cells are still unknown.

Although *NF2* is frequently inactivated in MMs, recent progress in *NF2*-targeted therapies has been limited [119]. To search for more effective drugs against *NF2*-deficient mesothelioma cells, we have to understand when, where, and how merlin exerts its tumor-suppressive effects, especially in mesothelial cells. Further, the roles of downstream signals that are activated by *NF2* loss in mesothelioma progression also remain incompletely defined. For example, is the activation of YAP via the inactivation of the Hippo signaling pathway enough for mesothelioma formation? If so, why are *YAP* gene mutations that constitutively activate their transcription activity undetected in MMs? The activation of *TAZ* in merlin-deficient MM cells should be evaluated as a potential key oncogene that drives tumor initiation and progression together with YAP. It is to be noted that *NF2* loss might be involved in drug resistance. A genome-wide CRISPR screen in human cells has identified *NF2* as the highest-ranking candidate whose loss is involved in the resistance to vemurafenib, a therapeutic RAF inhibitor [120]. Future studies focusing on defining the alteration of molecular networks caused by the loss of merlin expression would further foster the development of new therapeutic strategies in mesothelioma.

Acknowledgments: This work was supported in part by JSPS KAKENHI (15K19015, 16H04706, 17K19628), Takeda Science Foundation (for Tatsuhiro Sato), and AMED P-CREATE.

Author Contributions: Tatsuhiro Sato and Yoshitaka Sekido conceived, wrote, and critically proof-read the article.

Conflicts of Interest: Collaboration grant: Kyowa Hakko Kirin Co., Ltd., (Tokyo, Japan) and Eisai Co., Ltd., (Tokyo, Japan). The founding sponsors had no role in the writing of the manuscript. The founding sponsors had no role in the design of the study; in the collection, analyses, or interpretation of data; in the writing of the manuscript, and in the decision to publish the results.

Abbreviations

AMOT	angiomotin
BAP1	BRCA1 Associated Protein 1
CCND2	cyclin D2
CDK	cyclin-dependent kinase
COX-2	cyclooxygenase 2
CPI-17	protein kinase C-potentiated inhibitor protein of 17 kDa
CRL4	Cullin-RING ubiquitin ligase 4
CTD	C-terminal domain
CUL4	Cullin 4
DCAF1	DDB1- and CUL4-associated factor 1
DDB1	damaged DNA binding protein 1
ERM	ezrin/radixin/moesin
FAK	focal adhesion kinase
FDA	Food and Drug Administration
FERM	band 4.1/ezrin/radixin/moesin
FLIP	caspase-like apoptosis regulatory protein

FOXM1	forkhead box M1
LATS1/2	large tumor suppressor kinase 1/2
Lin28B	lin-28 homolog B
MAPK	mitogen-activated protein kinase
mLST8	mammalian lethal with Sec13 protein 8
MM	malignant mesothelioma
Mob	Mps one binder kinase activator-like protein
mTOR	mechanistic target of rapamycin
mTORC	mTOR complex
MST1/2	mammalian Ste20-like protein kinase 1/2
MYPT1-PP1δ	myosin phosphatase target subunit 1-protein phosphatase 1 δ
NAE	NEDD8-activating enzyme
NF2	neurofibromatosis type 2
PAK	p21-activated kinase
Pals1	protein associated with Lin7-1
Patj	Pals1-associated tight junction protein
PI3K	phosphatidylinositol 3-kinase
PKA	protein kinase A
PLCB4	phospholipase C beta 4
PTEN	phosphatase and tensin homolog
Rheb	Ras homolog enriched in brain
ROS	reactive oxygen species
SAV1	Salvador Family WW Domain Containing Protein 1
siRNA	small interfering RNA
TAZ	WW domain-containing transcription regulator 1
TEAD	TEA domain transcription factor
TNF	tumor necrosis factor
TRAF	TNF receptor-associated factor
TRAIL	TNF-related apoptosis inducing ligand
TSC1/2	tuberous sclerosis complex subunit 1/2
VprBP	viral protein R (VPR)-binding protein
YAP	Yes-associated protein 1
ZO-1	Zonula occludens-1

References

1. Trofatter, J.A.; MacCollin, M.M.; Rutter, J.L.; Murrell, J.R.; Duyao, M.P.; Parry, D.M.; Eldridge, R.; Kley, N.; Menon, A.G.; Pulaski, K.; et al. A novel moesin-, ezrin-, radixin-like gene is a candidate for the neurofibromatosis 2 tumor suppressor. *Cell* **1993**, *72*, 791–800. [CrossRef]

2. Altomare, D.A.; Vaslet, C.A.; Skele, K.L.; De Rienzo, A.; Devarajan, K.; Jhanwar, S.C.; McClatchey, A.I.; Kane, A.B.; Testa, J.R. A mouse model recapitulating molecular features of human mesothelioma. *Cancer Res.* **2005**, *65*, 8090–8095. [CrossRef] [PubMed]

3. Stemmer-Rachamimov, A.O.; Xu, L.; Gonzalez-Agosti, C.; Burwick, J.A.; Pinney, D.; Beauchamp, R.; Jacoby, L.B.; Gusella, J.F.; Ramesh, V.; Louis, D.N. Universal absence of merlin, but not other ERM family members, in schwannomas. *Am. J. Pathol.* **1997**, *151*, 1649–1654. [PubMed]

4. Ruttledge, M.H.; Sarrazin, J.; Rangaratnam, S.; Phelan, C.M.; Twist, E.; Merel, P.; Delattre, O.; Thomas, G.; Nordenskjold, M.; Collins, V.P.; et al. Evidence for the complete inactivation of the *NF2* gene in the majority of sporadic meningiomas. *Nat. Genet.* **1994**, *6*, 180–184. [CrossRef] [PubMed]

5. Sekido, Y.; Pass, H.I.; Bader, S.; Mew, D.J.; Christman, M.F.; Gazdar, A.F.; Minna, J.D. Neurofibromatosis type 2 (*NF2*) gene is somatically mutated in mesothelioma but not in lung cancer. *Cancer Res.* **1995**, *55*, 1227–1231. [PubMed]

6. Bianchi, A.B.; Mitsunaga, S.I.; Cheng, J.Q.; Klein, W.M.; Jhanwar, S.C.; Seizinger, B.; Kley, N.; Klein-Szanto, A.J.; Testa, J.R. High frequency of inactivating mutations in the neurofibromatosis type 2 gene (*NF2*) in primary malignant mesotheliomas. *Proc. Natl. Acad. Sci. USA* **1995**, *92*, 10854–10858. [CrossRef] [PubMed]

7. Li, W.; Cooper, J.; Karajannis, M.A.; Giancotti, F.G. Merlin: A tumour suppressor with functions at the cell cortex and in the nucleus. *EMBO Rep.* **2012**, *13*, 204–215. [CrossRef] [PubMed]

8. Petrilli, A.M.; Fernandez-Valle, C. Role of Merlin/NF2 inactivation in tumor biology. *Oncogene* **2016**, *35*, 537–548. [CrossRef] [PubMed]

9. McClatchey, A.I.; Saotome, I.; Ramesh, V.; Gusella, J.F.; Jacks, T. The *NF2* tumor suppressor gene product is essential for extraembryonic development immediately prior to gastrulation. *Genes Dev.* **1997**, *11*, 1253–1265. [CrossRef] [PubMed]

10. Giovannini, M.; Robanus-Maandag, E.; van der Valk, M.; Niwa-Kawakita, M.; Abramowski, V.; Goutebroze, L.; Woodruff, J.M.; Berns, A.; Thomas, G. Conditional biallelic NF2 mutation in the mouse promotes manifestations of human neurofibromatosis type 2. *Genes Dev.* **2000**, *14*, 1617–1630. [PubMed]

11. McClatchey, A.I.; Saotome, I.; Mercer, K.; Crowley, D.; Gusella, J.F.; Bronson, R.T.; Jacks, T. Mice heterozygous for a mutation at the *NF2* tumor suppressor locus develop a range of highly metastatic tumors. *Genes Dev.* **1998**, *12*, 1121–1133. [CrossRef] [PubMed]

12. Thurneysen, C.; Opitz, I.; Kurtz, S.; Weder, W.; Stahel, R.A.; Felley-Bosco, E. Functional inactivation of NF2/merlin in human mesothelioma. *Lung Cancer* **2009**, *64*, 140–147. [CrossRef] [PubMed]

13. Jongsma, J.; van Montfort, E.; Vooijs, M.; Zevenhoven, J.; Krimpenfort, P.; van der Valk, M.; van de Vijver, M.; Berns, A. A conditional mouse model for malignant mesothelioma. *Cancer Cell* **2008**, *13*, 261–271. [CrossRef] [PubMed]

14. Poulikakos, P.I.; Xiao, G.H.; Gallagher, R.; Jablonski, S.; Jhanwar, S.C.; Testa, J.R. Re-expression of the tumor suppressor NF2/merlin inhibits invasiveness in mesothelioma cells and negatively regulates FAK. *Oncogene* **2006**, *25*, 5960–5968. [CrossRef] [PubMed]

15. Kato, T.; Sato, T.; Yokoi, K.; Sekido, Y. E-cadherin expression is correlated with focal adhesion kinase inhibitor resistance in Merlin-negative malignant mesothelioma cells. *Oncogene* **2017**, *36*, 5522–5531. [CrossRef] [PubMed]

16. Xiao, G.H.; Gallagher, R.; Shetler, J.; Skele, K.; Altomare, D.A.; Pestell, R.G.; Jhanwar, S.; Testa, J.R. The *NF2* tumor suppressor gene product, merlin, inhibits cell proliferation and cell cycle progression by repressing cyclin D1 expression. *Mol. Cell. Biol.* **2005**, *25*, 2384–2394. [CrossRef] [PubMed]

17. Rouleau, G.A.; Merel, P.; Lutchman, M.; Sanson, M.; Zucman, J.; Marineau, C.; Hoang-Xuan, K.; Demczuk, S.; Desmaze, C.; Plougastel, B.; et al. Alteration in a new gene encoding a putative membrane-organizing protein causes neuro-fibromatosis type 2. *Nature* **1993**, *363*, 515–521. [CrossRef] [PubMed]

18. Chang, L.S.; Akhmametyeva, E.M.; Wu, Y.; Zhu, L.; Welling, D.B. Multiple transcription initiation sites, alternative splicing, and differential polyadenylation contribute to the complexity of human neurofibromatosis 2 transcripts. *Genomics* **2002**, *79*, 63–76. [CrossRef] [PubMed]

19. Gutmann, D.H.; Sherman, L.; Seftor, L.; Haipek, C.; Hoang Lu, K.; Hendrix, M. Increased expression of the *NF2* tumor suppressor gene product, merlin, impairs cell motility, adhesionand spreading. *Hum. Mol. Genet.* **1999**, *8*, 267–275. [CrossRef] [PubMed]

20. Sherman, L.; Xu, H.M.; Geist, R.T.; Saporito-Irwin, S.; Howells, N.; Ponta, H.; Herrlich, P.; Gutmann, D.H. Interdomain binding mediates tumor growth suppression by the *NF2* gene product. *Oncogene* **1997**, *15*, 2505–2509. [CrossRef] [PubMed]

21. Lallemand, D.; Saint-Amaux, A.L.; Giovannini, M. Tumor-suppression functions of merlin are independent of its role as an organizer of the actin cytoskeleton in Schwann cells. *J. Cell Sci.* **2009**, *122*, 4141–4149. [CrossRef] [PubMed]

22. Zoch, A.; Mayerl, S.; Schulz, A.; Greither, T.; Frappart, L.; Rubsam, J.; Heuer, H.; Giovannini, M.; Morrison, H. Merlin Isoforms 1 and 2 Both Act as Tumour Suppressors and Are Required for Optimal Sperm Maturation. *PLoS ONE* **2015**, *10*, e0129151. [CrossRef] [PubMed]

23. Sher, I.; Hanemann, C.O.; Karplus, P.A.; Bretscher, A. The tumor suppressor merlin controls growth in its open state, and phosphorylation converts it to a less-active more-closed state. *Dev. Cell* **2012**, *22*, 703–705. [CrossRef] [PubMed]

24. Bueno, R.; Stawiski, E.W.; Goldstein, L.D.; Durinck, S.; De Rienzo, A.; Modrusan, Z.; Gnad, F.; Nguyen, T.T.; Jaiswal, B.S.; Chirieac, L.R.; et al. Comprehensive genomic analysis of malignant pleural mesothelioma identifies recurrent mutations, gene fusions and splicing alterations. *Nat. Genet.* **2016**, *48*, 407–416. [CrossRef] [PubMed]

25. McClatchey, A.I.; Giovannini, M. Membrane organization and tumorigenesis—The NF2 tumor suppressor, Merlin. *Genes Dev.* **2005**, *19*, 2265–2277. [CrossRef] [PubMed]

26. Morrison, H.; Sherman, L.S.; Legg, J.; Banine, F.; Isacke, C.; Haipek, C.A.; Gutmann, D.H.; Ponta, H.; Herrlich, P. The *NF2* tumor suppressor gene product, merlin, mediates contact inhibition of growth through interactions with CD44. *Genes Dev.* **2001**, *15*, 968–980. [CrossRef] [PubMed]

27. Sainio, M.; Zhao, F.; Heiska, L.; Turunen, O.; den Bakker, M.; Zwarthoff, E.; Lutchman, M.; Rouleau, G.A.; Jaaskelainen, J.; Vaheri, A.; et al. Neurofibromatosis 2 tumor suppressor protein colocalizes with ezrin and CD44 and associates with actin-containing cytoskeleton. *J. Cell Sci.* **1997**, *110 Pt 18*, 2249–2260. [PubMed]

28. Murthy, A.; Gonzalez-Agosti, C.; Cordero, E.; Pinney, D.; Candia, C.; Solomon, F.; Gusella, J.; Ramesh, V. NHE-RF, a regulatory cofactor for Na^+-H^+ exchange, is a common interactor for merlin and ERM (MERM) proteins. *J. Biol. Chem.* **1998**, *273*, 1273–1276. [CrossRef] [PubMed]

29. Reczek, D.; Berryman, M.; Bretscher, A. Identification of EBP50: A PDZ-containing phosphoprotein that associates with members of the ezrin-radixin-moesin family. *J. Cell Biol.* **1997**, *139*, 169–179. [CrossRef] [PubMed]

30. Lallemand, D.; Curto, M.; Saotome, I.; Giovannini, M.; McClatchey, A.I. NF2 deficiency promotes tumorigenesis and metastasis by destabilizing adherens junctions. *Genes Dev.* **2003**, *17*, 1090–1100. [CrossRef] [PubMed]

31. Fehon, R.G.; McClatchey, A.I.; Bretscher, A. Organizing the cell cortex: The role of ERM proteins. *Nat. Rev. Mol. Cell Biol.* **2010**, *11*, 276–287. [CrossRef] [PubMed]

32. Mani, T.; Hennigan, R.F.; Foster, L.A.; Conrady, D.G.; Herr, A.B.; Ip, W. FERM domain phosphoinositide binding targets merlin to the membrane and is essential for its growth-suppressive function. *Mol. Cell. Biol.* **2011**, *31*, 1983–1996. [CrossRef] [PubMed]

33. Hamada, K.; Shimizu, T.; Matsui, T.; Tsukita, S.; Hakoshima, T. Structural basis of the membrane-targeting and unmasking mechanisms of the radixin FERM domain. *EMBO J.* **2000**, *19*, 4449–4462. [CrossRef] [PubMed]

34. Shimizu, T.; Seto, A.; Maita, N.; Hamada, K.; Tsukita, S.; Tsukita, S.; Hakoshima, T. Structural basis for neurofibromatosis type 2. Crystal structure of the merlin FERM domain. *J. Biol. Chem.* **2002**, *277*, 10332–10336. [CrossRef] [PubMed]

35. Xu, H.M.; Gutmann, D.H. Merlin differentially associates with the microtubule and actin cytoskeleton. *J. Neurosci. Res.* **1998**, *51*, 403–415. [CrossRef]

36. LaJeunesse, D.R.; McCartney, B.M.; Fehon, R.G. Structural analysis of *Drosophila* merlin reveals functional domains important for growth control and subcellular localization. *J. Cell Biol.* **1998**, *141*, 1589–1599. [CrossRef] [PubMed]

37. Johnson, K.C.; Kissil, J.L.; Fry, J.L.; Jacks, T. Cellular transformation by a FERM domain mutant of the *NF2* tumor suppressor gene. *Oncogene* **2002**, *21*, 5990–5997. [CrossRef] [PubMed]

38. Surace, E.I.; Haipek, C.A.; Gutmann, D.H. Effect of merlin phosphorylation on neurofibromatosis 2 (*NF2*) gene function. *Oncogene* **2004**, *23*, 580–587. [CrossRef] [PubMed]

39. Rong, R.; Surace, E.I.; Haipek, C.A.; Gutmann, D.H.; Ye, K. Serine 518 phosphorylation modulates merlin intramolecular association and binding to critical effectors important for NF2 growth suppression. *Oncogene* **2004**, *23*, 8447–8454. [CrossRef] [PubMed]

40. Jin, H.; Sperka, T.; Herrlich, P.; Morrison, H. Tumorigenic transformation by CPI-17 through inhibition of a merlin phosphatase. *Nature* **2006**, *442*, 576–579. [CrossRef] [PubMed]

41. Shaw, R.J.; Paez, J.G.; Curto, M.; Yaktine, A.; Pruitt, W.M.; Saotome, I.; O'Bryan, J.P.; Gupta, V.; Ratner, N.; Der, C.J.; et al. The *NF2* tumor suppressor, merlin, functions in Rac-dependent signaling. *Dev. Cell* **2001**, *1*, 63–72. [CrossRef]

42. Kissil, J.L.; Wilker, E.W.; Johnson, K.C.; Eckman, M.S.; Yaffe, M.B.; Jacks, T. Merlin, the product of the *NF2* tumor suppressor gene, is an inhibitor of the p21-activated kinase, Pak1. *Mol. Cell* **2003**, *12*, 841–849. [CrossRef]

43. Xiao, G.H.; Beeser, A.; Chernoff, J.; Testa, J.R. p21-activated kinase links Rac/Cdc42 signaling to merlin. *J. Biol. Chem.* **2002**, *277*, 883–886. [CrossRef] [PubMed]

44. Alfthan, K.; Heiska, L.; Gronholm, M.; Renkema, G.H.; Carpen, O. Cyclic AMP-dependent protein kinase phosphorylates merlin at serine 518 independently of p21-activated kinase and promotes merlin-ezrin heterodimerization. *J. Biol. Chem.* **2004**, *279*, 18559–18566. [CrossRef] [PubMed]

45. Okada, T.; López-Lago, M.; Giancotti, F.G. Merlin/*NF-2* mediates contact inhibition of growth by suppressing recruitment of Rac to the plasma membrane. *J. Cell Biol.* **2005**, *171*, 361–371. [CrossRef] [PubMed]

46. Laulajainen, M.; Muranen, T.; Carpen, O.; Gronholm, M. Protein kinase A-mediated phosphorylation of the NF2 tumor suppressor protein merlin at serine 10 affects the actin cytoskeleton. *Oncogene* **2008**, *27*, 3233–3243. [CrossRef] [PubMed]

47. Tang, X.; Jang, S.W.; Wang, X.; Liu, Z.; Bahr, S.M.; Sun, S.Y.; Brat, D.; Gutmann, D.H.; Ye, K. AKT phosphorylation regulates the tumour-suppressor merlin through ubiquitination and degradation. *Nat. Cell Biol.* **2007**, *9*, 1199–1207. [CrossRef] [PubMed]

48. Yi, C.; Troutman, S.; Fera, D.; Stemmer-Rachamimov, A.; Avila, J.L.; Christian, N.; Persson, N.L.; Shimono, A.; Speicher, D.W.; Marmorstein, R.; et al. A tight junction-associated Merlin-angiomotin complex mediates Merlin's regulation of mitogenic signaling and tumor suppressive functions. *Cancer Cell* **2011**, *19*, 527–540. [CrossRef] [PubMed]

49. Nguyen, R.; Reczek, D.; Bretscher, A. Hierarchy of merlin and ezrin N- and C-terminal domain interactions in homo- and heterotypic associations and their relationship to binding of scaffolding proteins EBP50 and E3KARP. *J. Biol. Chem.* **2001**, *276*, 7621–7629. [CrossRef] [PubMed]

50. Hennigan, R.F.; Foster, L.A.; Chaiken, M.F.; Mani, T.; Gomes, M.M.; Herr, A.B.; Ip, W. Fluorescence resonance energy transfer analysis of merlin conformational changes. *Mol. Cell. Biol.* **2010**, *30*, 54–67. [CrossRef] [PubMed]

51. Gutmann, D.H.; Hirbe, A.C.; Haipek, C.A. Functional analysis of neurofibromatosis 2 (NF2) missense mutations. *Hum. Mol. Genet.* **2001**, *10*, 1519–1529. [CrossRef] [PubMed]

52. Singhi, A.D.; Krasinskas, A.M.; Choudry, H.A.; Bartlett, D.L.; Pingpank, J.F.; Zeh, H.J.; Luvison, A.; Fuhrer, K.; Bahary, N.; Seethala, R.R.; et al. The prognostic significance of BAP1, NF2, and CDKN2A in malignant peritoneal mesothelioma. *Mod. Pathol.* **2016**, *29*, 14–24. [CrossRef] [PubMed]

53. Hanahan, D.; Weinberg, R.A. Hallmarks of cancer: The next generation. *Cell* **2011**, *144*, 646–674. [CrossRef] [PubMed]

54. Bosco, E.E.; Nakai, Y.; Hennigan, R.F.; Ratner, N.; Zheng, Y. NF2-deficient cells depend on the Rac1-canonical Wnt signaling pathway to promote the loss of contact inhibition of proliferation. *Oncogene* **2010**, *29*, 2540–2549. [CrossRef] [PubMed]

55. Gladden, A.B.; Hebert, A.M.; Schneeberger, E.E.; McClatchey, A.I. The NF2 tumor suppressor, Merlin, regulates epidermal development through the establishment of a junctional polarity complex. *Dev. Cell* **2010**, *19*, 727–739. [CrossRef] [PubMed]

56. Gonzalez-Agosti, C.; Xu, L.; Pinney, D.; Beauchamp, R.; Hobbs, W.; Gusella, J.; Ramesh, V. The merlin tumor suppressor localizes preferentially in membrane ruffles. *Oncogene* **1996**, *13*, 1239–1247. [PubMed]

57. Scherer, S.S.; Gutmann, D.H. Expression of the neurofibromatosis 2 tumor suppressor gene product, merlin, in Schwann cells. *J. Neurosci. Res.* **1996**, *46*, 595–605. [CrossRef]

58. Yokoyama, T.; Osada, H.; Murakami, H.; Tatematsu, Y.; Taniguchi, T.; Kondo, Y.; Yatabe, Y.; Hasegawa, Y.; Shimokata, K.; Horio, Y.; et al. YAP1 is involved in mesothelioma development and negatively regulated by Merlin through phosphorylation. *Carcinogenesis* **2008**, *29*, 2139–2146. [CrossRef] [PubMed]

59. Harvey, K.F.; Zhang, X.; Thomas, D.M. The Hippo pathway and human cancer. *Nat. Rev. Cancer* **2013**, *13*, 246–257. [CrossRef] [PubMed]

60. Zanconato, F.; Cordenonsi, M.; Piccolo, S. YAP/TAZ at the Roots of Cancer. *Cancer Cell* **2016**, *29*, 783–803. [CrossRef] [PubMed]

61. Tanaka, I.; Osada, H.; Fujii, M.; Fukatsu, A.; Hida, T.; Horio, Y.; Kondo, Y.; Sato, A.; Hasegawa, Y.; Tsujimura, T.; et al. LIM-domain protein AJUBA suppresses malignant mesothelioma cell proliferation via Hippo signaling cascade. *Oncogene* **2015**, *34*, 73–83. [CrossRef] [PubMed]

62. Murakami, H.; Mizuno, T.; Taniguchi, T.; Fujii, M.; Ishiguro, F.; Fukui, T.; Akatsuka, S.; Horio, Y.; Hida, T.; Kondo, Y.; et al. *LATS2* is a tumor suppressor gene of malignant mesothelioma. *Cancer Res.* **2011**, *71*, 873–883. [CrossRef] [PubMed]

63. Kakiuchi, T.; Takahara, T.; Kasugai, Y.; Arita, K.; Yoshida, N.; Karube, K.; Suguro, M.; Matsuo, K.; Nakanishi, H.; Kiyono, T.; et al. Modeling mesothelioma utilizing human mesothelial cells reveals involvement of phospholipase-C β4 in YAP-active mesothelioma cell proliferation. *Carcinogenesis* **2016**, *37*, 1098–1109. [CrossRef] [PubMed]

64. Mizuno, T.; Murakami, H.; Fujii, M.; Ishiguro, F.; Tanaka, I.; Kondo, Y.; Akatsuka, S.; Toyokuni, S.; Yokoi, K.; Osada, H.; et al. YAP induces malignant mesothelioma cell proliferation by upregulating transcription of cell cycle-promoting genes. *Oncogene* **2012**, *31*, 5117–5122. [CrossRef] [PubMed]

65. Li, W.; You, L.; Cooper, J.; Schiavon, G.; Pepe-Caprio, A.; Zhou, L.; Ishii, R.; Giovannini, M.; Hanemann, C.O.; Long, S.B.; et al. Merlin/NF2 suppresses tumorigenesis by inhibiting the E3 ubiquitin ligase CRL4(DCAF1) in the nucleus. *Cell* **2010**, *140*, 477–490. [CrossRef] [PubMed]

66. Li, W.; Cooper, J.; Zhou, L.; Yang, C.; Erdjument-Bromage, H.; Zagzag, D.; Snuderl, M.; Ladanyi, M.; Hanemann, C.O.; Zhou, P.; et al. Merlin/NF2 loss-driven tumorigenesis linked to CRL4(DCAF1)-mediated inhibition of the Hippo pathway kinases LATS1 and 2 in the nucleus. *Cancer Cell* **2014**, *26*, 48–60. [CrossRef] [PubMed]

67. Cooper, J.; Xu, Q.; Zhou, L.; Pavlovic, M.; Ojeda, V.; Moulick, K.; de Stanchina, E.; Poirier, J.T.; Zauderer, M.; Rudin, C.M.; et al. Combined Inhibition of NEDD8-Activating Enzyme and mTOR Suppresses NF2 Loss-Driven Tumorigenesis. *Mol. Cancer Ther.* **2017**, *16*, 1693–1704. [CrossRef] [PubMed]

68. Hikasa, H.; Sekido, Y.; Suzuki, A. Merlin/NF2-Lin28B-let-7 Is a Tumor-Suppressive Pathway that Is Cell-Density Dependent and Hippo Independent. *Cell Rep.* **2016**, *14*, 2950–2961. [CrossRef] [PubMed]

69. Saxton, R.A.; Sabatini, D.M. mTOR Signaling in Growth, Metabolism, and Disease. *Cell* **2017**, *168*, 960–976. [CrossRef] [PubMed]

70. Albert, V.; Hall, M.N. mTOR signaling in cellular and organismal energetics. *Curr. Opin. Cell Biol.* **2015**, *33*, 55–66. [CrossRef] [PubMed]

71. Yang, H.; Rudge, D.G.; Koos, J.D.; Vaidialingam, B.; Yang, H.J.; Pavletich, N.P. mTOR kinase structure, mechanism and regulation. *Nature* **2013**, *497*, 217–223. [CrossRef] [PubMed]

72. Sato, T.; Akasu, H.; Shimono, W.; Matsu, C.; Fujiwara, Y.; Shibagaki, Y.; Heard, J.J.; Tamanoi, F.; Hattori, S. Rheb protein binds CAD (carbamoyl-phosphate synthetase 2, aspartate transcarbamoylase, and dihydroorotase) protein in a GTP- and effector domain-dependent manner and influences its cellular localization and carbamoyl-phosphate synthetase (CPSase) activity. *J. Biol. Chem.* **2015**, *290*, 1096–1105. [CrossRef] [PubMed]

73. Robitaille, A.M.; Christen, S.; Shimobayashi, M.; Cornu, M.; Fava, L.L.; Moes, S.; Prescianotto-Baschong, C.; Sauer, U.; Jenoe, P.; Hall, M.N. Quantitative phosphoproteomics reveal mTORC1 activates de novo pyrimidine synthesis. *Science* **2013**, *339*, 1320–1323. [CrossRef] [PubMed]

74. Ben-Sahra, I.; Howell, J.J.; Asara, J.M.; Manning, B.D. Stimulation of de novo pyrimidine synthesis by growth signaling through mTOR and S6K1. *Science* **2013**, *339*, 1323–1328. [CrossRef] [PubMed]

75. López -Lago, M.A.; Okada, T.; Murillo, M.M.; Socci, N.; Giancotti, F.G. Loss of the tumor suppressor gene *NF2*, encoding merlin, constitutively activates integrin-dependent mTORC1 signaling. *Mol. Cell. Biol.* **2009**, *29*, 4235–4249. [CrossRef] [PubMed]

76. James, M.F.; Han, S.; Polizzano, C.; Plotkin, S.R.; Manning, B.D.; Stemmer-Rachamimov, A.O.; Gusella, J.F.; Ramesh, V. NF2/merlin is a novel negative regulator of mTOR complex 1, and activation of mTORC1 is associated with meningioma and schwannoma growth. *Mol. Cell. Biol.* **2009**, *29*, 4250–4261. [CrossRef] [PubMed]

77. Guo, Y.; Chirieac, L.R.; Bueno, R.; Pass, H.; Wu, W.; Malinowska, I.A.; Kwiatkowski, D.J. Tsc1-Tp53 loss induces mesothelioma in mice, and evidence for this mechanism in human mesothelioma. *Oncogene* **2014**, *33*, 3151–3160. [CrossRef] [PubMed]

78. Altomare, D.A.; You, H.; Xiao, G.H.; Ramos-Nino, M.E.; Skele, K.L.; De Rienzo, A.; Jhanwar, S.C.; Mossman, B.T.; Kane, A.B.; Testa, J.R. Human and mouse mesotheliomas exhibit elevated AKT/PKB activity, which can be targeted pharmacologically to inhibit tumor cell growth. *Oncogene* **2005**, *24*, 6080–6089. [CrossRef] [PubMed]

79. Suzuki, Y.; Murakami, H.; Kawaguchi, K.; Tanigushi, T.; Fujii, M.; Shinjo, K.; Kondo, Y.; Osada, H.; Shimokata, K.; Horio, Y.; et al. Activation of the PI3K-AKT pathway in human malignant mesothelioma cells. *Mol. Med. Rep.* **2009**, *2*, 181–188. [PubMed]

80. Sato, T.; Nakashima, A.; Guo, L.; Coffman, K.; Tamanoi, F. Single amino-acid changes that confer constitutive activation of mTOR are discovered in human cancer. *Oncogene* **2010**, *29*, 2746–2752. [CrossRef] [PubMed]

81. Grabiner, B.C.; Nardi, V.; Birsoy, K.; Possemato, R.; Shen, K.; Sinha, S.; Jordan, A.; Beck, A.H.; Sabatini, D.M. A diverse array of cancer-associated MTOR mutations are hyperactivating and can predict rapamycin sensitivity. *Cancer Discov.* **2014**, *4*, 554–563. [CrossRef] [PubMed]

82. Rehfeld, F.; Rohde, A.M.; Nguyen, D.T.; Wulczyn, F.G. Lin28 and let-7: Ancient milestones on the road from pluripotency to neurogenesis. *Cell Tissue Res.* **2015**, *359*, 145–160. [CrossRef] [PubMed]

83. Zhou, J.; Ng, S.B.; Chng, W.J. LIN28/LIN28B: An emerging oncogenic driver in cancer stem cells. *Int. J. Biochem. Cell Biol.* **2013**, *45*, 973–978. [CrossRef] [PubMed]

84. Johnson, S.M.; Grosshans, H.; Shingara, J.; Byrom, M.; Jarvis, R.; Cheng, A.; Labourier, E.; Reinert, K.L.; Brown, D.; Slack, F.J. RAS is regulated by the let-7 microRNA family. *Cell* **2005**, *120*, 635–647. [CrossRef] [PubMed]

85. Kumar, M.S.; Lu, J.; Mercer, K.L.; Golub, T.R.; Jacks, T. Impaired microRNA processing enhances cellular transformation and tumorigenesis. *Nat. Genet.* **2007**, *39*, 673–677. [CrossRef] [PubMed]

86. Bouwmeester, T.; Bauch, A.; Ruffner, H.; Angrand, P.O.; Bergamini, G.; Croughton, K.; Cruciat, C.; Eberhard, D.; Gagneur, J.; Ghidelli, S.; et al. A physical and functional map of the human TNF-α/NF-κB signal transduction pathway. *Nat. Cell Biol.* **2004**, *6*, 97–105. [CrossRef] [PubMed]

87. Scudiero, I.; Zotti, T.; Ferravante, A.; Vessichelli, M.; Reale, C.; Masone, M.C.; Leonardi, A.; Vito, P.; Stilo, R. Tumor necrosis factor (TNF) receptor-associated factor 7 is required for TNFα-induced Jun NH2-terminal kinase activation and promotes cell death by regulating polyubiquitination and lysosomal degradation of c-FLIP protein. *J. Biol. Chem.* **2012**, *287*, 6053–6061. [CrossRef] [PubMed]

88. Rippo, M.R.; Moretti, S.; Vescovi, S.; Tomasetti, M.; Orecchia, S.; Amici, G.; Catalano, A.; Procopio, A. FLIP overexpression inhibits death receptor-induced apoptosis in malignant mesothelial cells. *Oncogene* **2004**, *23*, 7753–7760. [CrossRef] [PubMed]

89. Clark, V.E.; Erson-Omay, E.Z.; Serin, A.; Yin, J.; Cotney, J.; Ozduman, K.; Avsar, T.; Li, J.; Murray, P.B.; Henegariu, O.; et al. Genomic analysis of non-NF2 meningiomas reveals mutations in TRAF7, KLF4, AKT1, and SMO. *Science* **2013**, *339*, 1077–1080. [CrossRef] [PubMed]

90. Shapiro, I.M.; Kolev, V.N.; Vidal, C.M.; Kadariya, Y.; Ring, J.E.; Wright, Q.; Weaver, D.T.; Menges, C.; Padval, M.; McClatchey, A.I.; et al. Merlin deficiency predicts FAK inhibitor sensitivity: A synthetic lethal relationship. *Sci. Transl. Med.* **2014**, *6*, 237ra68. [CrossRef] [PubMed]

91. Liu-Chittenden, Y.; Huang, B.; Shim, J.S.; Chen, Q.; Lee, S.J.; Anders, R.A.; Liu, J.O.; Pan, D. Genetic and pharmacological disruption of the TEAD-YAP complex suppresses the oncogenic activity of YAP. *Genes Dev.* **2012**, *26*, 1300–1305. [CrossRef] [PubMed]

92. Zhang, W.Q.; Dai, Y.Y.; Hsu, P.C.; Wang, H.; Cheng, L.; Yang, Y.L.; Wang, Y.C.; Xu, Z.D.; Liu, S.; Chan, G.; et al. Targeting YAP in malignant pleural mesothelioma. *J. Cell. Mol. Med.* **2017**, *21*, 2663–2676. [CrossRef] [PubMed]

93. Tranchant, R.; Quetel, L.; Tallet, A.; Meiller, C.; Renier, A.; de Koning, L.; de Reynies, A.; Le Pimpec-Barthes, F.; Zucman-Rossi, J.; Jaurand, M.C.; et al. Co-occurring Mutations of Tumor Suppressor Genes, *LATS2* and *NF2*, in Malignant Pleural Mesothelioma. *Clin. Cancer Res.* **2017**, *23*, 3191–3202. [CrossRef] [PubMed]

94. Shimada, K.; Skouta, R.; Kaplan, A.; Yang, W.S.; Hayano, M.; Dixon, S.J.; Brown, L.M.; Valenzuela, C.A.; Wolpaw, A.J.; Stockwell, B.R. Global survey of cell death mechanisms reveals metabolic regulation of ferroptosis. *Nat. Chem. Biol.* **2016**, *12*, 497–503. [CrossRef] [PubMed]

95. Song, S.; Xie, M.; Scott, A.W.; Jin, J.; Ma, L.; Dong, X.; Skinner, H.D.; Johnson, R.L.; Ding, S.; Ajani, J.A. A Novel YAP1 Inhibitor Targets CSC-Enriched Radiation-Resistant Cells and Exerts Strong Antitumor Activity in Esophageal Adenocarcinoma. *Mol. Cancer Ther.* **2018**, *17*, 443–454. [CrossRef] [PubMed]

96. Lin, K.C.; Moroishi, T.; Meng, Z.; Jeong, H.S.; Plouffe, S.W.; Sekido, Y.; Han, J.; Park, H.W.; Guan, K.L. Regulation of Hippo pathway transcription factor TEAD by p38 MAPK-induced cytoplasmic translocation. *Nat. Cell Biol.* **2017**, *19*, 996–1002. [CrossRef] [PubMed]

97. Ou, S.H.; Moon, J.; Garland, L.L.; Mack, P.C.; Testa, J.R.; Tsao, A.S.; Wozniak, A.J.; Gandara, D.R. SWOG S0722: Phase II study of mTOR inhibitor everolimus (RAD001) in advanced malignant pleural mesothelioma (MPM). *J. Thorac. Oncol.* **2015**, *10*, 387–391. [CrossRef] [PubMed]

98. Thoreen, C.C.; Kang, S.A.; Chang, J.W.; Liu, Q.; Zhang, J.; Gao, Y.; Reichling, L.J.; Sim, T.; Sabatini, D.M.; Gray, N.S. An ATP-competitive mammalian target of rapamycin inhibitor reveals rapamycin-resistant functions of mTORC1. *J. Biol. Chem.* **2009**, *284*, 8023–8032. [CrossRef] [PubMed]

99. Agarwal, V.; Campbell, A.; Beaumont, K.L.; Cawkwell, L.; Lind, M.J. PTEN protein expression in malignant pleural mesothelioma. *Tumour Biol.* **2013**, *34*, 847–851. [CrossRef] [PubMed]

100. Goel, S.; Wang, Q.; Watt, A.C.; Tolaney, S.M.; Dillon, D.A.; Li, W.; Ramm, S.; Palmer, A.C.; Yuzugullu, H.; Varadan, V.; et al. Overcoming Therapeutic Resistance in HER2-Positive Breast Cancers with CDK4/6 Inhibitors. *Cancer Cell* **2016**, *29*, 255–269. [CrossRef] [PubMed]

101. Franco, J.; Balaji, U.; Freinkman, E.; Witkiewicz, A.K.; Knudsen, E.S. Metabolic Reprogramming of Pancreatic Cancer Mediated by CDK4/6 Inhibition Elicits Unique Vulnerabilities. *Cell Rep.* **2016**, *14*, 979–990. [CrossRef] [PubMed]

102. Bonelli, M.A.; Digiacomo, G.; Fumarola, C.; Alfieri, R.; Quaini, F.; Falco, A.; Madeddu, D.; La Monica, S.; Cretella, D.; Ravelli, A.; et al. Combined Inhibition of CDK4/6 and PI3K/AKT/mTOR Pathways Induces a Synergistic Anti-Tumor Effect in Malignant Pleural Mesothelioma Cells. *Neoplasia* **2017**, *19*, 637–648. [CrossRef] [PubMed]

103. Rodrik-Outmezguine, V.S.; Okaniwa, M.; Yao, Z.; Novotny, C.J.; McWhirter, C.; Banaji, A.; Won, H.; Wong, W.; Berger, M.; de Stanchina, E.; et al. Overcoming mTOR resistance mutations with a new-generation mTOR inhibitor. *Nature* **2016**, *534*, 272–276. [CrossRef] [PubMed]

104. Fan, Q.; Aksoy, O.; Wong, R.A.; Ilkhanizadeh, S.; Novotny, C.J.; Gustafson, W.C.; Truong, A.Y.; Cayanan, G.; Simonds, E.F.; Haas-Kogan, D.; et al. A Kinase Inhibitor Targeted to mTORC1 Drives Regression in Glioblastoma. *Cancer Cell* **2017**, *31*, 424–435. [CrossRef] [PubMed]

105. Demierre, M.F.; Higgins, P.D.; Gruber, S.B.; Hawk, E.; Lippman, S.M. Statins and cancer prevention. *Nat. Rev. Cancer* **2005**, *5*, 930–942. [CrossRef] [PubMed]

106. Rubins, J.B.; Greatens, T.; Kratzke, R.A.; Tan, A.T.; Polunovsky, V.A.; Bitterman, P. Lovastatin induces apoptosis in malignant mesothelioma cells. *Am. J. Respir. Crit. Care Med.* **1998**, *157*, 1616–1622. [CrossRef] [PubMed]

107. Asakura, K.; Izumi, Y.; Yamamoto, M.; Yamauchi, Y.; Kawai, K.; Serizawa, A.; Mizushima, T.; Ohmura, M.; Kawamura, M.; Wakui, M.; et al. The cytostatic effects of lovastatin on ACC-MESO-1 cells. *J. Surg. Res.* **2011**, *170*, e197–e209. [CrossRef] [PubMed]

108. Yamauchi, Y.; Izumi, Y.; Asakura, K.; Fukutomi, T.; Serizawa, A.; Kawai, K.; Wakui, M.; Suematsu, M.; Nomori, H. Lovastatin and valproic acid additively attenuate cell invasion in ACC-MESO-1 cells. *Biochem. Biophys. Res. Commun.* **2011**, *410*, 328–332. [CrossRef] [PubMed]

109. Tuerdi, G.; Ichinomiya, S.; Sato, H.; Siddig, S.; Suwa, E.; Iwata, H.; Yano, T.; Ueno, K. Synergistic effect of combined treatment with gamma-tocotrienol and statin on human malignant mesothelioma cells. *Cancer Lett.* **2013**, *339*, 116–127. [CrossRef] [PubMed]

110. Hwang, K.E.; Kim, Y.S.; Hwang, Y.R.; Kwon, S.J.; Park, D.S.; Cha, B.K.; Kim, B.R.; Yoon, K.H.; Jeong, E.T.; Kim, H.R. Enhanced apoptosis by pemetrexed and simvastatin in malignant mesothelioma and lung cancer cells by reactive oxygen species-dependent mitochondrial dysfunction and Bim induction. *Int. J. Oncol.* **2014**, *45*, 1769–1777. [CrossRef] [PubMed]

111. Tanaka, K.; Osada, H.; Murakami-Tonami, Y.; Horio, Y.; Hida, T.; Sekido, Y. Statin suppresses Hippo pathway-inactivated malignant mesothelioma cells and blocks the YAP/CD44 growth stimulatory axis. *Cancer Lett.* **2017**, *385*, 215–224. [CrossRef] [PubMed]

112. Wang, Z.; Wu, Y.; Wang, H.; Zhang, Y.; Mei, L.; Fang, X.; Zhang, X.; Zhang, F.; Chen, H.; Liu, Y.; et al. Interplay of mevalonate and Hippo pathways regulates RHAMM transcription via YAP to modulate breast cancer cell motility. *Proc. Natl. Acad. Sci. USA* **2014**, *111*, E89–E98. [CrossRef] [PubMed]

113. Sorrentino, G.; Ruggeri, N.; Specchia, V.; Cordenonsi, M.; Mano, M.; Dupont, S.; Manfrin, A.; Ingallina, E.; Sommaggio, R.; Piazza, S.; et al. Metabolic control of YAP and TAZ by the mevalonate pathway. *Nat. Cell Biol.* **2014**, *16*, 357–366. [CrossRef] [PubMed]

114. Guerrant, W.; Kota, S.; Troutman, S.; Mandati, V.; Fallahi, M.; Stemmer-Rachamimov, A.; Kissil, J.L. YAP Mediates Tumorigenesis in Neurofibromatosis Type 2 by Promoting Cell Survival and Proliferation through a COX-2-EGFR Signaling Axis. *Cancer Res.* **2016**, *76*, 3507–3519. [CrossRef] [PubMed]

115. Wahle, B.M.; Hawley, E.T.; He, Y.; Smith, A.E.; Yuan, J.; Masters, A.R.; Jones, D.R.; Gehlhausen, J.R.; Park, S.J.; Conway, S.J.; et al. Chemopreventative celecoxib fails to prevent schwannoma formation or sensorineural hearing loss in genetically engineered murine model of neurofibromatosis type 2. *Oncotarget* **2018**, *9*, 718–725. [CrossRef] [PubMed]

116. Yap, T.A.; Aerts, J.G.; Popat, S.; Fennell, D.A. Novel insights into mesothelioma biology and implications for therapy. *Nat. Rev. Cancer* **2017**, *17*, 475–488. [CrossRef] [PubMed]

117. Guo, X.E.; Ngo, B.; Modrek, A.S.; Lee, W.H. Targeting tumor suppressor networks for cancer therapeutics. *Curr. Drug Targets* **2014**, *15*, 2–16. [CrossRef] [PubMed]

118. Cooper, J.; Giancotti, F.G. Molecular insights into NF2/Merlin tumor suppressor function. *FEBS Lett.* **2014**, *588*, 2743–2752. [CrossRef] [PubMed]

119. McCambridge, A.J.; Napolitano, A.; Mansfield, A.S.; Fennell, D.A.; Sekido, Y.; Nowak, A.K.; Reungwetwattana, T.; Mao, W.; Pass, H.I.; Carbone, M.; et al. State of the art: Advances in Malignant Pleural Mesothelioma in 2017. *J. Thorac. Oncol.* **2018**, in press. [CrossRef] [PubMed]

120. Shalem, O.; Sanjana, N.E.; Hartenian, E.; Shi, X.; Scott, D.A.; Mikkelson, T.; Heckl, D.; Ebert, B.L.; Root, D.E.; Doench, J.G.; et al. Genome-scale CRISPR-Cas9 knockout screening in human cells. *Science* **2014**, *343*, 84–87. [CrossRef] [PubMed]

International Journal of
Molecular Sciences

MDPI

Review

Heterogeneity in Immune Cell Content in Malignant Pleural Mesothelioma

Jorien Minnema-Luiting, Heleen Vroman, Joachim Aerts and Robin Cornelissen *

Erasmus MC Cancer Institute, Department of Pulmonary Medicine, 's-Gravendijkwal 230, 3015 CE Rotterdam,
The Netherlands; j.minnema-luiting@erasmusmc.nl (J.M.-L.); h.vroman@erasmusmc.nl (H.V.);
j.aerts@erasmusmc.nl (J.A.)
* Correspondence: r.cornelissen@erasmusmc.nl; Tel.: +31-107-040-704

Received: 27 February 2018; Accepted: 22 March 2018; Published: 30 March 2018

Abstract: Malignant pleural mesothelioma (MPM) is a highly aggressive cancer with limited therapy options and dismal prognosis. In recent years, the role of immune cells within the tumor microenvironment (TME) has become a major area of interest. In this review, we discuss the current knowledge of heterogeneity in immune cell content and checkpoint expression in MPM in relation to prognosis and prediction of treatment efficacy. Generally, immune-suppressive cells such as M2 macrophages, myeloid-derived suppressor cells and regulatory T cells are present within the TME, with extensive heterogeneity in cell numbers. Infiltration of effector cells such as cytotoxic T cells, natural killer cells and T helper cells is commonly found, also with substantial patient to patient heterogeneity. PD-L1 expression also varied greatly (16–65%). The infiltration of immune cells in tumor and associated stroma holds key prognostic and predictive implications. As such, there is a strong rationale for thoroughly mapping the TME to better target therapy in mesothelioma. Researchers should be aware of the extensive possibilities that exist for a tumor to evade the cytotoxic killing from the immune system. Therefore, no "one size fits all" treatment is likely to be found and focus should lie on the heterogeneity of the tumors and TME.

Keywords: malignant pleural mesothelioma (MPM); tumor microenvironment (TME); heterogeneity; immunotherapy; myeloid-derived suppressor cells (MDSCs); tumor-associated macrophages (TAMs); tumor-infiltrating lymphocytes (TIL); regulatory T cells (Tregs)

1. Introduction

Malignant pleural mesothelioma (MPM) is a rare and highly aggressive cancer arising from the mesothelial cells of the pleura with a median survival of 9 months. More than 70 percent of MPM results from exposure of asbestos [1]. The only licensed treatment is palliative antifolate and platinum combination chemotherapy which results in a moderate overall survival benefit of about three months [2]. MPM consists of three histological variants: (1) epithelioid (~60% of mesotheliomas); (2) sarcomatoid, characterized by spindle cell morphology (~20% of mesotheliomas); (3) biphasic, a mixture of epithelioid and sarcomatoid characteristics (~20% of mesotheliomas) [3,4]. Currently, accepted prognosticators include stage and histology of which sarcomatoid subtype results in the lowest survival rates [5]. It has been demonstrated that protumor and antitumor immune responses within the tumor and associated stroma also correlate with the clinical outcome of MPM [6,7]. This review discusses current knowledge of heterogeneity in immune cell content in MPM in relation to prognosis and prediction of treatment efficacy.

2. Tumor Microenvironment (TME) in Mesothelioma

The mesothelioma tumor microenvironment (TME) is a complex and heterogeneous mixture of stromal, endothelial and immune cells. This composition differs between individuals and histologic types, and can change upon administered anti-tumor therapies [8]. The role of immune cells within the TME has become a major area of interest, as these immune cells are capable of influencing tumor growth. In general, immune infiltration in tumors include natural killer (NK) cells, B and T lymphocytes, mast cells, neutrophils, myeloid-derived suppressor cells (MDSCs), macrophages and dendritic cells (DCs). NK cells, cytotoxic T cells, mature DCs and T helper cells are known to be anti-tumorgenic, while others, like regulatory T cells (Tregs), type 2 macrophages, and MDSCs suppress the immune response and therefore favor tumor growth and dissemination [9]. The TME in mesothelioma is unique as it arises from exposure of mesothelial cells to asbestos fibers [6,8]. It is known to be highly immunosuppressive, with higher numbers of immunosuppressive cells such as type 2 tumor associated macrophages and Tregs [10–13].

3. Macrophages

Macrophages are specialized phagocytic cells which play a dual role in cancer depending on their differentiation. Schematically, tumor-associated macrophages (TAMs) can be divided into classically activated (M1) macrophages and alternatively activated (M2) macrophages. M1 macrophages have pro-inflammatory, tissue destructive and anti-tumor activity. Whereas M2 macrophages can be seen as pro-tumorgenic by promoting the metastatic capacity of a tumor due to production of multiple cytokines (e.g., IL-1, IL-6, IL-10, VEGF and TGF-β). TAMs derive from circulating monocytic precursors. In tumors, chemokines play an important role in recruitment of monocytes. Once recruited, interleukins such as IL-4, IL-13 and IL-10 produced by tumor infiltrating lymphocytes (TILs) promote differentiation of macrophages towards an M2 phenotype [14,15]. Certain drugs can skew M2 macrophages into a more M1 phenotype [16,17]. Table 1 describes the antibodies and their associated immune cells. Burt et al. performed a CD68 staining on tissue microarray of 52 MPM patients. Macrophages were abundantly present in both epithelial (n = 34) and non-epithelial (n = 18) mesothelioma (tumor infiltrating macrophages in percentage of tumor area (%) 25.2 $\pm$ 9.3 and 29.7 $\pm$ 10.2, p = 0.11). The relatively high standard deviation indicates large heterogeneity in macrophage infiltration in MPM. In seven patients, three with epithelial and four with non-epithelial MPM, flow cytometry was performed displaying high levels of CD163 and CD206, characterizing them as M2 macrophages. The absolute number of macrophages was associated with worse prognosis in non-epithelioid mesothelioma after surgery, but not in epithelioid mesotheliomas [18]. Cornelissen et al. described expression of CD68 and CD163 in tumor specimens of sixteen patients with epithelial MPM, eight of them receiving induction chemotherapy and surgery and eight patients receiving chemotherapy only. In both groups macrophages were abundantly present, whereby a large spreading in actual number of macrophages was seen (surgery vs. non-surgery 211.3/0.025 cm^2 $\pm$ 80.2 and 213.9/0.025 cm^2 $\pm$ 100.4 p = 1.0). Most of these macrophages showed a M2 phenotype. A higher percentage of M2 macrophages was significantly negative correlated with overall survival [19]. In lung cancer, Cornelissen et al. described ten MPM patients with local tumor outgrowth after surgery and their matched controls without local tumor outgrowth. Two biphasic and eighteen epithelial MPM patients were included. Macrophage infiltration was characterized by large heterogeneity with a mean macrophage count of 202/0.025 cm^2, ranging from 45 to 408/0.025 cm^2. These macrophages show a M2 phenotype with a mean count of 153/0.025 cm^2 and a range of 42 to 422/0.025 cm^2 [20]. Marcq et al. found macrophages in stroma of all 54 MPM specimens, with a majority of samples having less than 50% CD68+ cells. Numbers of stromal macrophages were positively correlated to the number of stromal Tregs (R = 0.41, p = 0.002), suggesting that macrophages stimulate and recruit CD4+ cells by affecting the adaptive immune response [15,21]. Schürch et al. found heavy infiltration of M2 macrophages in all 40 MPM analyzed [22]. Table 2 summarizes the extent of macrophage infiltration found in these studies.

Table 1. Cell surface markers and correlating immune cell type.

Surface Marker	Present on
CD3	T lymphocytes
CD4	T helper cells
CD8	Cytotoxic T cells
CD11b	Monocytes, macrophages, MDSCs, NK cells, eosinophils, neutrophils, basophils, dendritic cells, mast cells, CD8+ T cells, B cells
CD16	Natural killer cells, myeloid cells, monocytes, neutrophils
CD19	B cells
CD20	B cells
CD33	Myeloid cells
CD45	Leucocytes
CD45RO	T effector and memory cells
CD56	Natural killer cells
CD68	Macrophages
CD163	M2 macrophages
CD206	M2 macrophages
Foxp3+CD4+CD25+	Regulatory T cells

Table 2. Infiltration of TAMs and M2 macrophages in mesothelioma.

Study	n	CD68+	Coefficient of Variation (CV) *	CD163+	Coefficient of Variation (CV) *
[18]	52	25.2 ± 9.3%, (epithelial) 29.7 ± 10.2%, (non-epithelial)	0.37, (epithelial) 0.34, (non-epithelial)	n.a. ***	n.a. ***
[19]	16	211.3 ** ± 80.2, (surgery) 213.9 ** ± 100.4, (non-surgery)	0.37, (surgery) 0.47, (non-surgery)	168.3 ** ± 80.2, (surgery) 164.1 ** ± 82.5, (non-surgery)	0.48, (surgery) 0.50, (non-surgery)
[20]	20	202 **	Range 45–408	153 **	Range 42–422
[21]	54	Present in all specimens	n.a. ***	n.a. ***	n.a. ***
[22]	40	Heavy infiltration	n.a. ***	Heavy infiltration	n.a. ***

* CV is defined as the ratio of the standard deviation to the mean; ** Cell count per field; *** n.a is not applicable.

4. Myeloid-Derived Suppressor Cells

Myeloid cells are abundantly present in stroma of MPM [8]. MDSCs are immature myeloid cells with immune suppressive capacities. MDSCs are generally characterized by being positive for CD33 and CD11b and low or negative for HLA-DR. They induce Tregs and produce nitric oxide and arginase, leading to loss of function of CD4+ and CD8+ T cells. These strongly immunosuppressive characteristics promote immune escape, tumor growth, invasion and angiogenesis [8,18]. Immune suppression by MDSCs was found to be one of the main factors for immunotherapy insufficiency [10]. MDSCs are induced by several tumor-derived factors, e.g., prostaglandins. Celecoxib reduces prostaglandin levels. Veltman et al. found celecoxib to improve dendritic cell-based immunotherapy by reducing numbers of MDSCs and suppressing function [10]. In mice, MDSCs are defined by IL-4Rα expression [23]. Burt et al. found IL-4Rα to be highly expressed on tumor cells of 52 MPM specimens, with presence of IL-4Rα in 97% of epithelial and 95% of non-epithelial tumors. Only a scattered and small fraction of stromal cells stained positive for IL-4Rα, conversely macrophages were predominantly found in stroma [24]. In another study of Burt et al., flow cytometry was performed on mononuclear cell suspensions from seven MPM patients; these macrophages displayed high levels of IL-4Rα [18].

Awad et al. found myeloid cells (CD33) to represent approximately 42% of CD45+ immune cells (range 5.7–86.1%); 0.6–31% of these myeloid cells were typed as MDSCs [25].

5. T Cells and Natural Killer Cells in Mesothelioma

TILs play an important role in the immune defense in cancer. They recognize tumor-specific antigens presented on HLA-1, to then kill the tumor cells via production of perforins and granzymes. In many cancers, T cell infiltration is associated with a good prognosis [26–28]. T helper CD4+ cells play an important role in the generation of a T cell-mediated antitumor response, via stimulation of CD8+ TILs and NK cells and via activation of antigen-presenting cells (APCs) [29–31]. NK cells are lymphoid cells of the innate immune system with strong immunostimulatory effector functions and efficient cytotoxic capacity [32]. In 1982, Leigh et al. were the first to describe a relation between presence of significant lymphoid infiltration and prolonged survival in 58 mesothelioma patients. Tumors were found without, with insignificant and with significant lymphocyte inflammation. Due to absence of modern immunohistochemical agents, no lymphocyte subsets could be identified [33]. Mudhar et al. performed immunohistochemical staining on fifteen cases of epithelioid MPM, scoring CD45, CD3, CD20 and CD56 with 0 (no significant infiltrate), 1 (non-brisk) or 2 (brisk infiltrate). In one patient, none of these immune cells were present. Specimens demonstrated some heterogeneity in numbers of T lymphocyte and NK cells. With brisk infiltration of T-lymphocytes and NK cells in one case, non-brisk infiltration in eleven and ten cases, respectively. The other three and four specimens showed absence of T lymphocytes and NK cells, respectively. No B cells were present in any specimens. No relation was found between the infiltration of immune cells and survival [34]. A comprehensive analysis by Hegmans et al. demonstrated leukocyte infiltration in all four MPM patients. Most inflammatory cells were identified as macrophages and NK cells (CD16). Some heterogeneity was noted. Eosinophils, mast cells, B cells and neutrophils were rarely detected. DCs were not found in the biopsies [35]. Immunohistochemical analysis of T cells of 32 extrapleural pneumonectomy specimens after induction chemotherapy was performed by Anraku et al. Results are summarized in Table 3. The distribution of T cells varies, with only CD3+ and CD45RO+ TILs showing normal distribution. The coefficient of variation ranges from 0.49 to 0.87, implying substantial heterogeneity. In multivariate data analyses, presence of CD8+ TILs was associated with better prognosis [36].

Table 3. Infiltration of T cell subtypes in 32 extrapeural pneumonectomy specimens.

Surface Marker	Mean (Cell Count per Field)	Standard Deviation	Coefficient of Variation (CV) *
CD3+	232.16	114.1	0.49
CD4+	119.9	94.2	0.79
CD8+	73.1	40.2	0.55
CD25+	17.5	12.6	0.72
FOXP3+	21.8	19.0	0.87
CD45RO+	115.7	56.2	0.49

* CV is defined as the ratio of the standard deviation to the mean.

Yamada et al. [37] analyzed presence of TILs and NK cells in 44 MPM cases, comprised of 26 epithelioid, fourteen biphasic and four sarcomatoid mesotheliomas. Results of T cell subtype counts are presented in Table 4. Again, the heterogeneity is substantial, indicated by wide ranges and CVs ranging from 0.82 to 1.54. Presence of CD4+ and CD8+ T cells was strongly correlated ($R = 0.74$, and $p = 0.001$). In multivariate data analysis high CD8+ TILs and epithelioid histology were independent favorable prognostic factors [37].

Awad et al. [25] performed flow cytometry with various leukocyte markers on 38 malignant mesothelioma, with all histologies. They found considerable variability in immune cell infiltration across tumors. Numbers of CD45+ leukocytes were increased in non-epithelioid mesothelioma compared to epithelioid mesothelioma (median 91.4% vs. 64.1%). Amount of T cells ranged from 5.2% to 81.2% of CD45+ cells, with a higher fraction of T cells in non-epithelioid mesothelioma.

There was considerable variability in numbers of leukocytes and in immune cell composition across cases [25]. Marcq et al. [21] found lymphocytic infiltration in all 54 tested mesotheliomas, ranging from 20% to 80% of stromal cells. The fourteen chemotherapy pretreated samples showed higher numbers of lymphocytes. CD8+ TILs were the predominant cell type of the immune infiltrate and were present in all samples. In 70% of the untreated and 57% of the pretreated samples, the majority of the lymphocytes were CD8+ TILs. High expression of CD45RO on stromal lymphocytes was associated with worse response to chemotherapy. T helper cells were found in 85% of untreated and 100% of pretreated samples. T helper cells in lymphoid infiltrates were associated with better survival in multivariate analysis [21]. Suzuki et al. [38] evaluated inflammatory responses in tumor and stroma of 175 chemotherapy naive epithelioid MPM specimens with H&E-stained slides. Acute response was represented by presence of neutrophils, while chronic inflammation was represented by lymphocytes and plasma cells. Acute inflammatory reaction was sparse in tumors and stroma, with high scores (>1% of total area) in 18% of specimens. The chronic reaction was more heterogenic, with high scores (>50% of total area) in 37% of tumors and 34% of stromal tissue. In multivariate analysis, chronic inflammation in stroma was an independent predictor of survival while other inflammatory responses were not significantly correlated with survival [38]. These studies suggest considerable infiltration of TILs in mesothelioma. Higher levels of TILs are associated with better survival in most studies.

Table 4. Infiltration of T cell subtypes in 44 MPM cases [37].

Surface Marker	Mean (Cell Count per Field)	Standard Deviation	Coefficient of Variation (CV) *	Range	Median
CD4+	51.1	41.8	0.82	0.2–159.7	37.3
CD8+	103.3	106.9	1.03	8.8–547.5	64.5
CD56+	5.4	8.3	1.54	0.0–41.8	1.8

* CV is defined as the ratio of the standard deviation to the mean.

6. Regulatory T Cells (Tregs)

FOXP3+CD25+CD4+ regulatory T cells maintain self-tolerance and prevent autoimmune disease. They are abundantly present in tumors, where they suppress activation and proliferation of effector T cells. High numbers of Tregs are associated with poor prognosis in many cancers [39]. Hegmans et al. demonstrated that human mesothelioma biopsies harbor significant numbers of Tregs at the rim of the tumor [35]. Marcq et al. [21] found Tregs to be present in 72% of samples, both chemotherapy pretreated and untreated. Lower numbers of Tregs were seen in samples pretreated with cisplatin and pemetrexed [21]. DeLong et al. performed flow cytometry on malignant pleural effusions from seven patients with mesothelioma; 7.8% ± 6.8% of T-lymphocytes were functionally suppressive CD4+CD25+ cells, which might be Tregs. This is a significant lower number of Tregs than seen in malignant effusions secondary to breast cancer or NSCLC. Some heterogeneity was noted, including two patients with <3% CD4+CD25+ T cells and one patient with 21% CD4+CD25+ T cells in pleural effusion. The latter was a sarcomatoid subtype [40].

7. B Lymphocytes

B lymphocytes contribute to humoral immunity as they can differentiate into antibody-secreting plasma cells. Also, B cells can stimulate T cells or serve as APCs. In several cancers, including mesothelioma, B lymphocyte infiltration is associated with better patient survival [41]. Two studies found low numbers of B lymphocytes (CD20) in mesothelioma [34,35]. A third study found low B lymphocyte (CD19) infiltration (median 3% of CD45+ cells), although some outliers with B cell infiltration up to 51.8% of CD45+ cells were seen [26]. Patil et al. [42] classified three molecular subgroups based on immune profiles; in one subgroup high numbers of B cells were found [42]. Generally, B cell infiltration in mesothelioma is sparse, although a subgroup with higher numbers of B cells is described. More research is needed for determining the clinical implications.

8. Cancer-Associated Fibroblasts (CAFs)

The major component of the TME are cancer-associated fibroblasts, also known as tumor-associated fibroblasts [43]. MPM recruit and activate CAFs by secreting fibroblast growth factor-2 (FGF-2) and platelet-derived growth factor-AA (PDGF-AA) [44]. CAFs can contribute to tumor growth by inhibiting cytotoxic T cell influx and by secreting several growth factors such as hepatocyte growth factor, thereby inducing angiogenesis [44,45]. In 1996 Harvey et al. demonstrated infiltration of CAFs in six of eight MPM samples [46]. Li et al. performed histological analyses on specimens from 51 MPM patients and revealed considerable CAF infiltration [44].

9. PD-L1 Expression and Other Immune Checkpoints

Programmed cell death 1 (PD-1) is an immune checkpoint receptor present on activated T cells. PD-1 and its ligands, PD-L1 and PD-L2, which are expressed by tumor cells and/or stromal cells share immunosuppressive capacities [47]. In several tumors, including NSCLC, PD-L1 enrichment is associated with higher response rates to PD-1 and PD-L1-blocking antibodies [47–49]. However, responses have also been observed in PD-L1-negative patients [50]. We found eight studies evaluating PD-L1 expression in mesothelioma. A summary of the results is displayed in Table 5. PD-L1 was found to be expressed in 16% to 65% of malignant mesothelioma. PD-L1 expression is higher in non-epithelioid mesothelioma compared to epithelioid mesothelioma (37.5–97.4% vs. 6.7–31%) [21,25,42,50–54]. Several studies found higher PD-L1 expression to be an independent prognostic indicator for worse overall survival in multivariate data analysis [50,52,53,55]. Khanna et al. [54] analyzed PD-L1 expression in peritoneal and pleural fluid of respectively six and three mesothelioma patients. PD-L1 expression was found in all samples, varying from 12% to 83%. Immune cells were evaluated for PD-1 expression in seven samples. PD-1 was expressed in 21.8% of CD4+ cells and 37.5% of CD8+ cells. Together, these data suggest that malignant effusions of mesothelioma patients have high PD-L1 expression on tumor cells as well as PD-L1 and PD-1 on infiltrating immune cells [54]. Staining for other checkpoint inhibitors such as TIM-3 and LAG-3 was performed by Marcq et al. [21] TIM-3 expression was found in 36 of 54 samples (both treated and untreated). LAG-3 expression was absent in all 54 MPM samples, pointing out the possible opportunities of TIM-3 as a promising immunotherapy target in mesothelioma. In multivariate analysis, TIM-3 expression in lymphoid aggregates was a prognosticator for better survival [21].

Table 5. PD-L1 expression in mesothelioma.

Study	PD-L1 Antibody	n	Positivity (%)	PD-L1 Positive (n (%))	PD-L1 Positive in Epithelioid (n (%))	PD-L1 Positive in Non-Epithelioid (n (%))	Survival in PD-L1+ (Months)	Survival in PD-L1- (Months)	p Value
[9]	5H1-A3	106	≥5	42 (40)	14/68 (21)	37/38 (97)	5	14.5	<0.0001
[1]	E1L3N	77	>1%	16 (21)	7/53 (13)	9/24 (38)	4.8	16.3	0.012
[20]	E1L3N	39	≥1%	18 (46)	8/26 (31)	10/13 (77)	shorter	longer	0.15
[30]	E1L3N	58	≥1%	17 (29)	8/34 (24)	9/24 (38)	n.a.*	n.a.*	n.a.*
[49]	SP142	58	≥1%	10 (17)	4/34 (12)	6/24 (25)	4	13	0.016
[14]	rabbit	65	≥5%	41 (63)	n.a.*	n.a.*	23.0	33.3	0.35
[21]	SP142	54	≥1%	35 (65)	n.a.*	More in sarcomatoid	n.a.*	n.a.*	n.a.*
[32]	E1L3N	175	≥5%	57 (33)	46/148 (31)	11/27 (41)	6	18	<0.01
[42]	SP142	99	>1%	16 (16)	5/75 (6.7)	9/24 (38)	shorter	longer	

* n.a is not applicable.

10. Discussion

We performed a comprehensive literature search focusing on the heterogeneity of immune cell infiltration, PD-L1 expression and other immune checkpoints in MPM. The composition of TME holds therapeutic and prognostic implications [6,7]. Stage and histology are currently accepted prognostic indicators [5], but evidence is accumulation that infiltrating immune cells and expression of immune checkpoints are of high prognostic value in MPM [7,50–53]. Infiltration of M2 macrophages seems to be associated with worse prognoses [18–20], as is PD-L1 expression [50–53]. Infiltration of cytotoxic T cells was associated with better prognosis in MPM in most studies [21,33,36–38].

TME composition differs between various histologic subtypes and individuals [25]. Macrophages are found to be abundantly present in all MPM, although the level of infiltration can vary significantly. Macrophages generally show an M2 phenotype [18–22,24]. Stroma of MPM is infiltrated by MDSCs [8,18,25]. Leukocyte infiltration was found in almost all mesothelioma, with higher numbers of leukocytes in non-epithelioid mesothelioma [25]. T cell subsets showed considerable heterogeneity with wide ranges and high coefficients of variation across all studies. Cytotoxic T cells, NK cells and T helper cells were most abundantly present [21,33–38]. B cell infiltration is sparse, although a (molecular) subgroup with an increased number B cells is described [25,34,35,42]. Significant numbers of Tregs were found in biopsies and pleural fluid of mesothelioma [21,35,40]. Tumor growth promoting CAFS are found in TME of most MPM [44,46]. PD-L1 expression is commonly found in MPM, with higher expression in non-epithelioid histologic subtypes [21,25,42,50–54].

Altogether, substantial heterogeneity in immune cell content in mesothelioma was found. MPM are highly infiltrated by immune effector cells, but also immune suppressive cells such as Tregs and M2 macrophages and PD-L1 expression are found. Apparently, the tumor finds several ways to bypass the immune system. Thoroughly mapping the composition of the TME is rational in targeting therapy in mesothelioma. For example, tumors with high amounts of T effector cells and Tregs might benefit from a combination of immunotherapy and drugs that control Tregs, to invigorate immunotherapy efficacy. Tumors highly infiltrated by MDSCs might benefit more from (dendritic cell-based) immunotherapy when this is combined with celecoxib, as this reduces the suppressive function and number of MDSCs [10]. In MPM expressing PD-L1 and cytotoxic T cells present in TME, treatment with PD-(L)1 inhibitors is more rational. Other rational treatment options include nintedanib or emactuzumab for the skewing of M2 macrophages to the M1 subtype in TME highly infiltrated with M2 macrophages [16,17], or OX40 for the stimulation of cytotoxic T cells when they are not already present in the TME [56]. Inhibition of the cytokines FGF-2, PDGF-AA, and HGF may be appropriate in MPM infiltrated with CAFs [44]. This opens up a whole new era of personalized immunotherapy in which we are just scratching the surface. Researchers should be aware of the extensive possibilities that exist for a tumor to evade the cytotoxic killing by the immune system. Therefore, no "one size fits all" treatment is likely to be found and focus should lie on the heterogeneity of the tumors and TME.

Author Contributions: Jorien Minnema-Luiting wrote the paper, contributed to the conception of the work, interpreted of data, drafted the work and has approved the submitted version. Heleen Vroman, Joachim Aerts and Robin Cornelissen contributed to the conception of the work, interpretation of data, have drafted the work and substantively revised it and have approved the submitted version.

Conflicts of Interest: Jorien Minnema-Luiting and Heleen Vroman declare no conflict of interest. Joachim Aerts: Speakers fee and consultancy Eli-Lilly, Boehringer Ingelheim, MSD, BMS, Astra Zeneca, Amphera, Roche. Stock owner Amphera b.v. Robin Cornelissen: Consultancy Roche, Boehringer Ingelheim. Speakers fee Roche, Pfizer, Boehringer Ingelheim, Novartis.

References

1. Carbone, M.; Ly, B.H.; Dodson, R.F.; Pagano, I.; Morris, P.T.; Dogan, U.A.; Gazdar, A.F.; Pass, H.I.; Yang, H. Malignant mesothelioma: Facts, myths, and hypotheses. *J. Cell. Physiol.* **2012**, *227*, 44–58. [CrossRef] [PubMed]

2. Vogelzang, N.J.; Rusthoven, J.J.; Symanowski, J.; Denham, C.; Kaukel, E.; Ruffie, P.; Gatzemeier, U.; Boyer, M.; Emri, S.; Manegold, C.; et al. Phase III study of pemetrexed in combination with cisplatin versus cisplatin alone in patients with malignant pleural mesothelioma. *J. Clin. Oncol.* **2003**, *21*, 2636–2644. [CrossRef] [PubMed]

3. Attanoos, R.L.; Gibbs, A.R. Pathology of malignant mesothelioma. *Histopathology* **1997**, *30*, 403–418. [CrossRef] [PubMed]

4. Travis, W.D.; Brambilla, E.; Burke, A.P.; Marx, A.; Nicholson, A.G. Introduction to The 2015 World Health Organization Classification of Tumors of the Lung, Pleura, Thymus, and Heart. *J. Thorac. Oncol.* **2015**, *10*, 1240–1242. [CrossRef] [PubMed]

5. Sugarbaker, D.J.; Flores, R.M.; Jaklitsch, M.T.; Richards, W.G.; Strauss, G.M.; Corson, J.M.; DeCamp, M.M.; Swanson, S.J.; Bueno, R.; Lukanich, J.M.; et al. Resection margins, extrapleural nodal status, and cell type determine postoperative long-term survival in trimodality therapy of malignant pleural mesothelioma: Results in 183 patients. *J. Thorac. Cardiovasc. Surg.* **1999**, *117*, 54–63. [CrossRef]

6. Mossman, B.T.; Shukla, A.; Heintz, N.H.; Verschraegen, C.F.; Thomas, A.; Hassan, R. New insights into understanding the mechanisms, pathogenesis, and management of malignant mesotheliomas. *Am. J. Pathol.* **2013**, *182*, 1065–1077. [CrossRef] [PubMed]

7. Bograd, A.J.; Suzuki, K.; Vertes, E.; Colovos, C.; Morales, E.A.; Sadelain, M.; Adusumilli, P.S. Immune responses and immunotherapeutic interventions in malignant pleural mesothelioma. *Cancer Immunol. Immunother.* **2011**, *60*, 1509–1527. [CrossRef] [PubMed]

8. Yap, T.A.; Aerts, J.G.; Popat, S.; Fennell, D.A. Novel insights into mesothelioma biology and implications for therapy. *Nat. Rev. Cancer* **2017**, *17*, 475–488. [CrossRef] [PubMed]

9. Hegmans, J.P.; Aerts, J.G. Immunomodulation in cancer. *Curr. Opin. Pharmacol.* **2014**, *17*, 17–21. [CrossRef] [PubMed]

10. Veltman, J.D.; Lambers, M.E.; van Nimwegen, M.; Hendriks, R.W.; Hoogsteden, H.C.; Aerts, J.G.; Hegmans, J.P. COX-2 inhibition improves immunotherapy and is associated with decreased numbers of myeloid-derived suppressor cells in mesothelioma. Celecoxib influences MDSC function. *BMC Cancer* **2010**, *10*, 464. [CrossRef] [PubMed]

11. Veltman, J.D.; Lambers, M.E.H.; van Nimwegen, M.; Hendriks, R.W.; Hoogsteden, H.C.; Hegmans, J.P.; Aerts, J.G. Zoledronic acid impairs myeloid differentiation to tumour-associated macrophages in mesothelioma. *Br. J. Cancer* **2010**, *103*, 629–641. [CrossRef] [PubMed]

12. Veltman, J.D.; Lambers, M.E.H.; van Nimwegen, M.; de Jong, S.; Hendriks, R.W.; Hoogsteden, H.C.; Aerts, J.G.; Hegmans, J.P. Low-dose cyclophosphamide synergizes with dendritic cell-based immunotherapy in antitumor activity. *J. Biomed. Biotechnol.* **2010**, *2010*, 798467. [CrossRef] [PubMed]

13. Van der Most, R.G.; Currie, A.J.; Mahendran, S.; Prosser, A.; Darabi, A.; Robinson, B.W.S.; Nowak, A.K.; Lake, R.A. Tumor eradication after cyclophosphamide depends on concurrent depletion of regulatory T cells: A role for cycling TNFR2-expressing effector-suppressor T cells in limiting effective chemotherapy. *Cancer Immunol. Immunother.* **2009**, *58*, 1219–1228. [CrossRef] [PubMed]

14. Mantovani, A.; Sozzani, S.; Locati, M.; Allavena, P.; Sica, A. Macrophage polarization: Tumor-associated macrophages as a paradigm for polarized M2 mononuclear phagocytes. *Trends Immunol.* **2002**, *23*, 549–555. [CrossRef]

15. Solinas, G.; Germano, G.; Mantovani, A.; Allavena, P. Tumor-associated macrophages (TAM) as major players of the cancer-related inflammation. *J. Leukoc. Biol.* **2009**, *86*, 1065–1073. [CrossRef] [PubMed]

16. Pradel, L.P.; Ooi, C.-H.; Romagnoli, S.; Cannarile, M.A.; Sade, H.; Rüttinger, D.; Ries, C.H. Macrophage Susceptibility to Emactuzumab (RG7155) Treatment. *Mol. Cancer Ther.* **2016**, *15*, 3077–3086. [CrossRef] [PubMed]

17. Herrmann, F.; Ayaub, E.; Parthasarathy, P.; Ackermann, M.; Inman, M.D.; Kolb, M.R.J.; Wollin, L.; Ask, K.; Tandon, K. Nintedanib attenuates the polarization of profibrotic macrophages through the inhibition of tyrosine phosphorylation on CSF1 receptor. *Am. J. Respir. Crit. Care Med.* **2017**, *195*, A2397.

18. Burt, B.M.; Rodig, S.J.; Tilleman, T.R.; Elbardissi, A.W.; Bueno, R.; Sugarbaker, D.J. Circulating and tumor-infiltrating myeloid cells predict survival in human pleural mesothelioma. *Cancer* **2011**, *117*, 5234–5244. [CrossRef] [PubMed]

19. Cornelissen, R.; Lievense, L.A.; Maat, A.P.; Hendriks, R.W.; Hoogsteden, H.C.; Bogers, A.J.; Hegmans, J.P.; Aerts, J.G. Ratio of intratumoral macrophage phenotypes is a prognostic factor in epithelioid malignant pleural mesothelioma. *PLoS ONE* **2014**, *9*, e106742. [CrossRef] [PubMed]

20. Cornelissen, R.; Lievense, L.A.; Robertus, J.-L.; Hendriks, R.W.; Hoogsteden, H.C.; Hegmans, J.P.; Aerts, J.G. Intratumoral macrophage phenotype and CD8+ T lymphocytes as potential tools to predict local tumor outgrowth at the intervention site in malignant pleural mesothelioma. *Lung Cancer* **2015**, *88*, 332–337. [CrossRef] [PubMed]

21. Marcq, E.; Siozopoulou, V.; De Waele, J.; van Audenaerde, J.; Zwaenepoel, K.; Santermans, E.; Hens, N.; Pauwels, P.; van Meerbeeck, J.P.; Smits, E.L.J. Prognostic and predictive aspects of the tumor immune microenvironment and immune checkpoints in malignant pleural mesothelioma. *Oncoimmunology* **2017**, *6*, e1261241. [CrossRef] [PubMed]

22. Schürch, C.M.; Forster, S.; Brühl, F.; Yang, S.H.; Felley-Bosco, E.; Hewer, E. The "don't eat me" signal CD47 is a novel diagnostic biomarker and potential therapeutic target for diffuse malignant mesothelioma. *Oncoimmunology* **2018**, *7*, e1373235. [CrossRef] [PubMed]

23. Mandruzzato, S.; Solito, S.; Falisi, E.; Francescato, S.; Chiarion-Sileni, V.; Mocellin, S.; Zanon, A.; Rossi, C.R.; Nitti, D.; Bronte, V.; et al. IL4Rα+ Myeloid-Derived Suppressor Cell Expansion in Cancer Patients. *J. Immunol.* **2009**, *182*, 6562–6568. [CrossRef] [PubMed]

24. Burt, B.M.; Bader, A.; Winter, D.; Rodig, S.J.; Bueno, R.; Sugarbaker, D.J. Expression of Interleukin-4 Receptor Alpha in Human Pleural Mesothelioma Is Associated with Poor Survival and Promotion of Tumor Inflammation. *Clin. Cancer Res.* **2012**, *18*, 1568–1577. [CrossRef] [PubMed]

25. Naito, Y.; Saito, K.; Shiiba, K.; Ohuchi, A.; Saigenji, K.; Nagura, H.; Ohtani, H. CD8+ T cells infiltrated within cancer cell nests as a prognostic factor in human colorectal cancer. *Cancer Res.* **1998**, *58*, 3491–3494. [PubMed]

26. Zhang, L.; Conejo-Garcia, J.R.; Katsaros, D.; Gimotty, P.A.; Massobrio, M.; Regnani, G.; Makrigiannakis, A.; Gray, H.; Schlienger, K.; Liebman, M.N.; et al. Intratumoral T cells, recurrence, and survival in epithelial ovarian cancer. *N. Engl. J. Med.* **2003**, *348*, 203–213. [CrossRef] [PubMed]

27. Schumacher, K.; Haensch, W.; Röefzaad, C.; Schlag, P.M. Prognostic significance of activated CD8(+) T cell infiltrations within esophageal carcinomas. *Cancer Res.* **2001**, *61*, 3932–3936. [PubMed]

28. Zhu, J.; Paul, W.E. CD4 T cells: Fates, functions, and faults. *Blood* **2008**, *112*, 1557–1569. [CrossRef] [PubMed]

29. Friedman, K.M.; Prieto, P.A.; Devillier, L.E.; Gross, C.A.; Yang, J.C.; Wunderlich, J.R.; Rosenberg, S.A.; Dudley, M.E. Tumor-specific CD4+ melanoma tumor-infiltrating lymphocytes. *J. Immunother.* **2012**, *35*, 400–408. [CrossRef] [PubMed]

30. Neurath, M.F.; Finotto, S. The emerging role of T cell cytokines in non-small cell lung cancer. *Cytokine Growth Factor Rev.* **2012**, *23*, 315–322. [CrossRef] [PubMed]

31. Van Acker, H.H.; Capsomidis, A.; Smits, E.L.; Van Tendeloo, V.F. CD56 in the Immune System: More Than a Marker for Cytotoxicity? *Front. Immunol.* **2017**, *8*, 892. [CrossRef] [PubMed]

32. Leigh, R.A.; Webster, I. Lymphocytic infiltration of pleural mesothelioma and its significance for survival. *S. Afr. Med. J.* **1982**, *61*, 1007–1009. [PubMed]

33. Mudhar, H.S.; Wallace, W.A.H. No relationship between tumour infiltrating lymphocytes and overall survival is seen in malignant mesothelioma of the pleura. *Eur. J. Surg. Oncol.* **2002**, *28*, 564–565. [CrossRef] [PubMed]

34. Hegmans, J.P.J.J. Mesothelioma environment comprises cytokines and T-regulatory cells that suppress immune responses. *Eur. Respir. J.* **2006**, *27*, 1086–1095. [CrossRef] [PubMed]

35. Anraku, M.; Cunningham, K.S.; Yun, Z.; Tsao, M.-S.; Zhang, L.; Keshavjee, S.; Johnston, M.R.; de Perrot, M. Impact of tumor-infiltrating T cells on survival in patients with malignant pleural mesothelioma. *J. Thorac. Cardiovasc. Surg.* **2008**, *135*, 823–829. [CrossRef] [PubMed]

36. Yamada, N.; Oizumi, S.; Kikuchi, E.; Shinagawa, N.; Konishi-Sakakibara, J.; Ishimine, A.; Aoe, K.; Gemba, K.; Kishimoto, T.; Torigoe, T.; et al. CD8+ tumor-infiltrating lymphocytes predict favorable prognosis in malignant pleural mesothelioma after resection. *Cancer Immunol. Immunother.* **2010**, *59*, 1543–1549. [CrossRef] [PubMed]

37. Awad, M.M.; Jones, R.E.; Liu, H.; Lizotte, P.H.; Ivanova, E.V.; Kulkarni, M.; Herter-Sprie, G.S.; Liao, X.; Santos, A.A.; Bittinger, M.A.; et al. Cytotoxic T Cells in PD-L1-Positive Malignant Pleural Mesotheliomas Are Counterbalanced by Distinct Immunosuppressive Factors. *Cancer Immunol. Res.* **2016**, *4*, 1038–1048. [CrossRef] [PubMed]

38. Suzuki, K.; Kadota, K.; Sima, C.S.; Sadelain, M.; Rusch, V.W.; Travis, W.D.; Adusumilli, P.S. Chronic inflammation in tumor stroma is an independent predictor of prolonged survival in epithelioid malignant pleural mesothelioma patients. *Cancer Immunol. Immunother.* **2011**, *60*, 1721–1728. [CrossRef] [PubMed]

39. Nishikawa, H.; Sakaguchi, S. Regulatory T cells in cancer immunotherapy. *Curr. Opin. Immunol.* **2014**, *27*, 1–7. [CrossRef] [PubMed]

40. DeLong, P.; Carroll, R.G.; Henry, A.C.; Tanaka, T.; Ahmad, S.; Leibowitz, M.S.; Sterman, D.H.; June, C.H.; Albelda, S.M.; Vonderheide, R.H. Regulatory T cells and cytokines in malignant pleural effusions secondary to mesothelioma and carcinoma. *Cancer Biol. Ther.* **2005**, *4*, 342–346. [CrossRef] [PubMed]

41. Ujiie, H.; Kadota, K.; Nitadori, J.; Aerts, J.G.; Woo, K.M.; Sima, C.S.; Travis, W.D.; Jones, D.R.; Krug, L.M.; Adusumilli, P.S. The tumoral and stromal immune microenvironment in malignant pleural mesothelioma: A comprehensive analysis reveals prognostic immune markers. *Oncoimmunology* **2015**, *4*, e1009285. [CrossRef] [PubMed]

42. Patil, N.S.; Righi, L.; Koeppen, H.; Zou, W.; Izzo, S.; Grosso, F.; Libener, R.; Loiacono, M.; Monica, V.; Buttigliero, C.; et al. Molecular and Histopathological Characterization of the Tumor Immune Microenvironment in Advanced Stage of Malignant Pleural Mesothelioma. *J. Thorac. Oncol.* **2018**, *13*, 124–133. [CrossRef] [PubMed]

43. Baglole, C.J.; Ray, D.M.; Bernstein, S.H.; Feldon, S.E.; Smith, T.J.; Sime, P.J.; Phipps, R.P. More than structural cells, fibroblasts create and orchestrate the tumor microenvironment. *Immunol. Investig.* **2006**, *35*, 297–325. [CrossRef] [PubMed]

44. Li, Q.; Wang, W.; Yamada, T.; Matsumoto, K.; Sakai, K.; Bando, Y.; Uehara, H.; Nishioka, Y.; Sone, S.; Iwakiri, S.; et al. Pleural Mesothelioma Instigates Tumor-Associated Fibroblasts To Promote Progression via a Malignant Cytokine Network. *Am. J. Pathol.* **2011**, *179*, 1483–1493. [CrossRef]

45. Lo, A.; Wang, L.-C.; Scholler, J.; Monslow, J.; Avery, D.; Newick, K.; O'Brien, S.; Evans, R.A.; Bajor, D.J.; Clendenin, C.; et al. Tumor-Promoting Desmoplasia Is Disrupted by Depleting FAP-Expressing Stromal Cells. *Cancer Res.* **2015**, *75*, 2800–2810. [CrossRef] [PubMed]

46. Harvey, P.; Warn, A.; Newman, P.; Perry, L.J.; Ball, R.Y.; Warn, R.M. Immunoreactivity for hepatocyte growth factor/scatter factor and its receptor, met, in human lung carcinomas and malignant mesotheliomas. *J. Pathol.* **1996**, *180*, 389–394. [CrossRef]

47. Gandini, S.; Massi, D.; Mandalà, M. PD-L1 expression in cancer patients receiving anti PD-1/PD-L1 antibodies: A systematic review and meta-analysis. *Crit. Rev. Oncol. Hematol.* **2016**, *100*, 88–98. [CrossRef] [PubMed]

48. Fehrenbacher, L.; Spira, A.; Ballinger, M.; Kowanetz, M.; Vansteenkiste, J.; Mazieres, J.; Park, K.; Smith, D.; Artal-Cortes, A.; Lewanski, C.; et al. Atezolizumab versus docetaxel for patients with previously treated non-small-cell lung cancer (POPLAR): A multicentre, open-label, phase 2 randomised controlled trial. *Lancet* **2016**, *387*, 1837–1846. [CrossRef]

49. Rittmeyer, A.; Barlesi, F.; Waterkamp, D.; Park, K.; Ciardiello, F.; von Pawel, J.; Gadgeel, S.M.; Hida, T.; Kowalski, D.M.; Dols, M.C.; et al. Atezolizumab versus docetaxel in patients with previously treated non-small-cell lung cancer (OAK): A phase 3, open-label, multicentre randomised controlled trial. *Lancet* **2017**, *389*, 255–265. [CrossRef]

50. Combaz-Lair, C.; Galateau-Sallé, F.; McLeer-Florin, A.; Le Stang, N.; David-Boudet, L.; Duruisseaux, M.; Ferretti, G.R.; Brambilla, E.; Lebecque, S.; Lantuejoul, S. Immune biomarkers PD-1/PD-L1 and TLR3 in malignant pleural mesotheliomas. *Hum. Pathol.* **2016**, *52*, 9–18. [CrossRef] [PubMed]

51. Cedrés, S.; Ponce-Aix, S.; Zugazagoitia, J.; Sansano, I.; Enguita, A.; Navarro-Mendivil, A.; Martinez-Marti, A.; Martinez, P.; Felip, E. Analysis of Expression of Programmed Cell Death 1 Ligand 1 (PD-L1) in Malignant Pleural Mesothelioma (MPM). *PLoS ONE* **2015**, *10*, e0121071. [CrossRef] [PubMed]

52. Inaguma, S.; Lasota, J.; Wang, Z.; Czapiewski, P.; Langfort, R.; Rys, J.; Szpor, J.; Waloszczyk, P.; Okoń, K.; Biernat, W.; et al. Expression of ALCAM (CD166) and PD-L1 (CD274) independently predicts shorter survival in malignant pleural mesothelioma. *Hum. Pathol.* **2018**, *71*, 1–7. [CrossRef] [PubMed]

53. Mansfield, A.S.; Roden, A.C.; Peikert, T.; Sheinin, Y.M.; Harrington, S.M.; Krco, C.J.; Dong, H.; Kwon, E.D. B7-H1 expression in malignant pleural mesothelioma is associated with sarcomatoid histology and poor prognosis. *J. Thorac. Oncol.* **2014**, *9*, 1036–1040. [CrossRef] [PubMed]

54. Khanna, S.; Thomas, A.; Abate-Daga, D.; Zhang, J.; Morrow, B.; Steinberg, S.M.; Orlandi, A.; Ferroni, P.; Schlom, J.; Guadagni, F.; et al. Malignant Mesothelioma Effusions Are Infiltrated by CD3+ T Cells Highly Expressing PD-L1 and the PD-L1+ Tumor Cells within These Effusions Are Susceptible to ADCC by the Anti-PD-L1 Antibody Avelumab. *J. Thorac. Oncol.* **2016**, *11*, 1993–2005. [CrossRef] [PubMed]

55. Cedrés, S.; Ponce-Aix, S.; Pardo-Aranda, N.; Navarro-Mendivil, A.; Martinez-Marti, A.; Zugazagoitia, J.; Sansano, I.; Montoro, M.A.; Enguita, A.; Felip, E. Analysis of expression of PTEN/PI3K pathway and programmed cell death ligand 1 (PD-L1) in malignant pleural mesothelioma (MPM). *Lung Cancer* **2016**, *96*, 1–6. [CrossRef] [PubMed]

56. Curti, B.D.; Kovacsovics-Bankowski, M.; Morris, N.; Walker, E.; Chisholm, L.; Floyd, K.; Walker, J.; Gonzalez, I.; Meeuwsen, T.; Fox, B.A.; et al. OX40 is a potent immune-stimulating target in late-stage cancer patients. *Cancer Res.* **2013**, *73*, 7189–7198. [CrossRef] [PubMed]

International Journal of
Molecular Sciences

MDPI

Review

Heterogeneous Contributing Factors in MPM Disease Development and Progression: Biological Advances and Clinical Implications

Bhairavi Tolani *, Luis A. Acevedo, Ngoc T. Hoang and Biao He *

Thoracic Oncology Program, Department of Surgery, Helen Diller Family Comprehensive Cancer Center, University of California, San Francisco, CA 94115, USA; Luis.Acevedo@ucsf.edu (L.A.A.); Ngoc.Hoang@ucsf.edu (N.T.H.)
* Correspondence: Bhairavi.Tolani@ucsf.edu (B.T.); Biao.He@ucsf.edu (B.H.); Tel.: +1-415-502-0555 (B.T. & B.H.)

Received: 21 November 2017; Accepted: 10 January 2018; Published: 13 January 2018

Abstract: Malignant pleural mesothelioma (MPM) tumors are remarkably aggressive and most patients only survive for 5–12 months; irrespective of stage; after primary symptoms appear. Compounding matters is that MPM remains unresponsive to conventional standards of care; including radiation and chemotherapy. Currently; instead of relying on molecular signatures and histological typing; MPM treatment options are guided by clinical stage and patient characteristics because the mechanism of carcinogenesis has not been fully elucidated; although about 80% of cases can be linked to asbestos exposure. Several molecular pathways have been implicated in the MPM tumor microenvironment; such as angiogenesis; apoptosis; cell-cycle regulation and several growth factor-related pathways predicted to be amenable to therapeutic intervention. Furthermore, the availability of genomic data has improved our understanding of the pathobiology of MPM. The MPM genomic landscape is dominated by inactivating mutations in several tumor suppressor genes; such as *CDKN2A*; *BAP1* and *NF2*. Given the complex heterogeneity of the tumor microenvironment in MPM; a better understanding of the interplay between stromal; endothelial and immune cells at the molecular level is required; to chaperone the development of improved personalized therapeutics. Many recent advances at the molecular level have been reported and several exciting new treatment options are under investigation. Here; we review the challenges and the most up-to-date biological advances in MPM pertaining to the molecular pathways implicated; progress at the genomic level; immunological progression of this fatal disease; and its link with developmental cell pathways; with an emphasis on prognostic and therapeutic treatment strategies.

Keywords: malignant pleural mesothelioma (MPM); tumor microenvironment heterogeneity; molecular pathways; tumor suppressors; immunotherapy; developmental cell pathways

1. Introduction

Malignant mesothelioma is an aggressive and universally lethal cancer that arises because of pathological transformations in the mesothelium, a protective serous membrane that lines several organs in the body, such as the lungs (pleural), the intestines (peritoneum), the heart (pericardium) and tunica vaginalis. Of these, malignant pleural mesothelioma (MPM) is the most common and accounts for the predominant subtype—80% of all cases. MPM has an unusually dismal prognosis; its 5-year survival rate is a mere 10%, making it the most fatal among rare cancers [1]. Although MPM is classified as a rare disease and an estimated 3000 cases are diagnosed in the United States each year, the incidence is expected to remain steady, or rise, until 2055, because of the developmental latency of the disease; correspondingly, two thirds of mesothelioma cases are diagnosed in patients over the age of 65 [1]. The established cause of carcinogenesis is direct workplace exposure to asbestos, a naturally-occurring silicate fibrous mineral used in insulating material, and approximately 80% of MPM cases can be linked to it; consequently, males have a four times higher incidence rate than females [1]. While asbestos use has declined in the United States since the 1980s, it has not been banned, as in other countries; thus, more clinical diagnoses are expected to emerge as a result of the disease's late manifestation (~20–40 years from exposure). Although the exact mechanism by which asbestos fibers lead to disease onset is still under debate, some hypotheses include the role of toxic oxygen radicals, elevated growth factor-induced cell signaling of kinases, and chronic inflammation, which ultimately leads to malignancy [2]. Other etiological factors include infection with simian virus 40 (SV40) and exposure to erionite and radiation, but their contributions remain controversial [3].

Current treatments for MPM are limited to surgery, radiation, and chemotherapy. However, since 80% of patients are diagnosed in stage III/IV, they are not candidates for surgical cure because their disease is no longer resectable [4]. Some hope lies in the recent advancement of intensity-modulated radiotherapy (IMRT), which precisely delivers radiation doses to the malignant tissues, while sparing normal counterparts [5]. Unfortunately, because of its low radio- and chemo-sensitivity, MPM remains unresponsive to systemic therapy. Currently, there are only two Food and Drug Administration (FDA)-approved chemotherapeutic drugs specifically for treating MPM: cisplatin and pemetrexed. Although combination therapy, consisting of cisplatin and pemetrexed, has shown promising prognostic outcomes and a response rate of 41.3%, making it the standard treatment for mesothelioma, overall survival has been a low 16.6 months [6].

Histologically, there are three main sub-types of malignant mesothelioma: epithelioid (~60% of cases), sarcomatous (~20%), and biphasic (combinations of the first two) [7]. Whereas patients diagnosed with epithelioid mesothelioma reportedly survive the longest (12–27 months), the appropriate course of treatment can be onerous to determine [8]. This is due to the inherent tumor heterogeneity and the difficulty in staging the disease, which underscores the need to better understand the disease pathobiology at the molecular level.

The goal of this review is to provide translational scientists with an up-to-date account of recent and potential therapeutics to address the treatment of MPM. We discuss current advances in mesothelioma biology and heterogeneity in the tumor microenvironment, starting with heterogeneity at (1) the molecular pathway level, followed by progress at (2) the genomic level, then (3) immunological progression of MPM, and finally the link with (4) developmental/stem cell pathways associated with the disease. In the first section on molecular pathways, we explicate translational implications of pathways, such as angiogenesis, apoptosis, and the cell cycle, along with their characteristics in mesothelioma. In the genomic landscape section, we explore the three main genetic alterations exclusive to this rare cancer. Next, we review the contribution of the immune environment, including the stromal compartment, immune population and tumor cells. The last section describes the development pathways, such as Hedgehog, Wnt/β-catenin, Notch and Hippo/yes-associated protein (YAP) in MPM. Our focus is to discuss promising new therapeutic strategies under investigative development, which could potentially permit a longer and better quality of life for those with mesothelioma.

2. Molecular Pathways in Malignant Pleural Mesothelioma

Therapeutic exploitation of the following vulnerabilities present in MPM, by harnessing overabundant growth signaling factors and cytokines, could help in the fight against this disease. A summary of the clinical study status and potential feasibility for the following are provided in Table 1.

Table 1. Malignant Pleural Mesothelioma (MPM) Targets, Potential Therapeutics, and Clinical Status.

Pathway	Target(s)	Prevalence of Target in MPM (%)	Therapeutic	Clinical Trial Status
Angiogenesis	VEGF	30% expression	Bevacizumab	Approved for colon cancer
	VEGF	30% expression	Bevacizumab + pemetrexed and cisplatin	Clinical trial phase III completed
	VEGFR	20% expression	Dovitinib; nintedanib; cediranib	Under clinical investigation
	VEGFR	20% expression	Vatalanib	Under clinical investigation
	VEGF, FGF	30%, 50% overexpression	Lenvatinib	Approved for thyroid/kidney Cancers
	HGF	85% overexpression	Adenovirus containing an HGF variant, NK-4	-
	?	?	Thalidomide	Approved for multiple myeloma
	VEGFR, PDGFR	20%, 30% overexpression	Sorafenib	Approved for thyroid, kidney, and liver cancer
	VEGFR, PDGFR	20%, 30% overexpression	Sunitinib	Approved for kidney and GI cancer
	p53	20–25% mutated	-	-
Apoptosis	*Survivin*	-	Antisense oligonucleotides + chemotherapy	-
	Bcl-xL	-	Small molecule HDAC inhibitors + antisense oligonucleotides	-
	Bcl-xL/Bcl-2	-	2-Methoxy antimycin A3 + chemotherapy	-
	Src	~50% expression	Dasatinib	Approved for leukemia
	Fas Ligand	Selective FasL-positive cells	Fas ligand + cisplatin	-
	Calcium Channels	Primary samples showed reduced calcium ion uptake	Exogenous calcium ions or mitochondrial calcium uniporter	-
	TRAIL	-	Administration of MSCs genetically engineered to express TRAIL	-
Cell Cycle	*CHEK1*	50% overexpression	*CHEK1* silencing + doxorubicin	-
	YAP	~70% expression	YAP silencing	-
Growth Factor	EGFR	~40% Expression	Gefitinib; Erlotinib	-
	PDGFR	20%, 30% overexpression	*PDGFR* silencing	-
	FGFR1	50% overexpression	FGFR-1 inhibitor PD-166866	-
	FGFR1	50% overexpression	Sorafenib	Approved for thyroid, kidney, and liver cancer
	FGFR1	50% overexpression	Ponatinib	Approved for thyroid, kidney, and liver cancer

Table 1. *Cont.*

Pathway	Target(s)	Prevalence of Target in MPM (%)	Therapeutic	Clinical Trial Status
DNA Replication	TERT	90% overexpression	Anti-telomerase drugs + other targeted therapies	-
	CDK2	70% homozygous deletion of *CDKN2A* known to inhibit CDK2	Milciclib/PHA-848125AC	Phase II for hepatocellular carcinoma
Tumor Suppressor-associated targets	Snail-p53	30–45% *NF2* deletion known to prevent p53 inhibition	GN25	-
	EZH2	60% *BAP1* inactivations known to modulate EZH2	Pinometostat (EPZ5676) and other methyltransferase inhibitors	-
Stromal Compartment	PD1	(%?) tumor microenvironment is immunosuppressive	Nivolumab	Phase II for MPM
Hedgehog	Smoothened	Inhibits Hh signaling (%?)	Vismodegib	Approved for basal cell carcinoma
	Gli	90% Gli1/Gli2 active	GANT61 and GLI-I	-
Wnt/β-catenin	PORCN	Inhibits Wnt signaling (5%)	LGK-974	Phase I for solid tumors
Notch	γ secretase	Inhibits Notch signaling (%?)	Semagacestat (LY450139)	Phase III for Alzheimer's disease
	CK2α	Down-regulates Notch1 signaling (%?)	Silmitasertib (CX-4945)	Phase I for solid tumors and multiple myeloma
Hippo/YAP	PI3K–AKT–mTOR	*LATS2* altered in 11%	PF-0469150 2	-
	Nedd8 activating enzyme (NAE)	Interferes with YAP (70%) activation	Pevonedistat (MLN4924)	Phase I for hematological malagnancies and melanoma
	YAP-TEAD	NF2 (40%) and YAP (70%) overactive	Verteporfin	Phase I for prostate cancer

Note: - indicates no data and ? indicates unknown. The above-mentioned therapeutics have not yet been approved for MPM but some of them could be promising in treating it in the near future.

2.1. Angiogenesis

The formation of new blood vessels via angiogenesis is prompted by the release of cellular cues, such as cytokines, and mesothelioma has been shown to express angiogenic factors along with their corresponding cellular receptors [9–12]. Approximately 30% of MPM cases express vascular endothelial growth factor (VEGF), and about 67% express the sub-type VEGF-C [11,12], one of the more abundant angiogenic factors found in any solid tumor patients [10]. Moreover, about 20% of cases show positive staining for the VEGF receptor (VEGFR-1), expressed almost exclusively in endothelial cells [11]. Inhibition of VEGF or its receptors can decrease proliferation in mesothelioma [10]. Other angiogenesis-related factors, such as fibroblast growth factor-2 (FGF-2) and hepatocyte growth factor (HGF), are hyper-expressed in mesothelioma, at 50% and 85%, respectively [12,13]. HGF, known to induce angiogenesis via activating endothelial cell migration and capillary tube formation, is significantly associated with epithelioid histology [13]. Given the strong correlation of elevated angiogenesis levels with diminished survival of MPM patients, several anti-angiogenic agents have been under clinical investigation. Bevacizumab, a neutralizing monoclonal antibody, targeted at VEGF-A, received FDA approval in 2004 for use in colon cancer [14]. Although it led to considerable toxicity in clinical trials, investigations have shown that the addition of bevacizumab to cisplatin and pemetrexed combinations was beneficial in MPM [15].

The strong rationale for angiogenesis inhibition has prompted the study of several alternative small molecules in MPM, such as vatalanib, lenvatinib, thalidomide, sorafenib and sunitinib, as well as other modalities, such as a competent adenovirus and NK-4, which is an HGF variant. A clinical trial was launched to investigate the efficacy of vatalanib, a dual pan-VEGFR and platelet derived growth factor receptor (PDGFR)-β inhibitor in mesothelioma, and although the results showed minimal benefits, its low toxicity could warrant further experimentation to explore synergistic effects if combined with other standard treatments [16]. Lenvatinib is a pan-tyrosine kinase inhibitor, aimed at multiple targets, such as VEGFR-2, fibroblast growth factor receptor (FGFR), and PDGFR, which inhibits endothelial cell growth, crucial for supplying blood to tumor cells [17]. Thalidomide, sorafenib and sunitinib are in various phases of clinical investigation [18–20] and a replicative competent adenovirus that targets the VEGF promoter is being tested preclinically [21]. NK-4, produced during inflammation, is a mimic fragment of HGF that can bind to its receptor, c-Met, without angiogenic stimulation, and therefore, when injected as an adenoviral vector into mesothelioma-bearing mice, there was tumor growth inhibition, caused by decreased blood vessel formation and apoptosis induction [22]. Finally, additional anti-angiogenesis agents under clinical investigation that target VEGFR, some of which are being combined with chemotherapy, include dovitinib, nintedanib, and cediranib and others [23].

2.2. Apoptosis

Programmed cell death, or apoptosis, is a carefully regulated process evaded by most cancers, partly because of inactivating mutations in the tumor suppressor gene, *TP53*, which is found to be mutated in about 50% of human tumors [24]. Distinct from other cancers, mesothelioma typically expresses functional p53 [25] and only 20–25% of cases harbor *TP53* mutations [26]; thus, both p53-dependent and independent mechanisms of apoptosis occur in MPM. Mesothelioma cell lines with normal p53 are sensitive to cisplatin treatment, show increased DNA binding and elevated phosphorylation, which prevents targeted degradation by MDM2, and thus, increased transcriptional activation of apoptotic genes ultimately leads to p53-dependent cell death [27,28]. Survivin, a negative regulator of p53, is an anti-apoptotic protein associated with unfavorable patient outcomes [28] and as a treatment strategy, introduction of inhibitory anti-sense *Survivin* oligonucleotides sensitized mesothelioma cells to chemotherapy and induced apoptosis [27,28].

Downstream of p53, many defects in the apoptotic core machinery have been reported in cancer. The Bcl-2 family of proteins is most well-known for orchestrating mitochondrial-mediated apoptosis and encodes both pro- and anti-apoptotic proteins that either protect tumor cells from chemotherapy-induced apoptosis or confer resistance to these drugs. One apoptotic repressor, Bcl-xL, which is strongly expressed in mesothelioma, prevents apoptosis induction by inhibiting mitochondrial permeabilization and activated caspase release, thus contributing to tumor growth [26]. Small molecule HDAC inhibitors and antisense oligonucleotides directed at Bcl-xL were shown to induce tumor cell death via apoptosis and enhance chemo-sensitivity in MPM [29,30]. A second anti-apoptotic protein, MCL1, is also expressed in mesothelioma, and together with Bcl-xL, can further inhibit apoptosis and confer resistance to mesothelioma [25]. Bcl-2 is not commonly expressed in mesothelioma, but down-regulating Bcl-2 in conjunction with Bcl-xL is more effective in inducing apoptosis [25]. One such small molecule, 2-methoxy antimycin A3, alone, or in chemotherapeutic combinations, exhibited a synergistic relationship in promoting tumor cell death via apoptosis, as well as promoting chemo-sensitivity [31].

Outlined below are several noteworthy proof-of-concept studies, which demonstrate that the induction of apoptosis in resistant mesothelioma cells could be used as a therapeutic approach. These comprise taking aim at the following targets: focal adhesion kinase (FAK) and Src, Fas receptor and modulation of reactive oxygen species (ROS), calcium ions, and tumor necrosis factor-related apoptosis inducing ligand (TRAIL).

2.2.1. FAK (Focal Adhesion Kinase) and Src, Fas Receptor and ROS (Reactive Oxygen Species)

FAK is an important regulator of cell proliferation, migration, and invasion expressed in primary mesothelioma cell lines. Dual pharmacological inhibition of FAK and MDM2 was synergistic and resulted in decreased cell viability and increased total phosphorylation of p53 and p21, along with a decrease in the proliferative marker, cyclin A [32]. Furthermore, about half of the mesothelioma samples express a protein similar to FAK called Src, whose presence is correlated with advanced tumor pathology given its role in cell migration and invasion. The use of the Src/Abl inhibitor, dasatinib, decreased activated Src in mesothelioma cell migration and invasion, which led to cell cycle arrest and apoptosis in sensitive cell lines [33]. The Fas family of receptors and ligands are part of many death-signaling pathways that can initiate apoptosis. One study showed that the combined effect of cisplatin-induced ROS levels with Fas ligands (FasL) and sensitized Fas receptor positive cells to apoptotic death by activating caspase 9, evidenced by apoptosis-specific DNA fragmentation [34]. These in vitro data were recapitulated in an in vivo mouse model, where tumor growth decreased significantly after treatment with cisplatin and FasL, thereby highlighting the possibility for personalized treatment of Fas-positive patients by combining ROS-mediated apoptosis induction with chemotherapy [34].

2.2.2. Calcium Ions

Mitochondrial calcium ion concentration can induce death via apoptosis in cancer cells and tissue. When exogenous calcium ions were added to primary mesothelioma culture, there was an increase in apoptosis via cleaved caspase 3 activities, which also occurred if the mitochondrial calcium ion uniporter was expressed exogenously, underscoring the utility of harnessing this mechanism for MPM treatment [35].

2.2.3. TRAIL

TRAIL can induce apoptosis by binding to death receptors expressed on cancer cells. Immune modulating human mesenchymal stromal cells (MSCs) have been genetically engineered to express TRAIL, which, when injected intraperitonially into mesothelioma-bearing mice, significantly reduced tumor burden [36]. This was accompanied by an increase in apoptotic, Terminal deoxynucleotidyl transferase dUTP nick end labeling (TUNEL)-positive cells, which potentially allows for personalized and autologous treatment of mesothelioma by harvest and modification of patients' own MSCs to express TRAIL without adverse immune reactions [36].

2.3. Cell Cycle Effectors

Uncontrolled cell growth results in cancer when mutations accumulate in several control mechanisms that cause cells to no longer respond to biological signals. The effectors of the cell cycle machinery play crucial roles in the progression of MPM and a study that compared patients by grouping long-term (>1 year) and short-term (<1 year) survivors reported that the majority of cell cycle genes, such as *AURKA*, *AURKB*, *CDC25C*, *PTTG3*, *CCND1*, and *KIF4*, are negatively correlated with survival [37]. Christensen et al. demonstrated that several cell cycle genes—*APC*, *CCND2*, *CDKN2A*, *CDKN2B*, *APPBP1*, and *RASSF1*—are methylated in older mesothelioma patients, and those with methylation of *RASSF1*, a gene that encodes an activator of G1/S cell cycle arrest, through sequestering of CCND1, is significantly associated with elevated asbestos body counts and therefore longer asbestos exposure [38]. These investigators also showed a significant correlation between asbestos body count/asbestos exposure and the number of cell cycle genes that are methylated, highlighting the importance of age and the duration of asbestos exposure in the development of mesothelioma [38].

Counterintuitively, an important regulator of DNA damage-mediated cell cycle arrest, checkpoint kinase 1 (CHEK1), is overexpressed in 50% of mesothelioma cases compared to normal pleural tissue, and *CHEK1* knockdown leads to synergistic apoptosis in mesothelioma cell lines when combined with the chemotherapeutic drug, doxorubicin [39]. In addition, over 70% of mesothelioma cases express YAP, which is known to up-regulate cell cycle genes in mesothelioma and subsequently decrease the Merlin–Hippo interaction. Given that YAP promotes TEA domain family member 1 (*TEAD*)-mediated transcription of cell cycle genes, such as *CCND1* and *FOXM1*, one study found that silencing *YAP* in mesothelioma resulted in suppression of these genes and inhibition of cell motility, invasion and anchorage-independent growth [40]. Concurrently, inhibition of YAP led to decreases in *E2F1*, *AURKB*, *PLK1*, *NEK2*, and anti-apoptotic *BIRC5*/*Survivin*, as well as an increase in pro-apoptotic *BCL2L11*/*BIM*, which ultimately resulted in broad decreases in anchorage properties, proliferation, migration, and invasion [40]. Thus, targeting YAP could potentially be feasible in treating this aggressive cancer.

2.4. TERT

In addition to cell cycle regulators and effectors, telomere production and regulation are also dysfunctional in mesothelioma [41]. For a tumor cell to continue to grow and proliferate, its telomere length must be elongated to ensure the basal integrity of its already unstable genome after many rounds of cell division. Tallet et al. established that telomerase activity is present in over 90% of mesothelioma cases, and its catalytic subunit *TERT* mRNA is expressed in over 80% of patient samples. While *TERT* can be mutated at its promoter region (>15% of cases present with C228T somatic mutations), the *TERT* locus (5p15.3) can also be amplified to induce higher *TERT* levels (around 50% of mesothelioma); however, the *TERT* promoter mutation is responsible for elevated *TERT* expression. Mutations in the *TERT* promoter along with *NF2* and *CDKN2A* alterations frequently occur concurrently and anti-telomerase drugs might therefore have clinical utility when combined with other targeted therapies [42,43].

2.5. Growth Factors Implicated in Mesothelioma

Several growth factors involved in MPM have been identified, such as epidermal growth factor receptor (EGFR), platelet derived growth factor receptor (PDGFR), and fibroblast growth factor (FGF), making them potential candidates for therapeutic intervention.

2.5.1. EGFR

Although mutations that lead to overexpression of EGFR have been associated with a wide range of malignancies, the role of elevated levels of EGFR in MPM appears limited. Okuda et al. showed that 0 out of 25 MPM samples expressed any of 13 known EGFR mutations, and only eight were positive for the presence of EGFR via immunohistochemistry (IHC) [44]. Although another study reported higher rates of EGFR expression in MPM tumors using IHC (38/83 positive for EGFR) [45], both studies confirmed by fluorescence in situ hybridization (FISH) analysis that there are only few occurrences of EGFR copy number gains, and thus EGFR overexpression did not appear to be due to gene amplification [44,45]. Whereas growth in EGFR-expressing mesothelioma cell lines is significantly deterred by EGFR small molecule inhibitors [45], clinical evaluations of gefitinib and erlotinib were dismal in MPM patients [46,47]. Only two mesothelioma patients out of 43 responded to gefitinib and 21 had stable disease, suggesting that a subset of patients might benefit from this treatment [46].

2.5.2. PDGFR

PDGFR has two subtypes: alpha (α) and beta (β); PDGFRα can bind three different PDGF ligand combinations (AA, AB, and BB), but PDGFRβ only binds to PDGF-BB [48,49]. Whereas PDGFRα functions in normal mesothelial cell growth, PDGF-BB and PDGFRβ are implicated in MPM growth [48, 49]. Like EGFR, PDGFRβ is overexpressed in 20–40% of MPM samples without frequent somatic mutations, and overexpression of PDGFRβ levels correlates with poor survival [44]. Additionally, human MPM cell lines have shown up to 70-fold increases in PDGFRB expression, when compared to a non-malignant mesothelial cell line [49]. The use of si*PDGFRB* to silence gene expression in these MPM cell lines caused a significant reduction in growth and clonogenicity [49]. Since non-malignant Met5A cells showed no noticeable adverse effects after *PDGFRβ* silencing, it could be a potential target gene for therapeutic intervention [49].

2.5.3. FGF

Hyper-expressed components of the FGF pathway contribute to MPM tumorigenicity and in particular, FGF2, FGF18, and FGFR1 have shown elevated levels in MPM cell lines and human tissue samples [50,51]. Schelch et al. demonstrated that FGF2-mediated stimulation increased MPM cell proliferation and motility but did not affect normal mesothelial controls [50], establishing the importance of FGF's role in tumorigenicity and thus the potential for therapeutic interference as a treatment route. Furthermore, high serum and pleural effusions of FGF levels are significantly associated with poor survival [52]. Use of the FGFR1 inhibitor, PD-166866, was particularly promising as it not only hindered in vitro MPM cell proliferation and caused an increase in death via apoptosis, but also reduced in vivo MPM tumor growth in mouse models; this was accompanied by synergism when coupled with cisplatin or radiotherapy treatment [50]. Moreover, Pattarozzi et al. reported that pharmacological inhibition with sorafenib reduced tumor growth in MPM primary cells, mainly by derailing FGFR1 activation [53]. Despite previous limited success with sorafenib in clinical trials, this study renews interest in sorafenib treatment in select MPM patients who have particularly highly active FGFR1. Similarly, the tyrosine kinase inhibitor (TKI) ponatinib has also been shown to selectively inhibit growth and clonogenicity of MPM cell lines that express FGFR1 [51], further suggesting its potential for targeted therapy in MPM.

3. Genomic Landscape Prevalent with Tumor Suppressor Inactivation

The genomic landscape of mesothelioma is characterized by frequent alterations in tumor suppressor genes resulting from mutational events involving copy number losses, single nucleotide variants (SNVs), gene fusions and splicing events. MPM exhibits a distinct genomic signature, including inactivating mutations in *cyclin-dependent kinase inhibitor 2A* (*CDKN2A*), and more recently, *ubiquitin carboxyl-terminal hydrolase* (*BAP1*) and *neurofibromin 2* (*NF2*), have been shown to be the most prevalent in this type of cancer (Figure 1A). Of particular interest, although the frequency of point mutations in cancer genes is low, several studies report global tumor suppressor inactivation, associated with CpG promoter methylation in MPM, which contrasts with the epigenetic landscape of normal pleura [54,55].

A

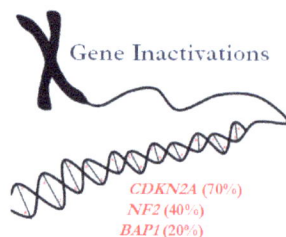

B

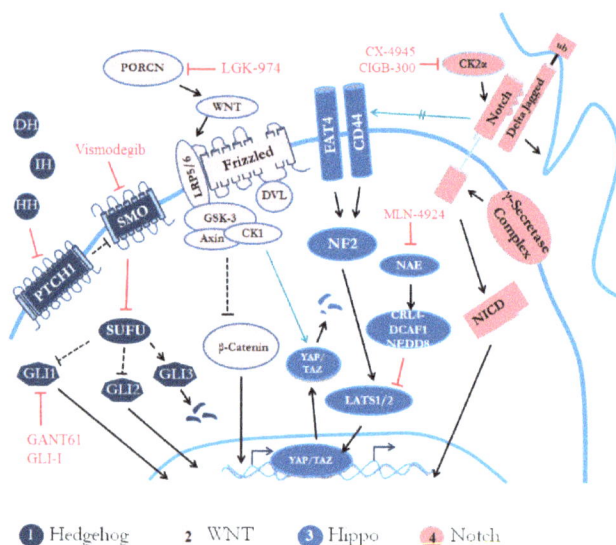

Figure 1. Heterogeneous contributing factors in MPM disease progression. (**A**) The MPM genomic landscape is dominated by frequent gene inactivation in *CDKN2A*, *NF2* and *BAP1*. (**B**) Several stem cellular signaling pathways, such as Hedgehog (Hh), Wnt/β-catenin, Hippo/YAP and Notch have been implicated in MPM pathobiology. (1) Hh ligands (DH, IH, and HH) bind patched (PTCH1) to relive inhibition of Smoothened (SMO) and this activates the Hh signaling pathway, via transcriptional activation of glioma-associated protein (Gli) factors; in the inactive state, Gli3 gets degraded when suppressor of fused homolog (Sufu) is inhibited. (2) Protein-serine O-palmitoleoyltransferase porcupine (PORCN) is required for efficient binding of Wnt ligands to cell-surface Frizzled (Fzd) receptors and to LRP5/6 which signals to the Dishevelled (Dvl) proteins. This causes an accumulation of β-catenin in the cytoplasm, followed by nuclear translocation and activation of transcription factors. (3) Hippo signaling consists of a cascade of kinases, which ultimately phosphorylate and activate serine/threonine-protein kinase (LATS)1/2 to inhibit two major downstream effectors of the Hippo pathway—YAP and Tafazzin (TAZ)—to stymie signaling (protein degradation). However, in *NF2*-mutant tumors, NAE activates CRL4^{DCAF1} and thus sustains tumor growth. (4) The Notch pathway can be activated by CK2α and when the Notch transmembrane receptors bind to a Notch ligand, a cascade of events promote gene transcription via γ secretase. Therapeutics (red) to target these pathway components have been developed for Smoothened (Vismodegib), Gli (GLI-I, GANT61), PORCN (LGK-974), NAE (MLN-4924) and CK2α (CX-4945 and CIGB-300). Arrows indicate interaction or activating effect and T-bars indicate inhibition. Green arrows indicate cross-talk between pathways.

Int. J. Mol. Sci. **2018**, *19*, 238

3.1. CDKN2A

Occurring at a frequency of over 70%, the most common homozygous deletion or incidence of gene inactivation in MPM is found in the 9p21 region of *CDKN2A*. As a result of alternative reading frames, this tumor suppressor gene acts as a negative regulator of cell proliferation, by encoding two distinct protein products: INK4A (p16, named by its molecular weight) and ADP-ribosylation factor (ARF) (human: p14, murine: p19). P16 is an inhibitory protein that prevents cyclin-dependent kinases, such as CDK4 and CDK6, from hyper-phosphorylating tumor suppressor protein RB, thereby activating RB, which blocks cell cycle progression from traversing the G1-S phase transition [26]. Inactivating mutations can thus disrupt this cell growth control pathway and contribute to carcinogenesis. Incidentally, P16 is also a key regulator of RB-mediated cellular senescence [56]. As part of the p53 pathway, P14 promotes ubiquitylation and degradation of MDM2, but loss of P14 increases MDM2 stability and activation, thereby repressing p53 and disrupting cell cycle control [26].

Using genetically engineered mouse models, Altomare et al. established the importance of losses in both $P16^{INK4a}$ and $P14^{ARF}$ in the latency and development of mesothelioma tumorigenesis. When exposed to asbestos, mice with mono-allelic deletions of $P16^{INK4a}$ or $P19^{ARF}$ (human: $P14^{ARF}$) developed mesothelioma earlier and more often than their wild type littermates, and those with deletions of both $P16^{INK4a}$ and $P19^{ARF}$ showed even faster progression into malignancy relative to the mice bearing single deletions [57]. Furthermore, adenoviral-based reintroduction of $P16^{INK4a}$ or $P14^{ARF}$ showed reversal of mesothelioma growth by decreasing hyper-phosphorylation of RB, increasing p53 levels, and ultimately resulted in cell cycle arrest [43]. Interestingly, over 75% of patients with mesothelioma have only $P16^{INK4a}$ deletions, and these patients generally have poor prognoses and shorter survival rates; almost 30% of primary tumors have been reported to have methylated $P16^{INK4a}$ [26,58]. Moreover, sustained $P16^{INK4a}$ expression is associated with better survival in patients after chemotherapy, regardless of histological subtypes, and deletion of $P16^{INK4a}$ correlates with poor outcomes [37,59].

Pharmacological intervention, through the use of CDK inhibitors, is an attractive therapeutic strategy and a multicenter phase II clinical trial as a second-line treatment for MPM is ongoing [60].

3.2. NF2

Deleted in around 35–40% of MPM cases and often functionally inactive if present due to somatic mutations, the tumor suppressor, Neurofibromin 2 (*NF2*) gene, has been implicated as a "gatekeeper" in asbestos-induced mesothelioma tumorigenicity via several mechanisms [61–64]. First, it is a putative regulator of the Hippo/SWH (Sav/Wts/Hpo) signaling pathway, known to be important for tumor suppression, and encodes the ezrin radixin and meosin (ERM) family protein, Merlin. *NF2*-mediated oncogenesis stems from the loss of Merlin and subsequent dysregulation of the Hippo pathway [62,63]. Whereas the Hippo pathway typically halts cell growth and inhibits the transcriptional regulator, YAP, the lack of Merlin leads to overactive YAP and thus uncontrolled cell growth and cancer progression. By transducing *NF2* into MM cells that had *NF2* mutations, YAP has been shown to be translocated from the nucleus to the cytoplasm [63], suggestive of NF2's role in YAP activity. In addition, YAP1 is negatively regulated by large tumor suppressor kinases (LATS1 and LATS2), which are commonly deleted in mesothelioma; in contrast, forced expression of LATS2 inactivates YAP1 and causes a reduction in cell growth [65].

Second, Merlin has also been suggested to negatively regulate the PI3K-AKT-mTOR signaling pathway, whose activation contributes to the pathogenesis of a number of malignancies [62,63]. In tumors that lack Merlin, the use of mTOR inhibitors led to reduced proliferation, yet were ineffective in Merlin-expressing tumors [62], suggestive of the role of the *NF2* deletion in MPM oncogenesis and its association with mTOR. *NF2*-depleted mesotheliomas are sensitive to the mTOR inhibitor molecule, rapamycin, and thus, a better understanding of the role of *NF2* in MPM, may hold pronounced potential for various targeted therapies.

Third, Cho et al. reported that NF2 typically prevents p53 inhibition caused by Snail-p53 interaction, but since it is often deleted in MPM, the tumor suppressive role of p53 is hindered, despite its low mutational frequency in MPM [25,26,61,66,67]. In MPM cell lines, a chemical inhibitor of Snail-p53 binding, GN25, induced p53 activity, followed by apoptosis, opening up potential therapeutic exploitation now that a clearer understanding of the molecular function of NF2 has been discerned [61]. GN25 is currently a drug candidate for MPM treatment, but there are no records of clinical trials at this time.

3.3. BAP1

The deubiquitinase enzyme encoded by the *BAP1* tumor suppressor gene is inactivated in over 60% of mesothelioma cases and is associated with global methylation via activation of PRC2. Rare germline mutations in *BAP1* contribute to predispositions in malignant mesothelioma, among other cancers, despite lack of exposure to high levels of asbestos or other carcinogens [68], suggesting that *BAP1* may serve as a predictive biomarker. Bott et al. reported that *BAP1* also harbored nonsynonymous mutations in 12/53 samples, was located in the center of the 3p21.1 locus, and also had somatic mutations and copy number losses [66]. Although they found no significant correlation between *BAP1* mutations and MPM subtype, Yoshikawa et al. found that *BAP1*-inactivating mutations occurred in epithelioid-type MM nearly exclusively, indicating that BAP1 may be more useful in diagnosing only epithelioid-type MM [67]. BAP1 plays a key role in chromatin biology by mediating deubiquitination of core histones and is recruited to double-strand breaks in DNA, given its role in DNA repair. Accordingly, *BAP1* inactivation results in impaired DNA repair functions, which can have dire consequences, consistent with its role as an epigenetic modulator and transcriptional regulator [69]. Thus, drugs designed to target the epigenome, and, in particular, enhancer of zeste homolog 2 (EZH2), may hold promise for MPM harboring *BAP1* mutations, but these have not been evaluated clinically.

4. Immune Microenvironment of MPM Tumors

The immune system plays a crucial role in tumor surveillance and attack; consequently, understanding the dynamic associations between pro-tumorigenic and anti-tumorigenic components of the MPM immune microenvironment is paramount to developing new treatments. Asbestos exposure by inhalation has become incontrovertible in MPM pathobiology and the physiological response of chronic inflammation as an immune reaction has been implicated in disease progression [70]. The secretion of VEGF and inflammatory cytokines by activated macrophages, coupled with the inability of mesothelial cells to expunge asbestos, along with the elevated presence of growth factors, results in malignant transformation. This leads to an influx of immune-suppressing cells like tumor-associated macrophages (TAMs), myeloid-derived suppressor cells (MDSCs) and regulatory T cells (T_{regs}) [71]. These events further propel tumor growth, due to immune surveillance escape, influx of stromal fibroblasts, and an increase in angiogenesis-sustaining endothelial cells [71]. Deciphering what differentiates one tumor from another could help in modulating the immune microenvironment to effectively induce an anti-tumor immune response, by strategic design and delivery of appropriate therapeutic agents.

Broadly, the microenvironment of the mesothelioma tumor contains a heterogeneous network of stromal cells, the immune population, tumor cells, fibroblasts, extracellular matrix and blood vessels [72]. Here, we focus on aspects of the tumor microenvironment where more information is available.

4.1. The Stromal Compartment

Connective tissue-derived stromal cells in the body secrete growth factors, like hepatocyte growth factor (HGF), that support regular cell division, but their interaction with cancer cells can be deleterious as they provide an extracellular matrix for tumor progression. Cancer-associated fibroblasts (CAFs) and myeloid cells are both stromal cells that help malignant cells to proliferate, migrate, and offer resistance to therapy, primarily by preventing immune T cells from entering the tumor [73]. In turn, the tumor cells themselves produce cytokines, such as fibroblast growth factor 2 (FGF-2) and platelet derived growth factor (PDGF), to continuously recruit more fibroblasts, and then these fibroblasts generate even greater amounts of HGF, thus contributing to the growth of the tumor mass [73]. This symbiosis between tumor cells and their stromal neighbors has been demonstrated by co-culturing lung fibroblasts with mesothelioma cell lines or patient-derived fibroblasts, because proliferation and migration of cancerous cells increased significantly due to activated MET-induced HGF; these effects were abrogated when the cells were treated with an anti-HGF antibody or FGFR/PDGFR inhibitors [70]. Thus, targeting fibroblast activation protein (FAP) in stromal cells has led to a decrease in mesothelioma tumor growth in preclinical animal studies [74]. Since chimeric antigen receptor (CAR)-T cell therapy is still quite new and exciting, it remains to be seen whether FAP targeting will become an effective therapeutic.

On the other hand, myeloid cells, which are highly abundant in mesothelioma, can be both immunosuppressive, as well as immunostimulatory, which can lead to tumor cell evasion or immune activation, depending on the environmental cues present. The tumor microenvironment has been demonstrated to be generally more immunosuppressive because of programmed death-ligand 1(PDL1) upregulation, which can inhibit T cell function [75]. Methods to target PD1/PDL1 using nivolumab are currently in phase II clinical trials for MPM use, and this may be adopted for clinical practice in the future. However, given variations in the tumor microenvironment from one mesothelioma patient to the next, an analysis of all individual cells and factors which make up this complex ecosystem is important to the pursuit of treatment.

4.2. The Immune Population

The presence of several different types of immune cells which support tumor formation can have a clinical impact on the immune microenvironment of MPM patients. $CD3^+$ pan T cells, $CD4^+$ helper T cells, and $CD8^+$ cytotoxic T cells are mainly found within the stromal compartment, $CD25^+CD4^+FOXP3^+$ regulatory T cells border the tumor cells, and $CD68^+$ macrophages and $CD16^+$ natural killer cells make up the majority of the tumor infiltrating population within the tumor [72]. The presence of $CD8^+$ cytotoxic T cells within the tumor has been associated with early stage disease, more apoptosis, and better survival. Conversely, in preclinical studies, the presence of immunosuppressive $CD4^+CD25^+FOXP3^+$ regulatory T cells and immunosuppressive soluble factors inside the tumor corresponded with poor prognosis in MPM, due to suppression of anti-tumor immune responses [76]. Accordingly, Hegmans et al. demonstrated that survival increased when $FoxP3^+CD4^+CD25^+$ T_{regs} were depleted in an in vivo model of MPM [72]. Adenosine and prostaglandin E2 (PGE2) immunosuppression has been shown to inhibit T cell function [77], and cyclooxygenase-2 (COX-2) suppression blocks the growth of mesothelioma tumors via an immunological mechanism which permits more effective cytotoxic T cell (CTL) build up in the tumors [78]. Kiyotani et al. demonstrated that different regions of the tumor yield different repertoires of tumor-infiltrating lymphocytes (TILs). This is due to the presence of distinct somatic mutations in different areas of the tumors, leading to discrete clonotypes of T cell receptors, and ultimate results in pronounced region-specific TILs [79]. The accumulation of $CD8^+$ tumor infiltrating lymphocytes (TILs) in tumors of MPM patients who underwent surgical resection correlated significantly with better survival [80]. Dendritic cells are rare, and interestingly, Cornwall et al. showed that there was a decline in the dendritic cell population, a reduction in expression of CD68, which can affect the cell's ability to become fully active, as well as a decrease in the cell's antigen-processing ability, in mesothelioma patients, compared to age-matched healthy donors [72,81]. Eosinophils, neutrophils, mast cells, and B cells are also rarely observed [72]. Thus, the presence of unique cell surface markers found on these various T cell populations could be exploited to selectively target those populations which are implicated in MPM progression, but progress has been limited thus far.

4.3. Tumor Cells

Cancer cells found within the tumor secrete high levels of VEGF to recruit endothelial cells for angiogenesis. The formation of new blood vessels has been established to contribute to tumor development by providing a continuous source of nutrients, growth factors and other related molecules described here. Tumor cells are reported to have elevated levels of chemokine (C-X-C motif) ligand 1 (CXCL1), a chemokine that plays a role in inflammation and tumorigenesis, as well as interleukin-6 (IL-6), a cytokine that prevents immune dendritic cell development [72]. Similar to their symbiotic relationship with stromal fibroblasts, tumor cells also have a beneficial relationship with the tumor-infiltrating macrophages. Cornelissen et al. demonstrated that factors secreted by the tumor cells transformed normal macrophages into malignant M2 macrophages, which reciprocally produce cytokines, such as interleukin-1 (IL-1), interleukin-6 (IL-6), interleukin-10 (IL-10), VEGF, and transforming growth factor beta (TGF-β), which contribute to tumor formation and development [82]. Chronic inflammation of the stromal compartment is associated with better patient survival in patients with epithelial histology, regardless of whether they received prior neoadjuvant therapy [83]. Cutting off the supply of one or more of the above-mentioned factors in the surrounding milieu is a therapeutic strategy being used to combat MPM tumors.

5. Developmental Pathways in Malignant Pleural Mesothelioma

MPM tumorigenesis is linked with asbestos exposure, chronic tissue inflammation, and subsequent tissue repair. Tissue regeneration and repair is associated with stem cell renewal genes and stem cell pathway activation. The activation of stem cell signaling is generally regulated and chronic stimulation and accompanying oncogenic events in these developmental pathways are correlated with poor prognosis in MPM patients [84]. Below we describe the role of four stem cell-associated developmental pathways (Figure 1B) and how they contribute to the heterogeneity in MPM tumors.

5.1. Hedgehog Pathway

Upon Hedgehog (Hh) ligand binding, the inhibition of transmembrane protein, Patched on Smoothened (Smo), is relieved and this activates the Hh signaling pathway via transcriptional activation of glioma-associated protein (Gli) factors [85]. Hedgehog-dependent Gli-mediated transactivation of target genes is negatively regulated by suppressor of fused homolog (Sufu); it is required for normal embryonic development and its loss leads to unfettered Hh activation and lethality [86]. This pathway is crucial for embryonic development, but also remains active in regulation of adult stem cells responsible for maintaining and regenerating adult tissues, post-injury [87]. Abnormal overactivation of the Hedgehog pathway has been suggested to lead to tumorigenesis, through the transformation of adult stem cells into the cancer stem cells that give rise to tumors. When aberrantly reactivated, this developmental pathway has been implicated in multiple human cancers, including MPM [88–90].

In a study of MPM specifically, in which quantitative PCR and in situ hybridization were used on 45 clinical samples, *GLI1*, *SHH*, and human hedgehog interacting protein (*HHIP*) gene expression were significantly increased in MPM tumors, compared with healthy pleural tissue, although other Hedgehog pathway genes, such as *PTCH1*, *IHH*, *DHH*, *SMO*, and *GLI2* did not differ [91]. In that study, smoothened inhibitors suppressed hedgehog signaling in primary MPM cell culture systems and decreased tumor growth in MPM xenografts, indicating a potential therapeutic approach [91]. In contrast, we showed that elevated *SMO* expression levels strongly correlated with poorer overall MPM patient survival (*n* = 46), underscoring the heterogeneity of Hh pathway gene expression profiles in these tumor microenvironments [92]. Furthermore, whereas autocrine Hh pathway up-regulation was initially described in MPM pathobiology, Meerang et al. detected paracrine activation of Hh signaling in MPM patients and reported heterogeneous expression of *GLI1* in both tumor and stroma fractions [93].

Their study provides evidence for targeting Hh signaling, as well as a number of MPM-specific pathway perturbations in the stromal compartment, as a treatment strategy for MPM [94].

Additional heterogeneity in MPM tumors is introduced by Hh-independent Gli activation and we showed aberrant *GLI1* and *GLI2* activation in about 90% of MPM tissue samples ($n = 46$) [95]. We also demonstrated that targeting Gli using siRNA and small molecule inhibitors (alone and in combination with chemotherapy) suppressed cell growth dramatically, both in vitro and in vivo, providing strong support for Gli being a novel clinical target for MPM treatment [95]. Evidence suggests that crosstalk between Gli and other cell signaling pathways that are independent of Hh, such as Kirsten rat sarcoma (KRAS), EGFR and mTOR, are linked to worse clinical outcomes in MPM and crosstalk between Gli and mTOR actually causes increased Gli activation [96]. Finally, loss of the tumor suppressor Numb, known to regulate Notch, Hedgehog and TP53 pathways, was reported in ~50% of tissue specimens and has been associated with poor prognosis in epithelioid MPM [97]. Forced expression of Numb conferred sensitivity to cisplatin and activated apoptotic pathway proteins, suggesting potential therapeutic options for MPM [97].

Pharmacological inhibition of the Hh pathway could prove extremely useful as a therapeutic strategy. Most experimental models report Hh antagonists that target Smo or Gli, such as cyclopamine, vismodegib, GANT61, arsenic trioxide or GLI-I, strongly suppressed MPM cell viability by blocking Gli activation. Several clinical trials of Hh inhibitors are underway; thus far, three mesothelioma patients who participated in the phase I study for vismodegib have not responded to treatment [94], but their tumors were not assessed for Hh activity before study enrollment. Thus, reliable biomarkers and patient stratification would significantly ameliorate unfavorable clinical outcomes so that appropriate individuals who can benefit most are selected [95,98].

5.2. Wnt/β-Catenin Pathway

The Wnt signal transduction pathway is activated by a secreted Wnt ligand, from a 19-member family, binding to one of 10 cell-surface Frizzled (Fzd) receptors and to LRP5/6 [99,100]. Through a series of intracellular events in the canonical pathway, Fzd signals to the Dishevelled (Dvl) proteins, which causes an accumulation of β-catenin in the cytoplasm, followed by nuclear translocation and activation of the T-cell factor/lymphoid enhancer factor (TCF/LEF) family of transcription factors in a context-dependent manner [99,100]. In the absence of Wnt, β-catenin is degraded in the cytoplasm via a destruction complex (Axin, APC, GSK3, PP2A, CK1α) and the pathway becomes inactive [99,100]. This conserved pathway orchestrates cell fate decisions during embryonic development, maintenance of self-renewing stem cells in adults, and integration of signaling cues from other pathways, like bone morphogenetic protein (BMP), FGF, TGF-β and retinoic acid [99]. Mutations in the Wnt pathway are frequently observed in cancer, including germline mutations in *APC*, responsible for familiar adenomatous polyposis and colorectal cancers [101]. Loss-of-function mutations in Axin lead to hepatocellular carcinomas, and mutations in β-catenin lead to melanoma and several solid tumors; in addition, studies have reported the expression of β-catenin, a Wnt stem cell pathway activation indicator, in the progression of mesothelioma [102,103].

In a comparative Wnt-specific microarray study between MPM and normal lung tissue, our laboratory reported that the most common event in MPM was up-regulation of Wnt2 [104]. Subsequent targeted-Wnt2 inhibition via siRNA and a monoclonal Ab has an anti-proliferative effect and prompted apoptosis in MPM cells, thereby validating Wnt2 as a therapeutic target [104,105]. Other members of the Wnt signal transduction pathway, such as secreted frizzled-related proteins (sFRP), which are negative Wnt modulators, were shown to be transcriptionally down-regulated in MPM primary tissues and cell lines [106].

To identify meaningful biomarkers in MPM, our group used qRT-PCR to evaluate the expression of Wnt7A in 50 tumor specimens from patients who underwent surgical resection at the University of California, San Francisco (UCSF) [107]. We found that Wnt7A expression predicted sensitivity to chemotherapy; in particular, low Wnt7A expression was significantly correlated with negative overall survival in a univariate analysis, and patients with epithelioid tumors and low Wnt7A had significantly worse prognoses [107]. Furthermore, patients who had low Wnt7A-expressing tumors and received neoadjuvant chemotherapy had a better prognosis than those who did not [107]. Finally, a set of microRNAs was identified, which antagonize Wnt signaling, and were all down-regulated in MPM, compared to lung adenocarcinoma; these biomarkers could facilitate differential diagnoses between these different, but related, cancers [108].

About two dozen small-molecule inhibitors have been designed to target the Wnt pathway, but the most effort has been aimed at the TCF/β-catenin complex, to try to block signaling at the transcriptional level [101]. Of these inhibitors, LGK-974, a protein-serine *O*-palmitoleoyltransferase porcupine (PORCN)-specific inhibitor that is a key regulator of the Wnt signaling pathway, is currently being evaluated in phase I clinical studies for solid malignancies with documented genetic alterations upstream in the Wnt pathway, but no MPM patients were reported to have enrolled (NCT01351103). Additional small molecules have been developed for other Wnt pathway targets, such as β-catenin, GSK-3β, KCNQ1/KCNE1, Wnt3A, TNKS1/2, TNIK, and Axin/β-catenin interaction. However, it is not yet known whether any of these or other therapies that target the Wnt pathway will be effective in curbing the growth of MPM tumors.

5.3. Notch Pathway

Yet another stem-cell signaling and/or developmental pathway that exerts its effect on MPM cell lines, is the Notch pathway, which maintains tissue homeostasis and regulates neural stem cells in adults [109,110]. Four single-pass transmembrane receptors—Notch 1–4—consisting of large extracellular domains, get bound to a Notch ligand, which induces proteolytic cleavage and release of the intracellular domain, which translocates to the nucleus to influence gene transcription [111]. The Notch family plays an important role in normal development, and when it goes awry, promotes tumorigenesis [112]. As described in the Hh pathway, Numb acts as a tumor suppressor and in a Notch-specific context, promotes neural differentiation and maintains stem cell compartments; loss of Numb expression is linked to worse clinical prognoses in epitheliod MPM, and its upregulation could be associated with chemosensitivity, making it an informative biomarker [97].

Paralog-specific heterogeneity in the Notch pathway has been reported, since Notch1 and Notch2 exhibit opposing effects, depending upon the disease context [113]. For instance, Notch2 suppresses the effects of Notch1 in mesothelioma, but in medulloblastoma, Notch2 actually stimulates tumorigenesis, whereas Notch1 inhibits it [110]. Furthermore, several Notch ligands exist, and some trigger pathway activation, whereas others have inhibitory effects; thus, the complexities and biological consequences need to be carefully considered in therapeutic targeting [114,115]. Several small molecules have been developed to target Notch and γ secretase and a few have advanced to phase II/III clinical trials for indications like solid cancers and Alzheimer's disease; testing these molecules in MPM might be worthwhile [116,117]. We have shown that CK2α inhibition down-regulates Notch1 signaling and reduces CSC-like populations in human lung cancer cells. Since there are no active trials of CK2 inhibitors in mesothelioma, small-molecules like CX-4945 or CIGB 300 anti-CK2 peptides may have therapeutic utility [118]. Interestingly, CK2 participates in the regulation of Hedgehog (Hh)/Gli1 signaling, underscoring the prevalence of cross-talk between crucial stem cell pathways implicated in cancer [119]. However, since much less investigation of MPM and Notch signaling has been reported, as compared to Hh and Wnt, further studies are warranted.

5.4. Hippo/YAP Pathway

Hippo signaling regulates organ size, cell contact inhibition, and stem cell self-renewal, but dysregulation can result in cancer development [120,121]. Central to Hippo signaling is a cascade of kinases, comprising Mst1/2, which complex with SAVI to phosphorylate and activate LATS1/2 [120]. These phosphorylation events inhibit two major downstream effectors of the Hippo pathway–YAP and TAZ—to stymie signaling [120,122]. However, in the dephosphorylated state, YAP/TAZ interact with TEAD 1–4 and other transcription factors in the nucleus to activate cell proliferation genes and inhibit apoptosis [120]. The Hippo pathway is regulated by several upstream factors, and two in particular, the *NF2/Merlin* and *YAP* oncogenes, are mutated or overactive in approximately 40% and 70% of MPM cases, respectively [96]. Furthermore, *NF2* regulates homeostasis in tissue repair and controls stem cell signaling, making it the most notorious perpetrator of MPM oncogenesis and an ideal therapeutic target for drug development [96]. In fact, Bueno et al. performed integrated analyses on 216 MPM samples and identified alterations in Hippo, among the most frequent signaling pathways in MPM [123].

In an effort to group MPM patients by molecular sub-type, to address intertumoral heterogeneity, Tranchant et al. reported that the tumor suppressor *LATS2* gene was altered in 11% of MPM [124]. They identified a new subgroup, C2LN, in which mutations in the *LATS2* and *NF2* genes co-occur within the same MPM and these patients showed enhanced sensitivity to PF-04691502, a PI3K-AKT-mTOR inhibitor [124]. Also, these investigators reported the *MOK* gene as a specific potential biomarker, noting that studies like these would be ideal for ameliorating clinical outcomes based on specialized gene signatures, but no clinical trial information is available [124]. To develop targeted treatment for *NF2*-mutant cancers, Cooper et al. demonstrated that MLN4924, an NEDD8-activating enzyme (NAE) inhibitor, suppresses CRL4DCAF1 (E3 ubiquitin ligase) and interferes with YAP activation in *NF2*-mutant cancers alone, and in combination with, the PI3K-mTOR inhibitor, GDC-0980 [125].

In order to investigate the association between *NF2/Merlin* and *YAP* gene alterations and clinical outcomes, Meerang et al. performed multivariate analyses in MPM patients [126]. They found that NF2/Merlin protein expression and the Survivin labeling index were prognosticators for poor clinical outcomes in two independent MPM cohorts, indicating that this data could guide treatment selection [126]. Interestingly, Felley-Bosco et al. reported an interaction between Hedgehog and YAP signaling, which makes Hh-specific modulating agents amenable to treating MPM [96]. Therapeutic strategies, aimed at disrupting YAP activity, include (1) targeting the Hh pathway as it down-regulates the YAP protein, (2) a small molecule called verteporfin, which inhibits YAP-TEAD transcription factor assembly, and (3) obstructions of lysophosphatidic acid (LPA) and thrombin receptor signaling, because upon agonist binding, they activate YAP [96].

We have outlined a few stem cell pathways which not only get deregulated in several types of cancer, but also in MPM. A number of therapeutics have been developed for components of these pathways, which are under clinical investigation and may be used for patient care in the near future.

6. Conclusions and Perspectives

Over the past 10 years our knowledge of MPM biology has advanced substantially, specifically regarding the molecular pathways involved, genetic and epigenetic associations, and the heterogeneous immune microenvironment of MPM tumors. This progress has led to the development of new and exciting therapeutic strategies and treatment options for patients diagnosed with this deadly malignancy. However, a "one-size fits all" approach will not suffice because heterogeneity in the tumor microenvironment exists from one malignancy to another, and one patient to another, and also evolves in response to therapies administered. In addition, stratification of patients based on genetic signature to guide targeted therapeutic interventions, such as the use of CDK and mTOR inhibitors, and epigenetic modulators, could be more effective for treating patients. Furthermore, several clinical investigations of therapies outlined in this review for MPM are underway.

Acknowledgments: No grants were used to fund this work or to publish it.

Author Contributions: Bhairavi Tolani conceived and designed the manuscript; Luis A. Acevedo and Ngoc T. Hoang performed the literature search; Bhairavi Tolani wrote the paper and Biao He provided publishing costs.

Conflicts of Interest: The authors declare no conflict of interest.

References

1. Tarver, T. *Cancer Facts and Figures*; American Cancer Society: Atlanta, GA, USA, 2012.
2. *Mortality and Morbidity Weekly Report*; Center for Disease Control: Atlanta, GA, USA, 2009.
3. Carbone, M.; Ly, B.H.; Dodson, R.F.; Pagano, I.; Morris, P.T.; Dogan, U.A.; Gazdar, A.F.; Pass, H.I.; Yang, H. Malignant mesothelioma: Facts, myths, and hypotheses. *J. Cell. Physiol.* **2012**, *227*, 44–58. [CrossRef] [PubMed]
4. Zucali, P.A.; Ceresoli, G.L.; Vincenzo, F.D.; Simonelli, M.; Lorenzi, E.; Gianoncelli, L.; Santoro, A. Advances in the biology of malignant pleural mesothelioma. *Cancer Treat. Rev.* **2011**, *37*, 543–558. [CrossRef] [PubMed]
5. Wu, L.; Perrot, M. Radio-immunotherapy and chemo-therapy as a novel treatment paradigm in malignant pleural mesothelioma. *Transl. Lung Cancer Res.* **2017**, *6*, 325–334. [CrossRef] [PubMed]
6. Tsao, A.; Wistuba, I.; Roth, J.; Kindler, H. Malignant Pleural Mesothelioma. *J. Clin. Oncol.* **2009**, *27*, 2081–2090. [CrossRef] [PubMed]
7. Travis, W.D.; Brambilla, E.; Burke, A.; Marx, A.; Nicholson, A.G. *WHO Classification of Tumours of the Lung, Pleura, Thymus and Heart*; WHO: Geneva, Switzerland, 2015.
8. National Lung Cancer Audit. *National Lung Cancer Audit Report 2014: Mesothelioma Report for the Period 2008–2012*; Health and Social Care Information Centre: Leeds, UK, 2014.
9. Strizzi, L.; Catalano, A.; Vianale, G.; Orecchia, S.; Casalini, A.; Tassi, G.; Puntoni, R.; Mutti, L.; Procopio, A. Vascular Endothelial Growth Factor is an Autocrine Growth Factor in Human Malignant Mesothelioma. *J. Pathol.* **2001**, *193*, 468–475. [CrossRef] [PubMed]
10. Masood, R.; Kundra, A.; Zhu, S.; Xia, G.; Scalia, P.; Smith, D.L.; Gill, P.S. Malignant Mesothelioma Growth Inhibition by Agents that Target the VEGF and VEGF-C Autocrine Loop. *Int. J. Cancer* **2003**, *104*, 603–610. [CrossRef] [PubMed]
11. Ohta, Y.; Shridhar, V.; Bright, R.K.; Kalemkerian, G.P.; Du, W.; Carbone, M.; Watanabe, Y.; Pass, H.I. VEGF and VEGF type C Play an Important Role in Angiogenesis and Lymphangiogenesis in Human Malignant Mesothelioma Tumors. *Br. J. Cancer* **1999**, *81*, 54–61. [CrossRef] [PubMed]
12. Kumar-Singh, S.; Weyler, J.; Martin, M.J.; Vermeulen, P.B.; Van MArck, E. Angiogenic Cytokines in Mesothelioma: A Study of VEGF, FGF-1 and -2, and TGF β Expression. *J. Pathol.* **1999**, *189*, 72–78. [CrossRef]
13. Tolnay, E.; Kuhnen, C.; Wiethege, T.; Konig, J.E.; Voss, B.; Muller, K.M. Hepatocyte Growth Factor/Scatter Factor and its Receptor c-Met are Overexpressed and Associated with an Increased Microvessel Density in Malignant Pleural Mesothelioma. *J. Cancer Res. Clin. Oncol.* **1998**, *124*, 291–296. [CrossRef] [PubMed]
14. Kindler, H.L.; Karrison, T.G.; Gandara, D.R.; Lu, C.; Krug, L.M.; Stevenson, J.P.; Janne, P.A.; Quinn, D.I.; Koczywas, M.N.; Brahmer, J.R.; et al. Multicenter, Double-blind, Placebo-controlled, Randomized Phase II Trial of Gemcitabine/Cispltain Plus Bevacizumab or Placebo in Patients with Malignant Mesothelioma. *J. Clin. Oncol.* **2012**, *30*, 2508–2515. [CrossRef] [PubMed]
15. Levin, P.A.; Dowell, J.E. Spotlight on bevacizumab and its potential in the treatment of malignant pleural mesothelioma: The evidence to date. *Onco Targets Ther.* **2017**, *10*, 2057–2066. [CrossRef] [PubMed]
16. Jahan, T.; Gu, L.; Kratzke, R.; Dudek, A.; Otterson, G.A.; Wang, X.; Green, M.; Vokes, E.E.; Kindler, H.L. Vatalanib in Malignant Mesothelioma: A Phase II Trial by the Cancer and Leukemia Group B (CALGB 30107). *Lung Cancer* **2012**, *76*, 393–396. [CrossRef] [PubMed]
17. Ikuta, I.; Yano, S.; Trung, V.T.; Hanibuchi, M.; Goto, H.; Li, Q.; Wang, W.; Yamada, T.; Ogino, H.; Kakiuchi, S.; et al. E7080, a Multi-tyrosine Kinase Inhibitor, Suppresses the Progression of Malignant Pleural Mesothelioma with Different Proangiogenic Cytokine Production Profiles. *Clin. Cancer Res.* **2009**, *15*, 7229–7237. [CrossRef] [PubMed]
18. Baas, P.; Boogerd, W.; Dalesio, O.; Haringhuizen, A.; Custers, F.; van Zandwijk, N. Thalidomide in patients with malignant pleural mesothelioma. *Lung Cancer* **2005**, *48*, 291–296. [CrossRef] [PubMed]
19. Janne, P.A.; Wang, X.F.; Krug, L.M.; Hodgson, L.; Vokes, E.E.; Kindler, H.L. Sorafenib in malignant pleural mesothelioma (MM): A phase II trial of the Cancer and Leukemia Group B (CALGB 30307). *J. Clin. Oncol.* **2007**, *25*, 2007–7707.
20. Nowak, A.K.; Millward, M.J.; Francis, R.; van der Schaaf, A.; Musk, A.W.; Byrne, M.J. Phase II study of sunitinib as second-line therapy in malignant pleural mesothelioma (MPM). *J. Clin. Oncol.* **2008**. [CrossRef]
21. Harada, A.; Unchino, J.; Harada, T.; Nakagaki, N.; Hisasue, J.; Fujita, M.; Takayama, K. Vascular Endothelial Growth Factor Promoter-based Conditional Replicative Adenoviruses Effectively Suppress Growth of Malignant Pleural Mesothelioma. *Cancer Sci.* **2017**, *108*, 116–123. [CrossRef] [PubMed]

22. Suzuki, Y.; Sakai, K.; Ueki, J.; Xu, Q.; Nakamura, T.; Shimada, H.; Nakamura, T.; Matsumoto, K. Inhibition of Met/HGF Receptor and Angiogenesis by NK4 Leads to Suppression of Tumor Growth and Migration in Malignant Pleural Mesothelioma. *Int. J. Cancer* **2010**, *127*, 1948–1957. [CrossRef] [PubMed]

23. *Identifier NCT00309946*; National Library of Medicine (US): Bethesda, MD, USA, 2017. Available online: https://clinicaltrials.gov/ct2/show/NCT00309946 (accessed on 10 September 2017).

24. Hollstein, M.; Sidransky, D.; Vogelstein, B.; Harris, C.C. p53 mutations in human cancers. *Science* **1991**, *253*, 49–53. [CrossRef] [PubMed]

25. Fennell, D.A.; Rudd, R.M. Defective Core-Apoptosis Signaling in Diffuse Malignant Pleural Mesothelioma: Opportunities for Effective Drug Development. *Lancet Oncol.* **2004**, *5*, 354–362. [CrossRef]

26. De Assis, L.V.; Locatelli, J.; Isoldi, M.C. The Role of Key Genes and Pathways involved in the Tumorigenesis of Malignant Mesothelioma. *Biochim. Biophys. Acta* **2014**, *1845*, 232–247. [CrossRef] [PubMed]

27. Hopkins-Donaldson, S.; Belyanskaya, L.L.; Simoes-Wust, A.P.; Sigris, B.; Kurtz, S.; Zangemeister-Wittke, U.; Stahel, R. p53-induced Apoptosis Occurs in the Absence of P14(ARF) in Malignant Pleural Mesothelioma. *Neoplasia* **2006**, *8*, 551–559. [CrossRef] [PubMed]

28. Gordon, G.J.; Mani, M.; Mukhopadhyay, L.; Dong, L.; Edenfield, H.R.; Glickman, J.N.; Yeap, B.Y.; Sugarbaker, D.J.; Bueno, R. Expression Patterns of Inhibitor of Apoptosis Proteins in Malignant Pleural Mesothelioma. *J. Pathol.* **2007**, *211*, 447–454. [CrossRef] [PubMed]

29. Cao, X.X.; Mohuiddin, I.; Ece, F.; McConkey, D.J.; Smythe, W.R. Histone deacetylase inhibitor downregulation of *bcl-xl* gene expression leads to apoptotic cell death in mesothelioma. *Am. J. Respir. Cell Mol. Biol.* **2001**, *25*, 562–568. [CrossRef] [PubMed]

30. Ozvaran, M.K.; Cao, X.X.; Miller, S.D.; Monia, B.A.; Hong, W.K.; Smythe, W.R. Antisense oligonucleotides directed at the *bcl-xl* gene product augment chemotherapy response in mesothelioma. *Mol. Cancer Ther.* **2004**, *3*, 545–550. [PubMed]

31. Cao, X.; Rodarte, C.; Zhang, L.; Morgan, C.D.; Littlejohn, J.; Smythe, W.R. Bcl2/bcl-xL inhibitor engenders apoptosis and increases chemosensitivity in Mesothelioma. *Cancer Biol. Ther.* **2007**, *6*, 1–7. [CrossRef]

32. Ou, W.B.; Lu, M.; Ellers, G.; Li, H.; Ding, J.; Meng, X.; Wu, Y.; He, Q.; Sheng, Q.; Zhou, H.M.; Fletcher, J.A. Co-targeting of FAK and MDM2 Triggers Additive Anti-proliferative Effects in Mesothelioma via a Coordinated Reactivation of P53. *Br. J. Cancer* **2016**, *115*, 1253–1263. [CrossRef] [PubMed]

33. Tsao, A.S.; He, D.; Saigal, B.; Liu, S.; Lee, J.J.; Bakkannagari, S.; Ordonez, N.G.; Hong, W.K.; Wistuba, I.; Johnson, F.M. Inhibition of c-Src Expression and Activation in Malignant Pleural Mesothelioma Tissues Leads to Apoptosis, Cell Cycle Arrest, and Decreased Migration and Invasion. *Mol. Cancer Ther.* **2007**, *6*, 1962–1972. [CrossRef] [PubMed]

34. Stewart, J.H.; Tran, T.L.; Levi, N.; Tsai, W.S.; Schrump, D.S.; Nguyen, D.M. The Essential Role of the Mitochondria and Reactive Oxygen Species in Cisplatin-mediated Enhancement of Fas Ligand-induced Apoptosis in Malignant Pleural Mesothelioma. *J. Surg. Res.* **2007**, *141*, 120–131. [CrossRef] [PubMed]

35. Patergnani, S.; Giorgi, C.; Maniero, S.; Missiroli, S.; Maniscalco, P.; Bononi, I.; Martini, F.; Cavallesco, G.; Tognon, M.; Pinton, P. The Endoplasmic Reticulum Mitochondrial Calcium Cross Talk is Downregulated in Malignant Pleural Mesothelioma Cells and Plays a Critical Role in Apoptosis Inhibition. *Oncotarget* **2015**, *6*, 23427–23444. [CrossRef] [PubMed]

36. Lathrop, M.J.; Sage, E.K.; Macura, S.L.; Brooks, E.M.; Cruz, F.; Bonenfant, N.R.; Sokocevic, D.; MacPherson, M.B.; Beuschel, S.L.; Dunaway, C.W.; et al. Antitumor Effects of TRAIL-expressing Mesenchymal Stromal Cells in a Mouse Xenograft Model of Human Mesothelioma. *Cancer Gene Ther.* **2015**, *22*, 44–54. [CrossRef] [PubMed]

37. Lopez-Rios, F.; Chuai, S.; Flores, R.; Shimizu, S.; Ohno, T.; Wakahara, K.; Illei, P.B.; Hussain, S.; Krug, L.; Zakowski, M.F.; et al. Global Gene Expression Profiling of Pleural Mesothelioma: Overexpression of Aurora Kinases and P16/CDKN2A Deletion as Prognostic Factors and Critical Evaluation of Microarray-based Prognostic Prediction. *Cancer Res.* **2006**, *66*, 2970–2979. [CrossRef] [PubMed]

38. Christensen, B.C.; Godleski, J.J.; Marsit, C.J.; Houseman, E.A.; Lopez-Fagundo, C.Y.; Longacker, J.L.; Bueno, R.; Sugarbaker, D.J.; Nelson, H.H.; Kelsey, K.T. Asbestos Exposure Predicts Cell Cycle Control Gene Promoter Methylation in Pleural Mesothelioma. *Carcinogenesis* **2008**, *29*, 1555–1559. [CrossRef] [PubMed]

39. Romagnoli, S.; Fasoli, E.; Vaira, V.; Falleni, M.; Pellegrini, C.; Catania, A.; Roncalli, M.; Marchetti, A.; Santambrogio, L.; Coggi, G.; et al. Identification of Potential Therapeutic Targets in Malignant Mesothelioma Using Cell-Cycle Gene Expression Analysis. *Am. J. Pathol.* **2009**, *174*, 762–770. [CrossRef] [PubMed]

40. Mizuno, T.; Murakami, H.; Fujii, M.; Ishiguro, F.; Tanaka, I.; Kondo, Y.; Akatsuka, S.; Toyokuni, S.; Yokoi, K.; Osada, H.; et al. YAP Induces Malignant Mesothelioma Cell Proliferation by Upregulating Transcription of Cell Cycle-promoting Genes. *Oncogene* **2012**, *31*, 5117–5122. [CrossRef] [PubMed]

41. Dhaene, K.; Hubner, R.; Kumar-Singh, S.; Weyn, B.; Van Marck, E. Telomerase activity in human pleural mesothelioma. *Thorax* **1998**, *53*, 915–918. [CrossRef] [PubMed]

42. Hanahan, D.; Weinberg, R.A. Hallmarks of Cancer: The Next Generation. *Cell* **2011**, *144*, 646–674. [CrossRef] [PubMed]

43. Whitson, B.A.; Kratzke, R.A. Molecular Pathways in Malignant Pleural Mesothelioma. *Cancer Lett.* **2006**, *239*, 183–189. [CrossRef] [PubMed]

44. Okuda, K.; Sasaki, H.; Kawano, O.; Yukiue, H.; Yokoyama, T.; Yano, M.; Fujii, Y. Epidermal growth factor receptor gene mutation, amplification and protein expression in malignant pleural mesothelioma. *J. Cancer Res. Clin. Oncol.* **2008**, *134*, 1105–1111. [CrossRef] [PubMed]

45. Rena, O.; Boldorini, L.R.; Gaudino, E.; Casadio, C. Epidermal growth factor receptor overexpression in malignant pleural mesothelioma: Prognostic correlations. *J. Surg. Oncol.* **2011**, *104*, 701–705. [CrossRef] [PubMed]

46. Govindan, R.; Kratzke, R.A.; Herndon, J.E., 2nd; Niehans, G.A.; Vollmer, R.; Watson, D.; Green, M.R.; Kindler, H.L.; Cancer and Leukemia Group B (CALGB 30101). Gefitinib in patients with malignant mesothelioma: A phase II study by the Cancer and Leukemia Group B. *Clin. Cancer. Res.* **2005**, *11*, 2300–2304. [CrossRef] [PubMed]

47. Garland, L.L.; Rankin, C.; Gandara, D.R.; Rivkin, S.E.; Scott, K.M.; Nagle, R.B.; Klein-Szanto, A.J.; Testa, J.R.; Altomare, D.A.; Borden, E.C. Phase II study of erlotinib in patients with malignant pleural mesothelioma: A Southwest Oncology Group Study. *J. Clin. Oncol.* **2007**, *25*, 2406–2413. [CrossRef] [PubMed]

48. Tsao, A.S.; Harun, N.; Fujimoto, J.; Devito, V.; Lee, J.J.; Kuhn, E.; Mehran, R.; Rice, D.; Moran, C.; Hong, W.K.; et al. Elevated *PDGFRB* gene copy number gain is prognostic for improved survival outcomes in resected malignant pleural mesothelioma. *Ann. Diagn. Pathol.* **2014**, *18*, 140–145. [CrossRef] [PubMed]

49. Melaiu, O.; Catalano, C.; de Santi, C.; Cipollini, M.; Figlioli, G.; Pelle, L.; Barone, E.; Evangelista, M.; Guazzelli, A.; Boldrini, L.; et al. Inhibition of the platelet-derived growth factor receptor β (PDGFRβ) using gene silencing, crenolanib besylate, or imatinib mesylate hampers the malignant phenotype of mesothelioma cell lines. *Genes Cancer* **2017**, *8*, 438–452. [PubMed]

50. Schelch, K.; Hoda, M.A.; Klikovits, T.; Munzker, J.; Ghanim, B.; Wagner, C.; Garay, T.; Laszlo, V.; Setinek, U.; Dome, B.; et al. Fibroblast growth factor receptor inhibition is active against mesothelioma and synergizes with radio- and chemotherapy. *Am. J. Respir. Crit. Care Med.* **2014**, *190*, 763–772. [CrossRef] [PubMed]

51. Marek, L.A.; Hinz, T.K.; von Massenhausen, A.; Olszewski, K.A.; Kleczko, E.K.; Boehm, D.; Weiser-Evans, M.C.; Nemenoff, R.A.; Hoffmann, H.; Warth, A.; et al. Nonamplified FGFR1 is a growth driver in malignant pleural mesothelioma. *Mol. Cancer Res.* **2014**, *12*, 1460–1469. [CrossRef] [PubMed]

52. Strizzi, L.; Vianele, G.; Catalano, A.; Muraro, R.; Mutti, L.; Procopio, A. Basic fibroblast growth factor in mesothelioma pleural effusions: Correlation with patient survival and angiogenesis. *Int. J. Oncol.* **2001**, *18*, 1093–1098. [CrossRef] [PubMed]

53. Pattarozzi, A.; Carra, E.; Favoni, R.E.; Wurth, R.; Marubbi, D.; Filiberti, R.A.; Mutti, L.; Florio, T.; Barbieri, F.; Daga, A. The inhibition of FGF receptor 1 activity mediates sorafenib antiproliferative effects in human malignant pleural mesothelioma tumor-initiating cells. *Stem Cell Res. Ther.* **2017**, *8*, 119. [CrossRef] [PubMed]

54. Christensen, B.C.; Houseman, E.A.; Poage, G.M.; Godleski, J.J.; Bueno, R.; Sugarbaker, D.J.; Wiencke, J.K.; Nelson, H.H.; Marsit, C.J.; Kelsey, K.T. Integrated profiling reveals a global correlation between epigenetic and genetic alterations in mesothelioma. *Cancer Res.* **2010**, *70*, 5686–5694. [CrossRef] [PubMed]

55. Christensen, B.C.; Houseman, E.A.; Godleski, J.J.; Marsit, C.J.; Longacker, J.L.; Roelofs, C.R.; Karagas, M.R.; Wrensch, M.R.; Yeh, R.F.; Nelson, H.H.; et al. Epigenetic profiles distinguish pleural mesothelioma from normal pleura and predict lung asbestos burden and clinical outcome. *Cancer Res.* **2009**, *69*, 227–234. [CrossRef] [PubMed]

56. Awad, M.M.; Jones, R.E.; Liu, H.; Lizotte, P.H.; Ivanova, M.; Kulkarni, M.; Herter-Sprie, G.S.; Lia, X.; Santos, A.A.; Bittinger, M.A.; et al. Cytotoxic T Cells in PD-L1-Positive Mesothelioma Are Counterbalanced by Distinct Immunosuppressive Factors. *Cancer Immunol. Res.* **2016**, *4*, 1038–1048. [CrossRef] [PubMed]

57. Altomare, D.A.; Menges, C.W.; Xu, J.; Pei, J.; Zhang, L.; Tadevosyan, A.; Neumann-Domer, E.; Liu, Z.; Carbone, M.; Chudoba, I.; et al. Losses of Both Products of the Cdkn2a/Arf Locus Contribute to Asbestos-induced Mesothelioma Development and Cooperate to Accelerate Tumorigenesis. *PLoS ONE* **2011**, *6*, e18828. [CrossRef] [PubMed]

58. Wong, L.; Zhou, J.; Anderson, D.; Kratzke, R.A. Inactivation of p16INK4a expression in malignant mesothelioma by methylation. *Lung Cancer* **2002**, *38*, 131–136. [CrossRef]

59. Jennings, C.J.; Murer, B.; O'Grady, A.; Hearn, L.M.; Harvey, B.J.; Kay, E.W.; Thomas, W. Differential p16/INK4A Cyclin-dependent Kinase Inhibitor Expression Correlates with Chemotherapy Efficacy in a Cohort of 88 Malignant Pleural Mesothelioma Patients. *Br. J. Cancer* **2015**, *113*, 69–75. [CrossRef] [PubMed]

60. Protocol for Study CDKO-125a-005, Version 29 September 2008. Nerviano Medical Sciences, 2008. Document No CDKO-125a-005-P. Available online: https://www.clinicaltrialsregister.eu/ctr-search/search?query=CDKO-125a-005 (accessed on 10 September 2017).

61. Cho, J.H.; Lee, S.J.; Oh, A.Y.; Yoon, M.H.; Woo, T.G.; Park, B.J. NF2 blocks Snail-mediated p53 suppression in mesothelioma. *Oncotarget* **2015**, *6*, 10073–10085. [CrossRef] [PubMed]

62. Ladanyi, M.; Zauderer, M.G.; Krug, L.M.; Ito, T.; McMillan, R.; Bott, M.; Giancotti, F. New strategies in pleural mesothelioma: BAP1 and NF2 as novel targets for therapeutic development and risk assessment. *Clin. Cancer Res.* **2012**, *18*, 4485–4490. [CrossRef] [PubMed]

63. Sekido, Y. Inactivation of Merlin in malignant mesothelioma cells and the Hippo signaling cascade dysregulation. *Pathol. Int.* **2011**, *61*, 331–344. [CrossRef] [PubMed]

64. Thurneysen, C.; Opitz, I.; Kurtz, S.; Weder, W; Stahel, R.A.; Felley-Bosco, E. Functional inactivation of NF2/merlin in human mesothelioma. *Lung Cancer* **2009**, *64*, 140–147. [CrossRef] [PubMed]

65. Miyanaga, A.; Masuda, M.; Tsuta, K.; Kawasaki, K.; Nakamura, Y.; Sakuma, T.; Asamura, H.; Gemma, A.; Yamada, T. Hippo pathway gene mutations in malignant mesothelioma: Revealed by RNA and targeted exon sequencing. *J. Thorac. Oncol.* **2015**, *5*, 844–851. [CrossRef] [PubMed]

66. Bott, M.; Brevet, M.; Taylor, B.S.; Shimizu, S.; Ito, T.; Wang, L.; Creaney, J.; Lake, R.A.; Zakowski, M.F.; Reva, B.; et al. The nuclear deubiquitinase BAP1 is commonly inactivated by somatic mutations and 3p21.1 losses in malignant pleural mesothelioma. *Nat. Genet.* **2011**, *43*, 668–672. [CrossRef] [PubMed]

67. Yoshikawa, Y.; Sato, A.; Tsujimura, T.; Emi, M.; Morinaga, T.; Fukuoka, K.; Yamada, S.; Murakami, A.; Kondo, N.; Matsumoto, S.; et al. Frequent inactivation of the *BAP1* gene in epithelioid-type malignant mesothelioma. *Cancer Sci.* **2012**, *103*, 868–874. [CrossRef] [PubMed]

68. Testa, J.R.; Cheung, M.; Pei, J.; Below, J.E.; Tan, Y.; Sementino, E.; Cox, N.J.; Dogan, A.U.; Pass, H.I.; Trusa, S.; et al. Germline BAP1 mutations predispose to malignant mesothelioma. *Nat. Genet.* **2011**, *43*, 1022–1025. [CrossRef] [PubMed]

69. Yap, T.A.; Aerts, J.G.; Popat, S.; Fennell, D.A. Novel insights into mesothelioma biology and implications for therapy. *Nat. Rev. Cancer* **2017**, *17*, 475–488. [CrossRef] [PubMed]

70. Li, Q.; Wang, W.; Yamada, T.; Matsumoto, K.; Sakai, K.; Bando, Y.; Uehara, H.; Nishioka, Y.; Sone, S.; Iwakiri, S.; et al. Pleural Mesothelioma Instigates Tumor-Associated Fibroblasts to Promote Progression via a Malignant Cytokine Network. *Am. J. Pathol.* **2011**, *179*, 1483–1493.

71. Sekido, Y. Molecular pathogenesis of malignant mesothelioma. *Carcinogenesis* **2013**, *34*, 1413–1419. [CrossRef] [PubMed]

72. Hegmans, J.P.; Hemmes, A.; Hammad, H.; Boon, L.; Hoogsteden, H.C.; Lambrecht, B.N. Mesothelioma Environment comprises cytokines and T-regulatory cells that suppress immune responses. *Eur. Respir. J.* **2006**, *27*, 1086–1095. [CrossRef] [PubMed]

73. Lo, A.; Wang, L.S.; Scholler, J.; Monslow, J.; Avery, D.; Newick, K.; O'Brien, S.; Evans, R.A.; Bajor, D.J.; Clendenin, C.; et al. Tumor-Promoting Desmoplasia Is Disrupted by Depleting FAP-Expressing Stromal Cells. *Cancer Res.* **2015**, *75*, 2800–2810. [CrossRef] [PubMed]

74. Wang, L.C.; Lo, A.; Scholler, J.; Sun, J.; Majumdar, R.S.; Kapoor, V.; Antzis, M.; Cotner, C.E.; Johnson, L.A. Targeting fibroblast activation protein in tumor stroma with chimeric antigen receptor T cells can inhibit tumor growth and augment host immunity without severe toxicity. *Cancer Immunol. Res.* **2014**, *2*, 154–166. [CrossRef] [PubMed]

75. Lievense, L.A.; Cornelissen, R.; Bezemer, K.; Kaijen-Lambers, M.E.; Hegmans, J.P.; Aerts, J.G. Pleural Effusion of Patients with Malignant Mesothelioma Induces Macrophage-Mediated T Cell Suppression. *J. Thorac. Oncol.* **2016**, *11*, 1755–1764. [CrossRef] [PubMed]

76. Anraku, M.; Cunningham, K.S.; Yun, Z.; Tsao, M.S.; Zhang, L.; Keshavjee, S.; Johnston, M.R.; de Perrot, M. Impact of Tumor-infiltrating T cells on Survival in Patients with Malignant Pleural Mesothelioma. *J. Thorac. Cardiovasc. Surg.* **2008**, *135*, 823–829. [CrossRef] [PubMed]

77. Newick, K.; O'Brien, S.; Sun, J.; Kapoor, V.; Maceyko, S.; Lo, A.; Pure, E.; Moon, E.; Albelda, S.M. Augmentation of CAR T-cell trafficking and antitumor efficacy by blocking protein kinase A localization. *Cancer Immunol. Res.* **2016**, *4*, 541–551. [CrossRef] [PubMed]

78. DeLong, P.; Tanaka, T.; Kruklitis, R.; Henry, A.C.; Kapoor, V.; Kaiser, L.R.; Sterman, D.H.; Albelda, S.M. Use of cyclooxygenase-2 inhibition to enhance the efficacy of immunotherapy. *Cancer Res.* **2003**, *63*, 7845–7852. [PubMed]

79. Kiyotani, K.; Park, J.H.; Inoue, H.; Husain, A.; Olugbile, S.; Zewde, M.; Nakamura, Y.; Vigneswaran, W.T. Integrated Analysis of Somatic Mutations and Immune Microenvironment in Malignant Pleural Mesothelioma. *Oncoimmunology* **2017**, *6*, e1278330. [CrossRef] [PubMed]

80. Yamada, N.; Oizumi, S.; Kikuchi, E.; Shinagawa, N.; Konishi-Sakakibara, J.; Ishimine, A.; Aoe, K.; Gemba, K.; Kishimoto, T.; Torigoe, T.; et al. CD8$^+$ tumor-infiltrating lymphocytes predict favorable prognosis in malignant pleural mesothelioma after resection. *Cancer Immunol. Immunother.* **2010**, *59*, 1543–1549. [CrossRef] [PubMed]

81. Cornwall, S.M.; Wikstrom, M.; Musk, A.W.; Alvarez, J.; Nowak, A.K.; Nelson, D.J. Human Mesothelioma induces Defects in Dendritic Cell Numbers and Antigen-processing Function which predicts Survival Outcomes. *Oncoimmunology* **2015**, *5*, e1082028. [CrossRef] [PubMed]

82. Cornelissen, R.; Lievense, L.A.; Maat, A.P.; Hendriks, R.W.; Hoogsteden, H.C.; Bogers, A.J.; Hegmans, J.P.; Aerts, J.G. Ratio of Intratumoral Macrophage Phenotypes is a Prognostic Factor in Epithelioid Malignant Pleural Mesothelioma. *PLoS ONE* **2014**, *9*, e106742. [CrossRef] [PubMed]

83. Suzuki, K.; Kadota, K.; Sima, C.S.; Sadelain, M.; Rusch, V.W.; Travis, W.D.; Adusumilli, P.S. Chronic Inflammation in Tumor Stroma is an Independent Predictor of Prolonged Survival in Epithelioid Malignant Pleural Mesothelioma Patients. *Cancer Immunol. Immunother.* **2011**, *60*, 1721–1728. [CrossRef] [PubMed]

84. Glinsky, G.V.; Berezovska, O.; Glinskii, A.B. Microarray analysis identifies a death-from-cancer signature predicting therapy failure in patients with multiple types of cancer. *J. Clin. Investig.* **2005**, *6*, 1503–1521. [CrossRef] [PubMed]

85. Beachy, P.A.; Hymowitz, S.G.; Lazarus, R.A.; Leahy, D.J.; Siebold, C. Interactions between Hedgehog proteins and their binding partners come into view. *Genes Dev.* **2010**, *18*, 2001–2012. [CrossRef] [PubMed]

86. Svard, J.; Heby-Henricson, K.; Persson-Lek, M.; Rozell, B.; Lauth, M.; Bergstrom, A.; Ericson, J.; Toftgard, R.; Teglund, S. Genetic elimination of suppressor of fused reveals an essential repressor function in the mammalian hedgehog signaling pathway. *Dev. Cell* **2006**, *10*, 187–197. [CrossRef] [PubMed]

87. Hui, C.C.; Angers, S. Gli proteins in development and disease. *Annu. Rev. Cell Dev. Biol.* **2011**, *27*, 513–537. [CrossRef] [PubMed]

88. Rubin, L.; de Sauvage, F.J. Targeting the Hedgehog pathway in cancer. *Nat. Rev. Drug Disc.* **2006**, *5*, 1026–1033. [CrossRef] [PubMed]

89. Nusslein-Volhard, C.; Wieschaus, E. Mutations affecting segment number and polarity in *Drosophila*. *Nature* **1980**, *287*, 795–801. [CrossRef] [PubMed]

90. Abe, Y.; Tanaka, N. The Hedgehog Signaling Networks in Lung Cancer: The Mechanisms and Roles in Tumor Progression and Implications for Cancer Therapy. *BioMed Res. Int.* **2016**, *2016*, 7969286. [CrossRef] [PubMed]

91. Shi, Y.; Noura, U.; Opitz, I.; Soltermann, A.; Rehrauer, H.; Thies, S.; Weder, W.; Stahel, R.A.; Felley-Bosco, E. Role of Hedgehog signaling in malignant pleural mesothelioma. *Clin. Cancer Res.* **2012**, *18*, 4646–4656. [CrossRef] [PubMed]

92. Zhang, Y.; He, J.; Zhang, F.; Li, H.; Yue, D.; Wang, C.; Jablons, D.M.; He, B.; Lui, N. SMO expression level correlates with overall survival in patients with malignant pleural mesothelioma. *J. Exp. Clin. Cancer Res.* **2013**, *32*, 7. [CrossRef] [PubMed]

93. Meerang, M.; Berard, K.; Felley-Bosco, E.; Lauk, O.; Vrugt, B.; Boss, A.; Kenkel, D.; Broggini-Tenzer, A.; Stahel, R.A.; Arni, S.; et al. Antagonizing the Hedgehog Pathway with Vismodegib Impairs Malignant Pleural Mesothelioma Growth In Vivo by Affecting Stroma. *Mol. Cancer Ther.* **2016**, *15*, 1095–1105. [CrossRef] [PubMed]

94. Felley-Bosco, E.; Opitz, I.; Meerang, M. Hedgehog Signaling in Malignant Pleural Mesothelioma. *Genes* **2015**, *6*, 500–511. [CrossRef] [PubMed]

95. Li, H.; Lui, N.; Cheng, T.; Tseng, H.H.; Yue, D.; Giroux-Leprieur, E.; Do, H.T.; Sheng, Q.; Jin, J.Q.; Luh, T.W.; et al. Gli as a novel therapeutic target in malignant pleural mesothelioma. *PLoS ONE* **2013**, *8*, e57346. [CrossRef] [PubMed]

96. Felley-Bosco, E.; Stahel, R. Hippo/YAP pathway for targeted therapy. *Transl. Lung Cancer Res.* **2014**, *3*, 75–83. [PubMed]

97. Kang, Y.; Ding, M.; Tian, G.; Guo, H.; Wan, Y.; Tao, Z.; Li, B.; Lin, D. Overexpression of Numb suppresses tumor cell growth and enhances sensitivity to cisplatin in epithelioid malignant pleural mesothelioma. *Oncol. Rep.* **2013**, *30*, 313–319. [CrossRef] [PubMed]

98. You, M.; Varona-Santos, J.; Singh, S.; Robbins, D.J.; Savaraj, N.; Nguyen, D.M. Targeting of the Hedghog signal transduction pathway suppresses survival of malignant pleural mesothelioma in vitro. *J. Thorac. Cardiovasc. Surg.* **2014**, *147*, 508–516. [CrossRef] [PubMed]

99. Angers, S.; Moon, R.T. Proximal events in Wnt signal transduction. *Nat. Rev. Mol. Cell Biol.* **2009**, *10*, 468–477. [CrossRef] [PubMed]

100. Cadigan, K.M.; Waternam, M.L. TCF/LEFs and Wnt signaling in the nucleus. *Cold Spring Harb. Perspect. Biol.* **2012**, *4*, a007906. [CrossRef] [PubMed]

101. Clevers, H.; Nusse, R. Wnt/B-catenin signaling and disease. *Cell* **2012**, *149*, 1192–1205. [CrossRef] [PubMed]

102. Uematsu, K.; Kanazawa, S.; You, L.; He, B.; Xu, Z.; Li, K.; Peterlin, B.M.; McCormick, F.; Jablons, D.M. Wnt pathway activation in mesothelioma: Evidence of Dishevelled overexpression and transcriptional activity of beta-catenin. *Cancer Res.* **2003**, *63*, 4547–4551. [PubMed]

103. Abutaily, A.S.; Collins, J.E.; Roche, W.R. Cadherins, catenins and PACE in pleural malignant mesothelioma. *J. Pathol.* **2003**, *201*, 355–362. [CrossRef] [PubMed]

104. Mazieres, J.; You, L.; He, B.; Xu, Z.; Twogood, S.; Lee, A.Y.; Reguart, N.; Batra, S.; Mikami, I.; Jablons, D.M. Wnt2 as a new therapeutic target in malignant pleural mesothelioma. *Int. J. Cancer* **2005**, *117*, 326–332. [CrossRef] [PubMed]

105. Kobayashi, M.; Huang, C.L.; Sonobe, M.; Kikuchi, R.; Ishikawa, M.; Kitamura, J.; Miyahara, R.; Menju, T.; Iwakiri, S.; Itoi, K.; et al. Intratumoral Wnt2B expression affects tumor proliferation and survival in malignant pleural mesothelioma patients. *Exp. Ther. Med.* **2012**, *3*, 952–958. [CrossRef] [PubMed]

106. Lee, A.Y.; He, B.; You, L.; Dadfarmay, S.; Xu, Z.; Mazieres, J.; Mikami, I.; McCormick, F.; Jablons, D.M. Expression of the secreted frizzled-related protein gene family is downregulated in human mesothelioma. *Oncogene* **2004**, *23*, 6672–6676. [CrossRef] [PubMed]

107. Hirata, T.; Zheng, Q.; Chen, Z.; Kinoshita, H.; Okamoto, J.; Kratz, J.; Li, H.; Lui, N.; Do, H.; Cheng, T.; et al. Wnt7a is a putative prognostic and chemosensitivity marker in human malignant pleural mesothelioma. *Oncol. Rep.* **2015**, *33*, 2052–2060. [CrossRef] [PubMed]

108. Gee, G.V.; Koestler, D.C.; Christensen, B.C.; Sugarbaker, D.J.; Ugolini, D.; Ivaldi, G.P.; Resnick, M.B.; Houseman, E.A.; Kelsey, K.T.; Marsit, C.J. Downregulated microRNAs in the differential diagnosis of malignant pleural mesothelioma. *Int. J. Cancer* **2010**, *127*, 2859–2869. [CrossRef] [PubMed]

109. Ables, J.L.; Breunig, J.J.; Eisch, A.J.; Rakic, P. Not(ch) just development: Notch signaling in the adult brain. *Nat. Rev. Neurosci.* **2011**, *12*, 269–283. [CrossRef] [PubMed]

110. Graziani, I.; Eliasz, S.; De Marco, M.A.; Chen, Y.; Pass, H.I.; De May, R.M.; Strack, P.R.; Miele, L.; Bocchetta, M. Opposite effects of Notch-1 and Notch-2 on mesothelioma cell survival under hypoxia are exerted through the Akt pathway. *Cancer Res.* **2008**, *68*, 9678–9685. [CrossRef] [PubMed]

111. Oswald, F.; Tauber, B.; Dobner, T.; Bourteele, S.; Kostezka, U.; Adler, G.; Liptay, S.; Schmid, R.M. p300 acts as a transcriptional coactivator for mammalian Notch-1. *Mol. Cell. Biol.* **2001**, *21*, 7761–7774. [CrossRef] [PubMed]

112. Artavanis-Tsakonas, S.; Matsuno, K.; Fortini, M.E. Notch signaling. *Science* **1995**, *268*, 225–232. [CrossRef] [PubMed]

113. Fan, X.; Mikolaenko, I.; Elhassan, I.; Ni, X.; Wang, Y.; Ball, D.; Brat, D.J.; Perry, A.; Eberhart, C.G. Notch1 and notch2 have opposite effects on embryonal brain tumor. *Cancer Res.* **2004**, *64*, 7787–7793. [CrossRef] [PubMed]

114. Pannuti, A.; Foreman, K.; Rizzo, P.; Osipo, C.; Golde, T.; Osborne, B.; Miele, L. Targeting Notch to target cancer stem cells. *Clin. Cancer Res.* **2010**, *16*, 3141–3152. [CrossRef] [PubMed]

115. Bocchetta, M.; Miele, L.; Pass, H.I.; Carbone, M. Notch-1 induction, a novel activity of S40 required for the growth of SV40-transformed human mesothelial cells. *Oncogene* **2003**, *22*, 81–89. [CrossRef] [PubMed]

116. Jung, K.H.; Zhang, J.; Zhou, C.; Shen, H.; Gagea, M.; Rodriguez-Aguayo, C.; Lopez-Berestein, G.; Sood, A.K.; Beretta, L. Differentiation therapy for hepatocellular carcinoma: Multifaceted effects of miR-148a on tumor growth and phenotype and liver fibrosis. *Hepatology* **2016**, *63*, 864–879. [CrossRef] [PubMed]

117. Saito, N.; Fu, J.; Zheng, S.; Yao, J.; Wang, S.; Liu, D.D.; Yuan, Y.; Sulman, E.P.; Lang, F.F.; Colman, H.; et al. A high Notch pathway activation predicts response to γ secretase inhibitors in proneural subtype of glioma tumor-initiating cells. *Stem Cells* **2014**, *32*, 301–312. [CrossRef] [PubMed]

118. Zhang, S.; Long, H.; Yang, Y.L.; Wang, Y.; Hsieh, D.; Li, W.; Au, A.; Stoppler, H.J.; Xu, Z.; Jablons, D.M.; et al. Inhibition of CK2a down-regulates Notch1 signaling in lung cancer cells. *J. Cell.Mol. Med.* **2013**, *17*, 854–862. [CrossRef] [PubMed]

119. Chua, M.M.; Ortega, C.E.; Sheikh, A.; Lee, M.; Abdul-Rassoul, H.; Hartshorn, K.L.; Dominguez, I. CK2 in Cancer: Cellular and Biochemical Mechanisms and Potential Therapeutic Target. *Pharmaceuticals* **2017**, *10*, 18. [CrossRef] [PubMed]

120. Badouel, C.; McNeill, H. SnapShot: The hippo signaling pathway. *Cell* **2011**, *145*, 484. [CrossRef] [PubMed]

121. Pan, D. The hippo signaling pathway in development and cancer. *Dev. Cell* **2010**, *19*, 491–505. [CrossRef] [PubMed]

122. Zhao, B.; Tumaneng, K.; Guan, K.L. The Hippo pathway in organ size control, tissue regeneration and stem cell self-renewal. *Nat. Cell Biol.* **2011**, *13*, 877–883. [CrossRef] [PubMed]

123. Bueno, R.; Stawiski, E.W.; Goldstein, L.D.; Durinck, S.; De Rienzo, A.; Modrusan, Z.; Gnad, F.; Nguyen, T.T.; Jaiswal, B.S.; Chirieac, L.R.; et al. Comprehensive Genome Analysis of Malignant Pleural Mesothelioma Identifies Recurrent Mutations, Gene Fusions and Splicing Alterations. *Nat. Genet* **2016**, *48*, 407–416. [CrossRef] [PubMed]

124. Tranchant, R.; Quetel, L.; Tallet, A.; Meiller, C.; Reiner, A.; de Koning, L.; de Reynies, A.; Le Pimpec-Barthes, F.; Zucman-Rosi, J.; Jaurand, M.C.; et al. Co-occurring Mutations of Tumor Suppressor Genes, LATS2 and NF2, in Malignant Pleural Mesothelioma. *Clin. Cancer Res.* **2017**, *23*, 3191–3202. [CrossRef] [PubMed]

125. Cooper, J.; Xu, Q.; Zhou, L.; Pavlovic, M.; Ojeda, V.; Moulick, K.; de Stanchina, E.; Poirier, J.T.; Zauderer, M.; Rudin, C.M.; et al. Combined Inhibition of NEDD8-Activating Enzyme and mTOR Suppresses NF2 Loss-Driven Tumorigensis. *Mol. Cancer Ther.* **2017**, *16*, 1693–1704. [CrossRef] [PubMed]

126. Meerang, M.; Berard, K.; Friess, M.; Bitanihirwe, B.K.; Soltermann, A.; Vrugt, B.; Felley-Bosco, E.; Bueno, R.; Richards, W.G.; Seifert, B.; et al. Low Merlin expression and high Survivin labeling index are indicators for poor prognosis in patients with malignant pleural mesothelioma. *Mol. Oncol.* **2016**, *10*, 1255–1265. [CrossRef] [PubMed]

International Journal of
Molecular Sciences

MDPI

Review

Secreted and Tissue miRNAs as Diagnosis Biomarkers of Malignant Pleural Mesothelioma

Vanessa Martínez-Rivera [1], María Cristina Negrete-García [1], Federico Ávila-Moreno [2] and Blanca Ortiz-Quintero [1,*]

[1] Research Unit, Instituto Nacional de Enfermedades Respiratorias "Ismael Cosio Villegas", Calzada de Tlalpan 4502, Colonia Sección XVI, 14080 Mexico City, Mexico; vanessa.m.r.0801@gmail.com (V.M.-R.); cristi.negrete@gmail.com (M.C.N.-G.)
[2] Unidad de Investigación en Biomedicina (UBIMED), Cancer Epigenomics and Lung Disease Laboratory 12, Facultad de Estudios Superiores (FES)-Iztacala, Universidad Nacional Autónoma de México, Avenida de los Barrios #1 Colonia los Reyes Iztacala, 54090 Mexico City, Mexico; f.avila@unam.mx
* Correspondence: boq@iner.gob.mx; Tel.: +52-55-54871705; Fax: +52-55-56654623

Received: 29 December 2017; Accepted: 30 January 2018; Published: 17 February 2018

Abstract: Malignant pleural mesothelioma (MPM) is a rare but aggressive tumor that originates in the pleura, is diagnosed in advanced stages and has a poor prognosis. Accurate diagnosis of MPM is often difficult and complex, and the gold standard diagnosis test is based on qualitative analysis of markers in pleural tissue by immunohistochemical staining. Therefore, it is necessary to develop quantitative and non-subjective alternative diagnostic tools. MicroRNAs are non-coding RNAs that regulate essential cellular mechanisms at the post-transcriptional level. Recent evidence indicates that miRNA expression in tissue and body fluids is aberrant in various tumors, revealing miRNAs as promising diagnostic biomarkers. This review summarizes evidence regarding secreted and tissue miRNAs as biomarkers of MPM and the biological characteristics associated with their potential diagnostic value. In addition to studies regarding miRNAs with potential diagnostic value for MPM, studies that aimed to identify the miRNAs involved in molecular mechanisms associated with MPM development are described with an emphasis on relevant aspects of the experimental designs that may influence the accuracy, consistency and real diagnostic value of currently reported data.

Keywords: malignant pleural mesothelioma; microRNAs; diagnosis biomarkers

1. Introduction

Malignant pleural mesothelioma (MPM) is an uncommon but aggressive tumor that originates in mesothelial cells of the pleural membrane [1,2]. The disease is associated with asbestos exposure in 80% of cases, and symptoms manifest after a prolonged period of latency after exposure (20–40 years) [2–4] with a survival of 9–12 months [1–3,5].

The World Health Organization (WHO) estimated 92,252 worldwide deaths in the period of 1994–2008 due to this disease [6]; however, this figure may be underestimated due to the lack of reliable records on MPM diagnosis [7]. Moreover, an increase in the global incidence of mesothelioma is predicted based on high exposure to asbestos in past years [2].

MPM is classified histologically into three types: epithelioid, sarcomatoid and biphasic [5,8]. The epithelioid type is the most frequent with approximately 50–60% of cases [9]. Overall, accurate diagnosis of MPM is considered difficult and complex. First, early clinical MPM symptomatology is not disease-specific; however, advanced stages are characterized by pleural effusion, chest pain and dyspnea [10], which usually lead to chest X-ray analysis. Broad chest X-ray analysis can detect the presence of diffuse pleural thickening and nodular prominences, which suggest mesothelioma [2,11,12]. Cytological examination of the pleural fluid can be performed; however, only 60% of true positive cases can be identified using this technique [2,13,14]. The confirmatory diagnosis test or gold standard is based on the detection of pleural tissue markers by immunohistochemistry [15], but it requires pleural tissue samples obtained by invasive techniques [16]. In addition, there is no known marker that is 100% diagnostic; therefore, a combination of antibodies that recognize several positive and negative markers in MPM is used [8]. A pathologist observes the tissue under a microscope and decides whether the test is positive or negative based on his/her criteria and experience; therefore, diagnosis is subjective and qualitative. For MPM diagnosis, at least two positive and two negative markers are recommended [17]. Typically, MPM is positive for calretinin and cytokeratin 5/6 but negative for thyroid transcription factor 1 (TTF-1) and carcinoembryonic antigen (CEA) [8,17]. In addition, epithelial MPM (tubulopapillary and acinar subtype) may be difficult to distinguish from metastatic lung adenocarcinoma (AD), due to the mesenchymal/epithelial pattern that is present in both [8,17]. Their differential diagnosis requires an additional panel of antibodies in immunohistochemical staining that is positive for epithelial MPM [17]. This differential diagnosis is relevant since the treatment regimen is different for each disease. Because of this complexity, the MPM diagnostic guide issued by the "International Mesothelioma Interest Group (IMIG)" recommends that diagnosis is based on the interpretation of clinical, radiological and pathological findings altogether [17] in order to increase the likelihood of an accurate MPM diagnosis.

Therefore, the development of alternative, quantitative diagnostic biomarkers would have significant clinical potential.

In recent years, microRNAs (miRNAs) have been the subject of intense studies as key regulators of gene expression at the post-transcriptional level. Furthermore, it was found that miRNA expression was altered in tumor tissue and body fluids from several neoplastic pathologies, pointing to miRNAs as potential diagnostic biomarkers. Moreover, the miRNAs found in biological samples have all been stable and quantifiable.

Alternative miRNA-based biomarker tests could add relevant adjunct information and increase the probability of reaching the right diagnosis.

This review aims to summarize evidence regarding the potential diagnostic value of secreted and tissue miRNAs as MPM biomarkers with an emphasis on the relevant aspects of experimental designs that may influence the accuracy, consistency and real diagnostic value of currently reported data.

2. MicroRNAs

miRNAs are small double-stranded RNAs of ~22 nt that regulate gene expression at the post-transcriptional level by blocking the translation of target messenger RNA [18]. miRNAs have been found in every organism analyzed to date and regulate essential cellular processes, such as differentiation, proliferation and apoptosis [18–20]. It is important to note that these cellular processes are deregulated in several neoplastic processes [21,22]. miRNAs are expressed in normal physiological conditions in a cell- and tissue-specific-manner, but their expression pattern was found to be aberrant in tumor tissue and could be distinguished from the normal expression pattern of healthy tissue. Cumulative evidence showed that altered miRNA expression profiles in tumor tissues could be associated with the diagnosis, prognosis and even histological classification of lung, breast, colorectal, and pancreatic cancer, hepatocellular carcinoma and malignant mesothelioma, among others [23–29]. Moreover, several of these tissue miRNAs have been associated with carcinogenesis per se: experimental manipulation of certain altered miRNAs in several cancers has been shown to regulate the tumorigenic properties of tumor cell lines and tumor growth in mouse models [30–33]. In addition, the expression of tissue miRNAs does not seem to depend on the age or race of the individual [34,35].

Nevertheless, tumor tissue sampling requires the use of invasive retrieval techniques. Favorably, cell-free miRNAs are also detected in peripheral blood (circulating miRNAs) and in other body fluids, such as tears, urine, and saliva. In healthy individuals, miRNAs are secreted by cells into body fluids with stable and constant concentrations [36]; however, similar to tissue miRNAs, alterations in their expression levels have been associated with several cancers [36]. These secreted cell-free miRNAs are resistant to endogenous RNases and external conditions, such as freeze-thaw and extreme pH [36–38]. These characteristics are desirable in potential diagnostic biomarkers due to their transport, storage and manipulation in the laboratory. Because obtaining peripheral blood and body fluids is less invasive than obtaining tissue samples, secreted miRNAs are considered to be potential non-invasive biomarkers of cancer.

The biogenesis of miRNAs (Figure 1) starts in the cell nucleus through the transcription of miRNA genes by RNA polymerase II (Pol II), which generates the primary miRNA precursors known as pri-miRNAs. Pri-miRNAs (60–100 nt) have a "hairpin" structure, consisting of a stem of 33–35 base pairs with a terminal loop [18,39]; pri-miRNAs are enzymatically cleaved by the Drosha-DGCR8 protein complex to produce a smaller precursor called pre-miRNA (~70 nt). These pre-miRNAs are transported to the cytoplasm through exportin-5-RanGTP. In the cytoplasm, pre-miRNAs are cleaved by DICER type III RNase, which produces mature ~22 nt miRNAs. Mature miRNAs are recruited by the protein complex called RISC or miRISC through its protein component Argonauta (AGO). The double strand of mature miRNA dissociates, and a single strand is retained (guide strand) in the miRISC complex. The guide strand recognizes the 3′ untranslated region (UTR) of its target mRNA, which has partial sequence complementary; however, this binding must exhibit perfect complementarity in a region of 2–8 nt called the seed region, which is located at the 5′ end of miRNAs [18,40]. The binding of miRNA to its target mRNA induces the blocking of translation by three possible mechanisms: translation repression, mRNA degradation and mRNA destabilization [41–43].

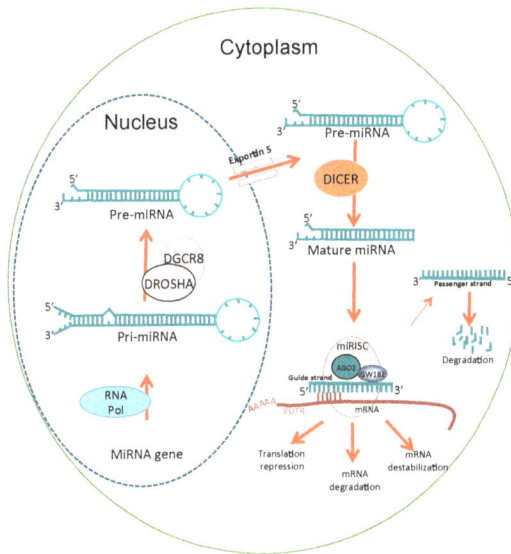

Figure 1. Biogenesis of miRNAs in the cell. miRNA precursors (pre-miRNAs) are transcribed in the nucleus and processed by the Drosha complex to generate pre-miRNAs. Pre-miRNAs are exported to the cytoplasm via exportin-5 and excised by DICER into a mature form of double-stranded RNA ~22 nt long. Double-stranded RNA is loaded onto AGO2 that is the catalytic component of the miRISC complex. One strand is removed from the duplex (the passenger strand) and the remaining RNA strand (the guide strand) binds to complementary sequences typically located in the 3′ untranslated region (UTR) of target mRNAs to repress translation or trigger mRNA cleavage.

Regarding miRNAs in body fluids, four mechanisms of secretion of miRNAs have been described to date: (a) inside exosomes; (b) associated with AGO2 protein; (c) associated with high-density lipoprotein (HDL); and (d) associated with nucleophosmin 1 protein (NPM1) [36]. As shown in

Figure 2, miRNAs are sorted into multivesicular bodies (MVBs) derived from early endosomes, which is a process that requires neutral sphingomyelinase 2 (nSMase2), endosomal sorting complex transport machinery (ESCRT) and sumoylated hnRNPA2B1 protein [44–46]. MVBs are enriched with two main components of the miRISC, GW182 and AGO2, which could be associated with miRNA functionality [36,47]. Finally, MVBs fuse with the plasma membrane and release exosomes into the extracellular medium. It has been shown that exosomes carrying miRNAs fuse with the plasma membrane of target cells and that the delivered miRNAs are functional inside the target cell [48,49]. In addition, miRNAs associated with Argonaute2 protein (AGO2) [50,51] and miRNAs bound to HDL [52,53] can be exported stably into human plasma samples. HLD-associated miRNAs are transferred actively to target cells in a functional form [52,54]; however, there is no experimental evidence for AGO-associated miRNAs; miRNAs released by these two mechanisms have not been reported in clinical samples from cancer patients. Lastly, one study reported that the RNA binding protein nucleophosmin (NPM1) binds miRNAs from the culture supernatants of tumor cell lines and fibroblasts while protecting them from RNase activity [55]. NPM1-associated miRNAs have not been reported in clinical samples or as part of active transfer into target cells.

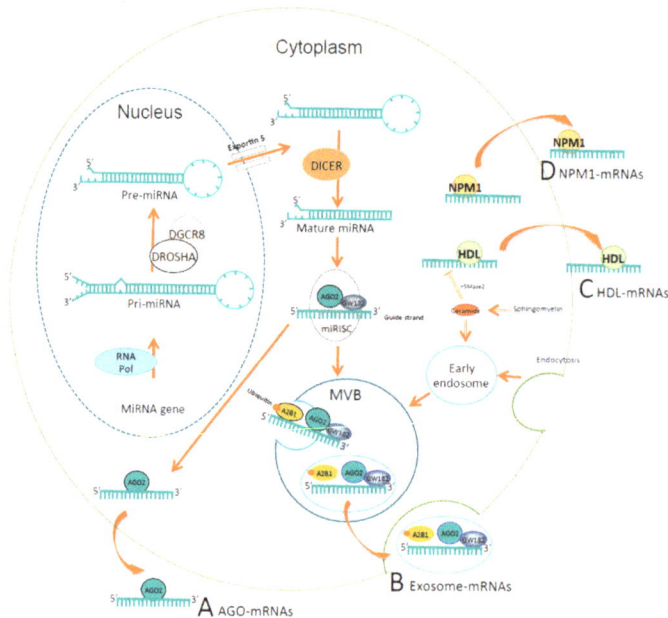

Figure 2. Mechanisms of secretion of miRNAs. (**A**) Secretion of the miRNAs associated with Argonaute2 protein (AGO2); (**B**) Secretion of the miRNAs by exosomes; (**C**) Secretion of the miRNAs associated with high-density lipoprotein (HDL); (**D**) Secretion of the miRNAs associated with the RNA binding protein nucleophosmin (NPM1). (**A**) miRNAs associated with AGO2, a main component of the RISC, can be stably exported into plasma samples. (**B**) miRNAs are sorted into multivesicular bodies (MVBs) derived from early endosomes. This mechanism requires ceramide production on the cytosolic side by neutral sphingomyelinase 2 (nSMase2), ESCRT machinery, and sumoylated hnRNPA2B1 protein, which specifically binds mature miRNAs and controls their loading into MVBs. MVBs are enriched with GW182 and AGO2, which are known to regulate the function of miRNAs. MVBs fuse with the plasma membrane and release exosomes into the extracellular medium. (**C**) miRNAs bound to HDL also can be stably exported into plasma samples via a mechanism repressed by nSMase2. (**D**) NPM1 binds miRNAs from the culture supernatants of tumor cell lines and fibroblasts while protecting them from RNase activity.

3. Studies Regarding microRNAs as Biomarkers for MPM

A biomarker in cancer can be any cellular, molecular or genetic component that can be measured and associated with the neoplastic process or the presence of disease [56,57]. Ideally, a biomarker for cancer diagnosis should distinguish between patients with a specific type of cancer and those who do not have the disease with high specificity and sensitivity. It also should have high stability in biological samples and should be measured with a simple, accurate and reproducible method in any laboratory [56,57].

Most studies with the main goal to identify candidate miRNAs with diagnostic value started by identifying miRNAs differentially expressed in tumor samples compared to non-tumor samples, which is known as the discovery phase. During the discovery phase, high-throughput techniques, such as semi-quantitative microarrays and deep sequencing, enable the analysis of an extensive number of miRNAs, but they usually have a limited number of samples due to the high cost of these methods. Alternatively, a limited number of miRNAs are tested as potential biomarkers based on a hypothesis-driven method. The following validation phase is usually performed using a quantitative technique, such as reverse transcription quantitative real-time PCR (RT-qPCR), preferably on a larger number of samples obtained from an independent set of patients.

In addition to studies with a main goal to identify candidate miRNAs with diagnostic value, studies that aimed to identify miRNAs involved in the oncogenic mechanisms of MPM and those that analyzed miRNA levels in biological samples are addressed in this section. Certainly, the potential association of miRNAs with the MPM carcinogenic process increases the likelihood of miRNAs having diagnostic value for this disease.

Here, studies regarding miRNAs with potential diagnostic value for MPM are described with an emphasis on relevant aspects of the experimental strategy, which includes the following: (i) source, size and preservation method of the tumor samples and the samples that they are compared against (normal controls, healthy controls or another cancers); (ii) specific assays used for miRNA analysis in the discovery and validation phase; (iii) number of miRNAs analyzed; (iv) sensitivity and specificity values (Receiver operating characteristic curve analysis); and (v) potential limitations of the experimental design.

The studies are listed by the year of publication and sectioned by the type of sample source for clarity. Table 1 summarizes the described studies.

3.1. Studies in Pleural Tissue

Guled et al. [58] reported the first study aimed to evaluate the expression of miRNAs in MPM in 2009. The authors evaluated the expression of 723 miRNAs in 17 frozen MPM tissue samples and in commercially available total RNA from normal human pericardium as a control (Ambion, Austin, TX, USA) using microarrays (Agilent human miRNAs V2). They reported 12 over-expressed miRNAs (let-7b*, miR-1228*, miR-195*, miR-30b*, miR-32*, miR-345, miR-483-3p, miR-584, miR-595, miR-615-3p, and miR-885-3p) and nine sub-expressed (let-7e*, miR-144*, miR-203, miR-340*, miR-34a*, miR-423, miR-582, miR-7-1* and miR-9) in MPM tissue compared to the single control. The authors also compared miRNA expression between MPM samples and reported seven miRNAs that were expressed exclusively in the epithelioid subtype, five in the biphasic subtype and three in the sarcomatoid subtype. Bioinformatic analysis was used to identify three suppressor genes (CDKN2A, NF2 and RB1) as putative targets of over-expressed miR-30b*, miR-32*, miR-483-3p, miR-584, and miR-885-3p, whereas oncogenes hepatocyte growth factor (HGF), Platelet Derived Growth Factor Subunit (PDGFA), Epidermal Growth Factor (EGF) and Jun proto-oncogene (JUN) were identified as putative targets of sub-expressed miR-9, miR-7-1* and miR-203. Nevertheless, experimental assays were not performed to verify these findings. There are major limitations in this study; first, only one normal control (total RNA from one donor) was used in the discovery phase, and second, quantitative validation of the microarray results was not performed. There was no follow up work for this study.

In 2010, Gee et al. [59], in order to identify biomarkers for the differential diagnosis of MPM and lung adenocarcinoma (AD), analyzed miRNA expression profiles of pleural tissue samples from MPM patients (*n* = 15) and AD patients (*n* = 10) using microarrays (Affymetrix). Microarray data were validated by RT-qPCR in 100 samples of MPM and 32 samples of AD. The authors reported seven miRNAs sub-expressed in MPM that could discriminate MPM from AD with greater than 80% specificity and sensitivity (miR-200c, miR-200b, miR-141, miR-429, miR-203 and miR-205), providing the first evidence of potential differential diagnosis biomarkers for these different tumors that are often difficult to differentiate histologically.

Also in 2010, Benjamin et al. [60] performed a study to identify biomarkers that could distinguish MPM from other carcinomas of epithelial origin that can invade pleura. In the discovery phase, seven pleural tissue samples of MPM and 97 tissue samples of epithelial carcinomas from various organs (lung, bladder, breast, and kidney, among others) were analyzed by microarray (Nexterion Slide E, Schott, Mainz, Germany). The results indicated that miR-193a-3p was over-expressed in MPM vs. all carcinomas, miR-200c was sub-expressed in MPM vs. renal cell carcinoma (RCC) and miR-192 was sub-expressed in MPM vs. non-RCC carcinomas. It should be noted that miR-200c was also reported by Gee et al. [59] as down-regulated in MPM compared to AD. These results were validated in 22 MPM and 43 carcinoma samples (new samples added: 15 MPM and 36 carcinomas) by RT-qPCR. The sensitivity and specificity of using these three miRNAs to diagnose MPM were analyzed in 32 MPM and 113 carcinoma samples (new samples added: 11 MPM and 77 carcinomas), resulting in a sensitivity of 97% and a specificity of 96%. Finally, a blind study of 63 new samples of pleural and lung tissue (14 MPM and 49 metastatic carcinomas) was performed. The results indicated that 14 MPM samples were correctly identified (100% sensitivity) and that 46 of the 49 carcinomas were correctly identified (94% specificity). This report was the first study that provided a quantitative diagnostic tool (RT-qPCR) to discriminate MPM vs. other epithelial carcinomas that may invade pleura and MPM vs. lung adenocarcinoma.

In 2011, Santarelli et al. [29] analyzed miRNA expression profiles in the pleural tissue of MPM patients and matched adjacent non-neoplastic pleural samples to identify candidate biomarkers for the diagnosis of MPM. They quantified the levels of 88 miRNAs that were previously reported to be associated with cancerous processes in ten samples of MPM and one sample of healthy mesothelial tissue from an RNA-pool of five individuals using a customized PCRArray (Array MAH-102A, SABiosciences). These samples were previously preserved in RNALater solution (Ambion) at −80 °C. Three sub-expressed miRNAs in MPM that were identified in PCRArray analysis (miR-335, miR-126 and miR-32) were subsequently analyzed by RT-qPCR in 27 formalin-fixed paraffin-embedded (FFPE) samples of MPM and 27 adjacent healthy pleural tissues. The data indicated that only hsa-miR-126 showed significant sub-expression in MPM compared to healthy tissues. It should be noted that this study was limited to the analysis of 88 microRNAs in the discovery phase and that samples were preserved with different methods. It has been reported that some miRNAs detected in frozen tissue samples may vary from those detected in FFPE samples because of the degradation of some miRNAs in the latter; however, these profiles were shown to be comparable [61]. Samples preserved in FFPE are frequently more available than frozen samples; therefore, this information is valuable for future studies regarding diagnostic purposes. Nevertheless, there were no follow up studies in malignant tissues from these authors.

In 2013, Xu et al. [62] performed miRNA profiling to identify altered miRNAs in MPM tissues that could be associated with the oncogenic transformation of mesothelial cells. They analyzed miRNA expression in 25 MPM specimens and six normal parietal pleural samples from patients without mesothelioma or other malignancies by microarray (BeadChips v2, Illumina). The results indicated that 49 miRNAs were over-expressed and 65 were under-expressed in MPM compared to controls. The authors reported the validation of four miRNAs by RT-qPCR (sub-expressed miR-206, miR-1 and miR-483-5p; over-expressed miR-155*), but they did not state clearly whether a new group of tissue samples was used for this validation. Because this study aimed to identify biologically relevant miRNAs in the development of MPM instead of miRNAs with diagnostic value, under-expressed miR-1 was transfected into two MPM cell lines (H513, epithelioid type and H2052, sarcomatoid type), which caused cycle arrest and apoptosis. This effect was associated with increased mRNA expression of the tumor suppressors p53, Bcl2-associated X protein (BAX) and cyclin-dependent kinase inhibitors p16/p21 and decreased mRNA expression of anti-apoptotic Bcl-2 and surviving. However, direct experimental evidence of this association was not provided in the paper. None of the putative target genes of the altered miRNAs identified by informatics analysis were tested experimentally.

Reid et al. [63] in 2013 reported that miR-16, miR15a, miR-15b, and miR-195 were markedly sub-expressed in FFPE tissue samples of MPM patients ($n = 60$) compared to normal pleural tissue samples of cardiac surgery patients ($n = 23$). Four-fold to 22-fold (miR-16) under-expression was verified in four MPM cell lines relative to the normal mesothelial cell line MeT-5A (two-fold to five-fold). The expression of miR16/15 was restored in cell lines to elucidate their biological function in MPM to find a new potential target for therapy. Restoring miR-16 using synthetic mimics resulted in growth inhibition, cell cycle arrest in G0/G1, increased apoptosis and reduced colony formation in the MPM cell line MSTO-211H but not in MeT-5A cells. These effects correlated with down-regulation of the miR-16 known target anti-apoptotic gene Bcl-2 and cyclin D1-encoding gene (CCND1) in H28 and MSTO-211H cell lines. Further, intravenous injection of miR-16-containing minicells to nude mice already xenografted with MSTO-211H cells led to tumor growth inhibition that was dose-dependent.

In 2014, Cioce et al. [64] performed a screening test (887 miRNAs) on 29 pleural tissue samples of MPM vs. 12 tissue samples of peritoneal mesothelial cysts (preserved in FFPE) with microarrays (Human miRNA Microarray V2, Agilent). Among the 19 miRNAs differentially expressed in MPM, sub-expressed miR-145 was chosen for validation by RT-qPCR in fresh pleural tissue samples of MPM ($n = 6$) and normal tissue ($n = 14$) and subsequently in frozen samples of pleural tissue with MPM ($n = 36$) and normal peritoneal mesothelial tissue ($n = 36$, same patients). They proceeded with over-expression of miR-145 in three MPM cell lines and a normal mesothelial cell line and reported reductions in proliferation and migration in two of the MPM cell lines compared to the control. They further performed a xenotransplant of MSTO-211H cells that over-expressed miR-145 in SCID mice and observed an inhibition of tumor growth in six of eight treated mice compared to controls. The authors concluded that the results suggested that miR-145 functions as a tumor suppressor; however, the number of experimental animals and cell lines were rather limited to provide reliable evidence. In addition, peritoneal mesothelial cyst tissue was used as a comparative control group in the discovery phase. Cystic mesothelial lesions are benign; therefore, they may not be comparable to the validation phase when normal peritoneal mesothelial tissue was used as a control. Like the study reported by Xu [62], this paper does not focus on the diagnostic value of miR-145 in MPM but on its potential association with a carcinogenic process. Additionally, this study also reported the sub-expression of miR-200c, which coincides with the results reported by Gee, et al. in 2011 [59].

Ramirez-Salazar et al. [65] in 2014 analyzed miRNA expression profiles in pleural tissue with epithelioid MPM ($n = 5$), pachypleuritis (PP) ($n = 4$) and atypical mesothelial hyperplasia (HP) ($n = 5$) and in non-cancerous/non-inflammatory tissue ($n = 5$) as a control using PCRArray (TaqMan Array, Applied Biosystems). The aim of this study was to elucidate mechanisms associated with the development of MPM since pleural chronic inflammation is considered to be a detonating factor in MPM pathogenesis. Different from most studies on mesothelioma, this study provided a description of the histological diagnosis of all tissue samples. Moreover, only tumor samples containing >80% neoplastic cells were used, which provided better tumor representativeness. The authors reported 19 miRNAs that were differentially expressed in MPM samples compared to control samples, 11 that were sub-expressed (miR-517b-3p, miR-627, miR-766-3p, miR-101-3p, miR-501-3p, miR-212-3p, miR-596, miR-145-5p, miR-671-3p, miR-181a-5p and miR-18a-3p), and eight that were over-expressed (miR-30e-3p, miR-34a-3p, miR-622, let-7a-5p, miR-196b-5p, miR-135b-5p, miR-18a-5p and miR-302-3p). Bioinformatic analysis revealed that the targets of four under-expressed miRNAs in MPM (miR-181a-5p, miR-101-3p, miR-145-5p and miR-212-3p), one in PP (miR-101-3p) and one in HP (miR-494) were significantly associated with "cancer pathways". Nevertheless, the authors did not perform any experimental studies to assess the predictive results. Coincidently, Cioce et al. also reported the sub-expression of miR-145 in MPM [64].

Andersen et al. [66] sought to identify miRNA candidates for diagnostic biomarkers by analyzing miRNA expression in samples preserved in FFPE of five MPM specimens previously treated with chemotherapy (MPM), five preoperative diagnostic biopsy samples of MPM (DB) and five non-neoplastic pleural tissue samples of a patient with MPM previously treated with chemotherapy (NNP) using PCRArray (miRCURYLNA Universal RT microRNA Ready-to-Use, Human Panels I + II v2). The authors chose four sub-expressed (miR-126, miR-143, miR-145 and miR-653) and two over-expressed (miR-193a-3p, miR-193b) miRNAs found in either the DB or MPM samples compared to NNP samples for RT-qPCR validation using 40 MPM, 12 DB and 14 NNP samples. The results indicated statistically significant sub-expression of miR-126, miR-652, miR-145 and miR-143 in both DB and MMP compared to NNP. It was reported that miR-145 and miR-652 had a sensitivity or specificity >80%, whereas miR-143 and miR-126 had a sensitivity or specificity <80% (Receiver operating characteristic curve or ROC curve). It is important to note the main potential design limitations of this study: first, the samples analyzed at the screening phase contained 40–85% neoplastic tissue, but it was not clear how many samples contained specific percentages in that range. This point could be relevant if we consider the representativeness of each sample as a tumor whose non-neoplastic content was 60% versus 15%. Second, the authors stated that in order to test for any chemotherapy-induced changes in miRNA expression profiles, they had to compare diagnostic biopsy samples without treatment (DB) to NNP; however, the latter samples were previously treated with chemotherapy. Chemotherapy affects normal and tumor tissues, potentially inducing changes in the miRNA expression profiles of both; therefore, this aim cannot be achieved with the stated comparison. Nevertheless, this study was the third to report the sub-expression of miR-145 in MPM pleural tissue.

In 2015, Ak et al. [67] investigated miRNA expression levels in frozen pleural tissues from MPM and benign asbestos-related pleural effusion (BAPE) patients using PCRArray. BAPE tissue samples showed non-specific pleuritis/fibrosis. The authors performed PCRArray (384 miRNAs, Applied Biosystems) on 18 MPM (11 with chemotherapy treatment) and six BAPE samples and reported 11 over-expressed miRNAs in MPM samples compared to BAPE (miR-484, miR-320, let-7a, miR-744, miR-20a, miR-193b, let-7d, miR-125a-5p, miR-92a, miR-155, and miR-152). They evaluated the diagnostic value of these miRNAs to differentiate MPM from BAPE using ROC and area under the curve (AUC) analysis. The results showed that four miRNAs had AUC values ≥0.90 (miR-484, miR-320, let-7a and miR-125a-5p). Meanwhile, miR-484 had a sensitivity and specificity of 100%, miR-320 had a sensitivity of 78% and a specificity of 100%, let-7a had a sensitivity of 94% and a specificity of 83% and miR-125a-5p had a sensitivity of 89% and a specificity of 100%. This study had some limitations: using a mixture of tissue samples from MPM patients treated with chemotherapy and without treatment and a limited number of samples without validation in an independent cohort of patients. Potential chemotherapy-induced changes in the miRNA expression profiles may have induced bias in the analysis.

Also in 2015, Birnie et al. [68] investigated the role of miR-223 in MPM based on evidence that suggested that miR-223 might be a tumor suppressor in other hematopoietic and solid tumors and on their own initial findings that indicated that miR-223 was down-regulated in three MPM cell lines compared to one human mesothelial cell control. They examined the expression levels of miR-223 in 17 MPM pleural tissue samples and six normal pleural tissue samples from non-cancer patients undergoing cardiac or aortic surgery by RT-qPCR. They enriched the tumor content of the FFPE-conserved specimens by performing laser-capture micro-dissection. In addition, they examined miR-223 expression in cells obtained from the pleural effusion of 26 MPM and ten benign pleural disease patients. Down-regulation of miR-223 was confirmed in MPM tissues and MPM effusion cells. After over-expression of miR-223 in MPM cell lines (Human JU77 and CRL2081), STMN1 levels were reduced, cell motility was inhibited, and tubulin acetylation was induced. Migration of both cell lines was significantly reduced following the knockdown of STMN1 expression. In addition, miR-223 levels increased, whereas STMN1 was reduced following the re-expression of c-Jun N-terminal kinase (JNK) isoforms in JNK-null murine embryonic fibroblasts, suggesting that miR-223 and its target STMN1 are involved in the regulation of MPM cell motility, which may be associated with their carcinogenic properties.

In 2016, Cappellesso et al. [69] searched for miRNAs that could be used as a complementary tool for the diagnosis of MPM in pleural effusion cytology. For this study, the authors decided to test 15 miRNAs previously reported by three publications as potential candidates for MPM biomarkers using RT-qPCR. First, they analyzed miRNA expression levels in two MPM cell lines (H2052 and H28) and one normal mesothelium cell line (MET-5A) and reported the over-expression of miR-19a, miR-19b, miR-21 and miR-25 and sub-expression of miR-126. These miRNAs were further analyzed in pleural tissue samples preserved in FFPE of 51 MPM and 40 benign/reactive pleurae with the same results. Likewise, these five miRNAs were evaluated in 29 cytological samples of MPM and 24 cytological samples with reactive mesothelial cells (RMCs). It was indicated that 31 samples were air-dried and stained with May–Grunwald–Giemsa and 22 samples were fixed in 95% alcohol and stained with Papanicolaou, but the authors did not clarify which type of samples were stained (MPM or RMCs). They found over-expression of miR-19a and miR-21 and sub-expression of miR-126 in MPM compared to RMCs. ROC analysis suggested that miR-19a, miR-19b, miR-21 and miR-126 could be diagnostic biomarkers of MPM in cytological samples because they showed a sensitivity or specificity >0.80. The results showed that the five analyzed miRNAs were detectable in these samples. One striking detail is that this study is the first to report the quantification of miRNAs from stained cytological samples. The authors stated that "staining, fixation, and the length of time in storage did not markedly affect final RNA quality or yield among smears", but they did not provide evidence to support this statement. It would have been relevant to this field and to other researchers to report this evidence as new findings.

In a subsequent study, Cappellesso et al. (2017) [70] searched for candidate biomarker miRNAs for differential diagnosis of MPM from AD in histologic and cytological specimens. First, a bioinformatic analysis of three data sets regarding the expression of miRNAs in MPM and AD was performed to select candidate miRNAs. Three upregulated miRNAs (miR-130a, miR-193a, and miR-675) and three downregulated miRNAs (miR-141, miR-205, and miR-375) in MPM vs. AD were selected. Their expression was tested in 41 epithelioid MPM and 40 AD histologic specimens (FFPE) and 26 MPM and 27 AD cytological specimens, by RT-qPCR. In this study, tumor cells were microdissected manually from histologic samples and the cells were scraped from each slide from cytological samples to ensure a tumor cell content >70%. Results indicated that only miR-130a was significantly overexpressed in both histologic and cytological MPM specimens compared to AD. Finally, the ROC analysis of miR-130a in cytological samples demonstrated a low sensitivity of 77%, a specificity of 67%.

3.2. Studies in the Cellular Fraction of Peripheral Blood

Tissue miRNAs have promising diagnostic value for several tumors, including MPM. However, obtaining pleural tissue samples requires invasive procedures. An alternative is using biological samples than can be obtained with less invasive procedures, such as peripheral blood.

In 2012, Weber et al. [71] published the first study that analyzed the diagnostic value of miRNAs in the cell fraction of human peripheral blood of MPM patients, cancer-free asbestos-exposed individuals (AEC) and cancer-free individuals from the general population (CGP). They first analyzed 328 miRNAs in 23 MPM and 17 AEC using microarrays (miRVana miRNA Probe Set v3.3, Ambion, TX, USA) and found that miR-20a and miR-103 were under-expressed in MPM. These two miRNAs were quantified in 23 MPM, 17 AEC and 25 CGP by RT-qPCR, and the results showed that only miR-103 was significantly under-expressed in MPM. ROC analysis showed that miR-103 could discriminate MPM from AEC with a sensitivity of 83% and specificity of 71% and could discriminate MPM from CGP with a sensitivity of 78% and specificity of 76%.

In 2014, Weber et al. published a follow up paper [72] in which they analyzed the diagnostic value of a combination of miR-103a-3p levels in the cell fraction of peripheral blood (by RT-qPCR) and the mesothelin concentration in plasma (ELISA test). The analysis of 43 male MPM patients and 52 male controls formerly exposed to asbestos revealed that the combination of mesothelin and miR-103a-3p showed a sensitivity of 95% and a specificity of 81% for MPM diagnosis. For individual determinations, mesothelin showed a sensitivity of 74% and specificity of 85%, whereas miR-103a-3p showed a sensitivity of 89% and a specificity of 63%.

These two studies did not provide details on how the healthy status of normal controls and asbestos-exposed subjects were assessed.

3.3. Studies in Serum and Plasma

MicroRNAs are secreted into the liquid fraction of peripheral blood (serum and plasma) and into all body fluids analyzed to date. These studies described here focus on miRNAs found in serum and plasma (also known as circulating miRNAs) as potential biomarkers for MPM.

In their study performed in 2011, Santarelli et al. [29] also evaluated the levels of miR-126 in serum samples of 44 MPM patients, 196 asbestos-exposed subjects and 50 healthy subjects together with the levels of soluble mesothelin-related peptides (SMRPs) using RT-qPCR and ELISA, respectively. ROC analysis showed that cut-off values of miR-126 discriminated asbestos-exposed subjects from controls with a sensitivity of 60% and specificity of 74%, and from MPM patients with a sensitivity of 73% and specificity of 74%. These values are promising, although the recommended values for good biomarkers are a sensitivity and specificity >80%. One advantage is the large number of samples analyzed. In addition, asbestos-exposed subjects were very well classified as a control group in this study. Chest radiography and high-resolution computed tomography were performed to verify the absence of tumors, and detailed questionnaires on asbestos exposure were administered. The authors also reported that a combination of decreased levels of miR-126 and increased levels of SMRPs correlated with a higher risk of developing MPM, but they did not report sensitivity or specificity values for that determination.

In a subsequent study (2012), Tomasetti et al. [73] aimed to validate an optimized method for the detection of miR-126 in serum. In this new contribution, endogenous and exogenous controls were used for the normalization of RT-qPCR data, the accuracy and precision of the method were evaluated, and relative plus absolute RT-qPCR quantification was performed. The authors quantified miR-126 in diluted serum samples of 45 MPM, 20 non-small cells lung cancer (NSCLC) patients and 56 healthy controls using RT-qPCR. The results showed an under-expression of miR-126 in MPM and NSCLC compared to healthy controls, which significantly discriminated MPM patients from healthy controls and from NSCLC patients, but did not differentiate NSCLC patients from healthy controls. ROC analysis indicated that miR-126 in serum is a candidate biomarker for MPM with high sensitivity (80%) but low specificity (60%).

Also in 2012, Kirschner et al. [74] analyzed miRNA expression profiles in the plasma of five MPM patients (three epithelioid and two sarcomatoid types) and three healthy controls using microarrays (V3, miRBase V12.0, Agilent Technologies, Santa Clara, CA, USA). They found 15 over-expressed miRNAs in MPM compared to controls. The authors further validated the microarray results of 12 candidate miRNAs with the most significant elevation levels in the plasma of 15 MPM patients and 14 control subjects (eight patients with coronary artery disease and six healthy subjects) using RT-qPCR. The results indicated that only miR-625-3p was significantly over-expressed in MPM. ROC analysis showed that plasma miR-625-3p levels discriminated between MPM patients and control subjects with an accuracy of 82.4%, a sensitivity of 73.3% and a specificity of 78.5%. Instead of testing a larger cohort of plasma samples, the levels of miR-625-3p were quantified in serum from a new cohort of 30 MPM patients and ten subjects with asbestosis. Levels of miR-625-3p were significantly elevated in MPM patients compared with asbestosis patients with an accuracy of 79.3%, a sensitivity of 70% and a specificity of 90%. Then, the authors decided to evaluate the levels of 12 miRNAs identified in microarray analysis of 18 tissue samples from MPM patients and seven pericardial tissue samples used as controls. Unlike previous results in plasma and serum, the data indicated an over-expression of miR-625-3p and an under-expression of miR-29c*, miR-16, miR-196b, miR-26a-2-3p and miR-1914-3p in MPM tumor samples compared to controls. These mismatched data are not surprising; there are studies that suggest that miRNA expression profiles in cells/tissues do not necessarily reflect the profiles of their secreted miRNAs [75,76]. The disadvantages of this study include a limited number of samples for each type of biological sample used. Perhaps choosing a larger number of plasma samples instead of analyzing a very limited number of serum and tissue samples would have been more useful. In addition, the authors used plasma from eight patients with coronary artery disease (together with six healthy subjects) as normal controls in the validation phase. These samples should not be considered controls because they are not from healthy subjects and because miRNA profiles can be altered due to coronary artery conditions [77,78]. Details on how the healthy status of control subjects was assessed were not provided. Later, levels of miR-16 were tested in a larger number of tissue samples (60 MPM samples) in a paper published in 2013 [63], but they were not tested in serum or plasma samples.

On the other hand, Gayosso-Gómez (2014) [79] searched for candidate biomarker miRNAs for differential diagnosis of MPM from AD in serum. They analyzed the miRNA profiles of pooled serum samples of 22 MPM (epithelioid), 36 AD patients and 45 healthy controls using deep sequencing (Illumina). The results indicated over-expression of seven miRNAs in MPM and 12 miRNAs in AD patients compared to healthy controls. Among these miRNAs, four were common to both neoplasms (miR-4791, miR-185-5p, miR-96-5p and miR-1271-5p), whereas miR-1292, miR-409-5 and miR-92b -5p were over-expressed exclusively in MPM. Comparative analysis of MPM vs. AD patients showed 13 miRNAs over-expressed and five miRNAs sub-expressed in MPM patients. The disadvantages of this study are the lack of quantitative validation of the sequencing results and the lack of a follow up study. An advantage was that the respiratory function of healthy subjects was verified.

In 2015, Lamberti et al. [80] reported the identification of two serum miRNA signatures that correlate with the clinical outcome and histological subtype of MPM. They quantified 384 miRNAs in the serum of 14 MPM patients (seven epithelioid, three sarcomatoid and four biphasic types) and ten patients affected by non-cancer-related pleural effusions as normal controls using PCRArray (Microfluidic card A, Applied Biosystems). The results indicated over-expression of miR-101, miR-25, miR-26b, miR-335 and miR-433 and sub-expression of miR-191 and miR-223 in MPM. Additionally, miR-29a and miR-516 were detected exclusively in MPM patients. Notably, it was stated that RT-qPCR was performed to evaluate miRNAs in "an extended group of patients", but clear information about these patients was not provided in the study. Patients were subdivided into two groups: group A, which was composed of patients with over-expression of $\geq$3/9 miRNAs and miR-516 undetectable or unchanged, and group B, which contained patients with at least 3/9 miRNAs sub-expressed or without change and/or miR-29a sub-expressed. Based on these criteria, patients in group A (n = 5) had a significantly shorter mean survival than patients in group B (n = 9) (7 months vs. 17 months, p = 0.0021). They reported that in patients with signature A, 2/5 had sarcomatoid and 3/5 had biphasic MPM, but statistical significance was not provided in this study. Therefore, two important pieces of information are missing. In addition, MPM patients were compared to non-cancer related pleural effusion patients as normal controls. Pleural effusion patients should not be considered normal or healthy controls. Pleural effusion does not occur in healthy subjects. In addition to tumor-related conditions, etiologies of pleural effusions are diverse and range from cardiopulmonary disorders to systemic inflammatory conditions that could be infectious (viral or bacterial) or non-infectious. More importantly, the authors did not report the diagnosis or etiology of the ten pleural effusion patients; therefore, there is no information on a potential bias in case of different etiologies. In addition to differences in pathogenesis and clinical characteristics, for example, viral or bacterial pneumonitis vs. cardiac failure, it has been suggested that miRNA expression profiles are distinctive for different diseases [81–84]. Perhaps the initial approach to this study should have considered the discovery of miRNAs in MPM patients that can distinguish them from patients with other non-cancer related diseases that induce pleural effusion, which also would have clinical value.

In a new contribution, Santarelli et al. [85] analyzed the diagnostic value of a combination of three markers (miR-126, methylated thrombomodulin promoter or Met-TM and SMRPs) in serum samples of 45 MPM patients, 99 asbestos-exposed subjects and 44 healthy controls to detect MPM. They further evaluated the three biomarkers in 18 MPM, 50 asbestos-exposed subjects, 20 healthy controls and 42 lung cancer (LC) patients. The population of LC patients was included for cancer specificity evaluation. The data indicated that the risk of MPM significantly increased at high SMRP levels with at least one or both altered epigenetic biomarkers (low miR-16 or high Met-TM), whereas the disease risk was maximum when all three biomarkers were altered. Conversely, the LC patients showed low miR-126 and high Met-TM levels but were associated with low levels of SMRPs. The combination of SMRPs, miR-126 and Met-TM improves the differential diagnosis of MPM up to an AUC of 0.857 (95% CI, 0.767–0.927) compared to SMRP alone at 0.818 (95% CI, 0.723–0.914). Importantly, the authors reported that the expression of these biomarkers was independent of gender, age, smoking and duration of asbestos exposure, which is a new contribution for MPM-related studies.

In 2016, Bononi et al. [86] investigated miRNA expression profiles in the serum of ten MPM patients, ten subjects exposed to asbestos (AE) and ten healthy subjects (HC) by using microarrays (Agilent Technologies, Human miRNA G4470A) in the discovery phase. Out of 37 differentially expressed miRNAs in MPM, three were validated in 30 sera previously used for microarray analysis and in additional 19 serum samples (ten MPM, five AE and four HC) by RT-qPCR. The results indicated that miR-197-3p, miR-1281 and miR-32-3p were up-regulated in MPM compared to HC; miR-197-3p and miR-32-3p were up-regulated in MPM compared to AE; and miR-1281 was up-regulated in MPM and AE compared to HC. AUC in all cases were a little less than 0.8. This work was one of few studies that identified endogenous stable miRNA that could be used as suitable normalizer. On the other hand, some relevant information was missing, such as the histological subtypes of MPM patients and how the healthy status of the controls was verified. This study did not find the down-regulation of miR-126 in MPM as previously reported by Santarelli et al. [29,85], who also used serum from asbestos-exposed subjects as comparative controls.

Cavalleri et al. [87] in 2017 aimed to identify a specific miRNA signature in plasmatic extracellular vesicles (EV) that discriminates MPM patients from past asbestos-exposed subjects (PAE). They analyzed 754 miRNAs in plasmatic EVs of 23 MPM patients and 19 cancer-free subjects exposed to asbestos in the past using an OpenArray method. Among 62 miRNAs differentially expressed in MPM compared to PAE (sub-expression), 16 out of 20 analyzed miRNAs were quantitatively confirmed by RT-qPCR. The authors found a signature of the five best discriminating miRNAs of miR-103a-3p, miR-98, miR-148b, miR-744 and miR-30e-3p with an AUC of 0.864, 0.864, 0.852, 0.845 and 0.827, respectively. They further simplified the signature with miR-30e-3p and miR-103a-3p, which generated an AUC of 0.942 with a sensitivity of 95.5% and specificity of 80%. Down-regulation of miR-103a-3p was also found in the cellular fraction of peripheral blood of MPM patients in two previous reports [71,72]. In addition, the authors tested other miRNAs reported in the literature as potential biomarkers of MPM (miR-126, miR-625-3p, miR-25, miR-29 and miR-433) and did not find significant differences between the study groups. Nevertheless, the evidence indicates that the type and levels of miRNAs may vary in the exosomal fraction compared to those found in exosome-free fraction and may vary between different body fluids, such as serum vs. plasma [36]. This work is the first study that exclusively analyzed the miRNAs in exosomes isolated from plasma samples of MPM patients, which may provide relevant information about miRNA release mechanisms associated with neoplastic processes for future studies. However, a potential disadvantage of this approach for future clinical applications is that ultracentrifugation requires a specialized heavy apparatus (ultracentrifuge) that is not common in clinical laboratories and two extra hours of sample processing.

Table 1. Studies regarding tissue and secreted miRNAs differentially expressed in MPM with potential diagnosis biomarker value.

Studies in	Sample Source	Study Design and Sample Size	Assay (and Number of miRNAs Analyzed)	miRNAs Differentially Expressed in MPM	ROC Analysis	Study Aim	Selection of Endogenous Stable Normalizer (Validation Phase)	Reference
Tissue	Pleural tissue (frozen)	Discovery: 17 MPM pleural tissue vs. 1 total RNA from normal human pericardium	Microarray (723)	let-7b*↑, MiR-1228*↑, miR-195*↑, miR-30b*↑, miR-32*↑, miR-345↑, miR-483-3p↑, miR-584↑, miR-595↑, miR-615-3p↑, and miR-885-3p↑; let-7e*↓, miR-144*↓, miR-203↓, miR-340*↓, miR-34a*↓, miR-423↓, miR-582↓, miR-7-1*↓, and miR-9↓	Not performed (NP)	Oncogenic mechanisms	NA	[58]
	Pleural and lung tissue	Discovery: 15 MPM pleural tissues vs. 10 AD pleural tissues. Validation: 100 MPM pleural tissues vs. 32 AD lung tissues	Microarray (2564), RT-qPCR (7)	RT-qPCR: miR-200c↓, miR-200b↓, miR-203↓, miR-141↓, miR-429↓ and miR-205↓	Specificity and sensitivity >80%	Diagnosis	NP. Use of RNU44 and RNU48	[59]
	Pleural and several tumor tissue (FFPE)	Discovery: 7 MPM pleural tissues vs. 97 epithelial carcinomas. Validation: (1) 32 pleura tissues MPM vs. 113 epithelial carcinomas. (2) 16 MPM pleural tissues vs. 23 epithelial carcinomas. (3) 14 pleural tissues MPM vs. 46 epithelial carcinomas	Microarray (747), RT-qPCR (3)	RT-qPCR: miR-193a-3p↑, miR-200c↓ and miR-192↓	Specificity 94%, sensitivity 100%	Diagnosis	NP. Use of U6 snoRNA	[60]
	Pleural tissue (frozen), pleural tissue (FFPE).	Discovery: 10 MPM vs. 5 Healthy controls (frozen). Validation: 27 MPM vs. 27 adjacent normal pleural tissues (FFPE)	PCRArray (88). RT-qPCR (3)	miR-126↓	NP	Diagnosis	NP. Use of U6 small nuclear RNA	[29]
	Pleural tissue (frozen)	Discovery: 25 MPM vs. 6 normal parietal pleura (patients without cancer). Validation: Same cohort? Not specified	Microarray (1145), RT-qPCR (4)	RT-qPCR: miR-206↓, miR-1↓, miR-483-5p↓, and miR-155↑	NP	Oncogenic mechanisms	NP. Use of RNU44	[62]
	Pleural tissue (FFPE). Also MPM cell lines.	Discovery: [?]. Validation: 60 MPM vs. 23 normal pleural tissues	RT-qPCR (4)	miR-16↓, miR15a↓, miR-15b↓, and miR-195↓	NP	New therapy targets	NP. Use of RNU6B	[63]
	Pleural tissue (fresh, frozen & FFPE). Peritoneal tissue (frozen and FFPE).	Discovery: 29 MPM pleural tissues vs. 12 peritoneal mesothelial cysts (FFPE). Validation: (1) 6 MPM pleural tissues vs. 14 benign pleural tissues (Fresh). (2) 36 pleural tissues MPM vs. 36 peritoneal mesothelium (frozen)	Microarray (887), RT-qPCR (1)	RT-qPCR: miR-145↓	NP	Oncogenic mechanisms	NP. Use of RNU6B and RNU49	[64]

Table 1. *Cont.*

Studies in	Sample Source	Study Design and Sample Size	Assay (and Number of miRNAs Analyzed)	miRNAs Differentially Expressed in MPM	ROC Analysis	Study Aim	Selection of Endogenous Stable Normalizer (Validation Phase)	Reference
	Pleural tissue (FFPE)	Discovery: 5 MPM pleural tissues vs. 5 non-cancerous/non-inflammatory pleural tissues vs. 4 pleural chronic inflammation tissues vs. 5 mesothelial hyperplasia tissue	PCRArray (667)	miR-517b-3p↓, miR-627↓, miR-766-3p↓, miR-101-3p↓, miR-501-3p↓, miR-212-3p↓, miR-596↓, miR-145-5p↓, miR-671-3p↓, miR-181a-5p↓, miR-18a-3p↓, miR-30e-3p↑, miR-34a-3p↑, miR-622↑, let-7-g-5p↑, miR-196b-5p↑, miR-135b-5p↑, miR-18a-5p↑, miR-302b-3p↑	NP	Oncogenic mechanisms	Normalization factor: the global mean expression value	[65]
	Pleural tissue (FFPE)	Discovery: 5 preoperative pleural tissues with MPM (before Cth = DB) and 5 pleural tissues MPM (after Cth = MPMc) vs. 5 non-neoplastic pleura tissues (after Cth = NNP). Validation: 40 MPMc vs. 12 DB vs. 14 NNP	PCRArray (742), RT-qPCR (4)	RT-qPCR: miR-126↓*, miR-143↓, miR-145↓, miR-652↓	Specificity and sensitivity close to or >80%	Diagnosis	NP. Use of snord49A	[66]
	Pleural tissue (frozen)	Discovery: 18 MPM pleural tissues vs. 6 pleural tissues from benign asbestos-related pleural effusion patients (BAPE) (tissue with unspecific pleuritis/fibrosis)	PCRArray (384)	miR-484↑, miR-320↑, let-7a↑, miR-125a-5p↑	Specificity and sensitivity close to or >80%	Diagnosis	NP. Use of U6 snoRNA	[67]
	Pleural tissue (FFPE) and cells from pleural effusion.	Discovery and Validation: 17 MPM pleural tissues vs. 6 normal pleural tissues patients without cancer undergoing cardiac or aortic surgery. Cells from pleural effusion of 26 MPM patients vs. 10 benign pleural diseases	RT-qPCR (1)	miR-223↓	NP	Oncogenic mechanisms	NP. Use of RNU6B for tumor and RNU48, RNU44, or SNOR202 for cells	[68]
	Cell lines, pleural tissue (FFPE), pleural citology	Discovery: 2 MPM cell lines vs. 1 mesothelium cell line. Validation: (1) 51 MPM pleural tissues vs. 40 benign/reactive pleurae. (2) 29 MPM cytologic specimens vs. 24 reactive mesothelial cells	RT-qPCR (15)	miR-19a↑, miR-19b↑, miR-25↑, miR-21↑, miR-126↓	Specificity and sensitivity >80%	Diagnosis	NP. Use of RNA U6B	[69]
	Pleural tissue (FFPE), pleural citology	Discovery: Bioinformatic analysis 3 database. Validation: 41 epithelioid MPM vs. 40 AD and 26 cytologic specimen epithelioid MPM vs. 26 AD	RT-qPCR (6)	miR-130a↑ (histological and cytological specimens)	Specificity 67% and sensitivity 77%	Differential diagnosis MPM vs. AD	NP. Use of RNU6B	[70]

Table 1. *Cont.*

Studies in	Sample Source	Study Design and Sample Size	Assay (and Number of miRNAs Analyzed)	miRNAs Differentially Expressed in MPM	ROC Analysis	Study Aim	Selection of Endogenous Stable Normalizer (Validation Phase)	Reference
Peripheral blood	Cellular fraction of peripheral blood	Discovery: 23 MPM vs. 17 asbestos-exposed controls (AE). Validation: 23 MPM vs. 17 AE vs. 25 healthy controls	Microarray (328). RT-qPCR (2)	RT-qPCR: miR-103↓	Specificity and sensitivity >80%	Diagnosis	Yes: miR-125a	[71]
	Cellular fraction of peripheral blood	Discovery: [7]. Validation: 43 MPM vs. 52 asbestos-exposed controls	RT-qPCR (1)	miR-103a-3p↓ (plus mesothelin↑ in plasma)	Specificity and sensitivity >80%	Diagnosis	Yes: miR-125a	[72]
	Serum	Discovery: In pleural tissue (miR-126↓ same paper). Validation: 44 MPM vs. 196 asbestos-exposed controls vs. 50 Healthy controls	RT-qPCR (1)	miR-126↓	Specificity 74% and sensitivity 73%.	Diagnosis	NP. Use of U6 snoRNA	[29]
	Serum	Pre-Validation: [29]. Validation: 45 MPM vs. 20 NSCLC vs. 56 healthy controls	RT-qPCR (1)	miR-126↓	Specificity 60% and sensitivity 80%	Diagnosis	Yes: U6 snoRNA and use of exogenous control cel-miR-39	[73]
Serum and plasma	Plasma, serum. Also pleural tissue (FFPE)	Discovery: 5 MPM (plasma) vs. 3 healthy controls (HC). Validation: 15 MPM (plasma) vs. 14 HC. Validation serum: 30 MPM (serum) vs. 10 asbestosis (serum). Validation tissue: 18 MPM pleural tissues vs. 7 pericardial tissues	Microarray (854), RT-qPCR	RT-qPCR: miR-625-3p↑ (plasma & serum). miR-625-3p↑, miR-29c*↓, miR-16↓, miR-196b↓, miR-26a-2-3p↓ and miR-1914-3p↓ (tissue)	Specificity & sensitivity close to or >80 % (plasma & serum miR-625-3p)	Diagnosis	Only SD of Cq range values without specified clearly which samples were used. Previous work (plasma): miR-16	[74]
	Serum	Discovery: 11 MPM (epithelial) vs. 45 healthy controls vs. 36 AD	Deep sequencing (Ilumina)	MPM vs. control: miR-479↑, miR-185-5p↑, miR-96-5p↑, miR-1271-5p, miR-1292-5p↑, miR-409-5p↑y miR-92b-5p↑	NP	Diagnosis	NA	[79]
	Serum	Discovery: 14 MPM vs. 10 non-cancer related effusions patients. Validation: Not specified	PCRArray (384). RT-qPCR (7)	RT-qPCR: miR-101↑, miR-25↑, miR-26b↑, miR-335↑, miR-29a↑, miR-516↑, miR-433↑, miR-191↓, miR-223↓	Not performed	Prognosis	NP. Use of miR-16	[80]
	Serum	Discovery: [29]. Validation 1: 45 MPM vs. 99 asbestos-exposed subjects (AE) vs. 44 healthy subjects. Validation 2:18 MPM vs. 50 (AE) vs. 20 healthy controls and 42 lung cancer (LC) patients	RT-qPCR (1)	Combination SMRPs↑, miR-126↓, and Met-TM↑	AUC of 0.857 (95% CI, 0.767–0.927)	Diagnosis	NP. Use of U6 snoRNA. Use of exogenous control cel-miR-39	[85]

Table 1. *Cont.*

Studies in	Sample Source	Study Design and Sample Size	Assay (and Number of miRNAs Analyzed)	miRNAs Differentially Expressed in MPM	ROC Analysis	Study Aim	Selection of Endogenous Stable Normalizer (Validation Phase)	Reference
	Serum	Discovery: 10 MPM vs. 10 asbestos-exposed subjects (AE) vs. 10 healthy controls (HC). Validation: 20 MPM vs. 15 AE vs. 14 HC	Microarray (1201), RT-qPCR (3)	RT-qPCR: miR-197-3p↑, miR-1281↑, miR-32-3p↑ (MPM vs. HS and MPM vs. AE)	AUC 95% CI, 0.5398-0.8959 (miR-197-3p)	Diagnosis	Yes: miR-3665	[86]
	Plasma (exosomal fraction)	Discovery: 23 MPM vs. 19 past asbestos-exposed subjects. Validation: Same samples minus 4	OpenArray (754). RT-qPCR (20)	2-miRNAs signatures: miR-103a-3p↓, miR-30e-3p↓	Specificity 80% and sensitivity 95.5%	Diagnosis	RNU48. It is not clear	[87]
	Plasma	Discovery: 21 MPM vs. 21 asbestos-exposed controls. Validation: 22 MPM vs. 44 asbestos-exposed controls	PCRArray (377), RT-qPCR (2)	RT-qPCR: miR-132-3p↓	Specificity 61% and sensitivity 86%	Diagnosis	Yes. Use of miR-146b-5p. Another untested normalizer was used too	[88]

AD = Lung Adenocarcinoma; Cth = chemotherapy; Met-TM = methylated thrombomodulin promoter; SMRPs = soluble mesothelin-related peptides. snoRNAs: small nuclear RNA; SD: standard desviation; NA = does not apply; NP = not performed. ROC = Receiver operating characteristic. ↑ = upregulated expression; ↓ = downregulated expression.

In 2017, Weber et al. [88] aimed to identify candidate biomarker miRNAs in plasma for diagnosis of MPM. For the discovery phase, they analyzed 377 miRNAs in plasma of 21 MPM patients and 21 asbestos-exposed controls, using PCRArray (TaqMan Low density Array Human MicroRNA CardA v2.0). The authors reported the identification of three stable reference miRNAs (miR-20b, miR-28-3p and miR-146b-5p), and also reported that they normalized the raw Ct values from the PCRArray with combinations of the three reference miRNAs to identify the candidate miRNAs. Then, miR-24 (miR-146b-5p as reference) and miR-132-3p (miR-146b-5p as reference), miR-24 (miR-146b-5pMiR-28-3p as reference) and miR-132-3p (miR-28-3p as reference) showed statistically significant down-regulation in MPM, and they were analyzed in the subsequent validation phase (22 MPM patients and 44 asbestos-exposed controls) using RT-qPCR. Results indicated that only miR-132-3p (and miR-146b-5p as reference) reached a significant difference in MPM patients compared to controls.

The authors additionally measured the expression of miR-126 (using U6 snoRNA as reference) and miR-625-3p (miR-16 as reference) in the verification phase to evaluate the discrimination potential of biomarker combinations. These miRNAs were previously described as candidate MPM biomarkers by Santarelli [29] and Kirschner et al. [74]. Results indicated that miR-126 was statistically significantly downregulated in MPM compared to controls. However, there are confusing details in this experimental design and results. First, miR-126 is included in the Human MicroRNA Card A v2.0 that was used for this study (discovery phase), and the authors did not report that the miR-126 expression was altered in MPM compared to controls. Therefore, it is not clear how the levels of miR-126 were found downregulated in the validation phase but not in the discovery phase in this study. Second, Santarelli et al., 2011 [29] used U6 snoRNA as a normalizer, but U6 snoRNA was not discovered as a suitable normalizer in this study, but it was used as reference anyway. Perhaps this inconsistency in miR-126 levels can be explained by the use of inadequate normalizer for these samples. Finally, the authors also reported that the combination of miR-132-3p and miR-126 within a panel (with two different references for normalization) implicate a less robust diagnosis method.

3.4. Studies in Other Body Fluids

To date, there are no publications that analyze the diagnostic value of secreted miRNAs in other biological fluids in MPM. Samples such as pleural effusion fluid could be a good option. This condition is a common clinical manifestation of late MPM, which is when most patients seek medical attention. Moreover, recent evidence suggests that secreted miRNAs in pleural effusion may have diagnostic value in other neoplasms, such as lung cancer [89]. As mentioned before, Birnie et al. analyzed the miRNA expression patterns of cells from pleural effusion [68]. Although they were not secreted miRNAs, miRNAs from pleural effusion cells would be a good option for potential non-invasive biomarkers.

4. Relevant Aspects of the Experimental Designs That May Influence the Accuracy, Consistency and Real Diagnostic Value of Currently Reported Data

4.1. Number of Samples

Most of the MPM-related studies analyzed a limited number of biological samples. Even for the validation phase, less than 100 samples were analyzed with a few exceptions. In fact, power and sample size calculations were not presented in any of the published studies. There is no doubt that data obtained from large-scale studies are considered the most reliable; however, it is relevant to notice that MPM is a low frequency disease that is difficult to diagnose and therefore samples may not be available in great number to the investigators at the time of research. For this particularly rare but aggressive tumor, information provided by well-designed studies will be relevant even with a limited number of samples. One strategy to overcome this scenario is to validate findings in follow up studies with larger, independent sets of patients when samples are available.

4.2. Follow up Studies of Promising Candidate miRNAs Biomarkers

Unfortunately, most of the studies reported data that were never confirmed in subsequent independent analysis with a few exceptions. Down-regulation of miR-126 was assessed in the serum of MPM patients in three subsequent studies by Santarelli et al. in 2011 [29], Tomasetti et al. in 2012 [73] and Santarelli et al. in 2015 [85]. Weber et al. reported the sub-expression of miR-103 in the cellular fraction of peripheral blood of MPM patients in two subsequent studies [71,72].

On the other hand, independent studies identified some common miRNAs as candidate biomarkers for MPM: in addition to Weber et al., an independent research group reported the sub-expression of miR-103 in the plasma of MPM patients [87]. Three independent authors found that miR-145 was downregulated in the pleural tissue of MPM patients [64–66].

Consistently reported miRNAs in different publications are more likely to have clinical relevance; therefore, a simple vote-counting method could be applied to choose promising candidate miRNAs that will be evaluated in statistically well-powered prospective studies. Table 2 summarizes the candidate miRNAs biomarkers of MPM reported by at least two independent studies and includes only RT-qPCR-validated miRNAs.

Table 2. miRNAs with potential diagnosis value for MPM reported by at least 2 independent studies.

miRNA	Number of Studies	Sample Source	Comparative Analysis Design	References
miR-200c↓	2	Pleural tissue	(1) MPM vs. AD. (2) MPM vs. epithelial carcinoma	[59,60]
miR-126↓	3	Pleural tissue	(1) MPM vs. normal pleura. (2) MPM vs. normal pleura (with Cth). (3) MPM vs. benign/reactive pleurae	[29,66,69]
miR-145↓	3	Pleural tissue	(1) MPM vs. benign pleural tissue. (2) MPM vs. normal pleura. (3) MPM vs. normal pleura (with Cth)	[64–66]
miR-16↓	2	Pleural tissue	(1) MPM vs. pericardial tissues. (2) MPM vs. normal pleural tissue	[63,74]
miR-103↓	2	Cellular fraction of peripheral blood	Two subsequent studies: MPM vs. asbestos-exposed controls	[71,72]
miR-126↓	3	Serum	Three subsequent studies: (a) MPM vs. asbestos-exposed controls vs. healthy controls. (b) MPM vs. Healthy controls. (c) MPM vs. asbestos-exposed controls vs. healthy controls	[29,73,69]

Only validated miRNAs (RT-qPCR) are included in this list. ↑ = upregulated expression; ↓ = downregulated expression.

On the other hand, a meta-analysis is a better approach because it provides statistical analysis of multiple data, increasing the likelihood of finding good candidate miRNAs. However, performing a meta-analysis requires an investigator to decide how to search for studies, how to select those studies, which criteria to use and which data to include. These choices affect the results of this analysis.

Therefore, both vote-counting methods and meta-analysis should be performed carefully by choosing reliable data for the analysis, for example only RT-qPCR-validated miRNAs.

In this regard, Micolucci et al. (2016) [90] performed a systematic review and meta-analysis in order to identify high-confidence miRNAs that could serve as biomarkers of MPM compared to asbestos exposure subjects. First, they listed the most frequently reported miRNAs that had been described in 2–5 papers in their table 1 and supplementary table 1 by using a vote-counting method. Nevertheless, the authors listed all miRNAs reported in those papers without analyzing the reliability of the studies. For example, both tables included miRNAs that were identified by a single microarray analysis, without verification with RT-qPCR. Moreover, Table 1 included miR-20a (reported by Weber 2012 [71]), which was not significantly down-regulated in MPM after the verification test performed by Weber et al. themselves. Another example is miR-101, which was reported by Kemp et al. [91] as downregulated in only six tumor samples vs. three normal pleural tissues. Next, the authors conducted a qualitative meta-analysis using only RT-qPCR-validated miRNAs to improve the results of the vote-counting method, but details of other criteria were not provided. This method identified nine miRNAs as the most significant in tissue (miR-145-5p, miR-126-3p, miR-16-5p, miR-192-5p, miR-193a-3p, miR-200b-3p, miR-203-3p, miR-143-3p and miR-652-3p) and three circulating miRNAs (miR126-3p, miR-103a-3p and miR-625-3p). In addition, authors analyzed the biological function of these promising miRNAs to estimate their potential value as biomarkers. In spite of the heterogeneity of MPM studies, these qualitative meta-analyses and functional research provide an undeniably useful list of candidate miRNAs that should be analyzed in a large-scale study in order to assess their clinical relevance.

4.3. Relevance of Tumor Representativeness in Tissue Samples

It is notable that only two studies assessed tumor representativeness in the analyzed tissue samples by using only MPM samples with >80% tumor content and by using laser micro-dissection [65,68]. Few other studies mentioned the percentage of tumor content in MPM tissue samples, which ranged from 40–85% [58,60,62,66]. As previously mentioned, this information could be relevant if we consider the representative tumor content in a sample with a non-neoplastic content of 60% vs. 15%, for example. Remember that a potential diagnostic value is based on the hypothesis that tumor tissue expresses miRNA profiles that are distinct from normal tissue; therefore, the experimental design should ensure that the obtained data corresponds to each type of tissue.

4.4. Relevance of the Normal/Healthy Controls Used in These Studies

Only three studies that analyzed miRNAs in peripheral blood, serum or plasma of healthy subjects as controls provided details on how the "healthy" status of those subjects was verified [73,79,85]. This relevant information should have been included in the other studies. On the other hand, a few studies used subjects who suffered from disease as "normal controls". For example, patients with pleural effusion were used as controls, which implies that these subjects suffered from a disease that was not disclosed in the study. Another example was the use of patients with coronary artery disease as normal controls. This scenario may be considered an important flaw in the design if the study aim was to compare MPM patients to "healthy" subjects. Evidence indicates that miRNAs are altered in several diseases; therefore, there is bias in analysis when the presence of another disease is not acknowledged.

4.5. Relevance of Proper Normalizers for Quantitative RT-qPCR Analysis (Validation Phase)

Currently, the relevance of proper and rigorous normalization of quantitative RT-qPCR data is well known. The accuracy of expression measurements requires the identification and validation of appropriate reference miRNA for each type of biological sample used [92–94]. It is notable that only a few studies identified and experimentally validated the most stable miRNAs to normalize qPCR expression data (Table 1). Inappropriate normalization can result in statistical confidence in the wrong conclusion [92] and can lead to false discovery.

4.6. Analysis of Different Histological Subtypes of MPM

In addition to different histological characteristics, there are differences in the clinical behavior, malignity and outcome between the three main histological subtypes of MPM. Consequently, analysis of miRNA expression may be separately performed for each histological subtype of MPM in order to maximize discoveries with clinical usefulness. Unfortunately, most studies did not report such analysis despite using different histological subtypes of MPM, or they omitted this analysis from the report it if it was actually performed. Moreover, some studies did not report the histological subtype of the samples used.

4.7. More Than One Biomarker Used for Diagnosis

Cancer, including MPM, is a multifactorial disease that involves multiple genetic/epigenetic alterations and environmental risk factors. Thus, it is unlikely that a single biomarker will provide a method of detection with the sensitivity and specificity required to reach an accurate diagnosis. Accordingly, evidence showed that the diagnostic value of serum levels of miR-126 increased when it was measured together with SMRPs and Met-TM [73,85], and the diagnostic value of levels of miR-103a-3p in the cellular fraction of peripheral blood increased when it was measured together with mesothelin in plasma [71,72]. In addition, signatures of two or more miRNAs may increase diagnostic value in future clinical use [87].

4.8. Complete and Accurate Reporting

Complete and accurate reporting allows readers to critically identify the strengths and weaknesses of the research study and therefore evaluate the validity and potential applicability of the reported data. However, critical information is often missing or unclear in most of the reviewed publications. This scenario is not unique for studies regarding MPM; lack of relevant information in design, conduct and analysis of diagnostic studies has been detected previously [95,96]. Because of this lack of information, the Standards for Reporting of Diagnostic Accuracy (STARD) initiative emerged to improve the quality of reporting of studies of diagnostic accuracy, among other guidelines. Recommendations of guidelines, such as the Standards for Reporting of Diagnostic Accuracy (STARD) for diagnostic studies [97], the Reporting recommendations for tumour Marker prognostic Studies (REMARK) for prognostic studies [98], or the Strengthening the Reporting of Observational Studies in Epidemiology (STROBE) for observational studies [99], are useful to determine the reliability and quality of biomarkers in the initial discovery phase. These guidelines have been available since 2003, 2007 and 2012, respectively. Following the recommendations of these guidelines may be helpful to standardize which vital information is published.

5. miRNAs Associated with Neoplastic Mechanisms of MPM and Their Potential Diagnostic Value

In addition, increasing evidence suggests that aberrant levels of miRNAs contribute to oncogenesis, progression and metastasis of several cancers, such as tumor suppressor or oncomiR [100]. As previously mentioned, the potential association of altered miRNAs with MPM carcinogenic mechanisms increases the likelihood of having a diagnostic value for this disease. This feature, together with their tumor tissue-specific expression, may facilitate the identification of diagnostic and prognostic miRNA biomarkers that can be applied for clinical use. Table 3 summarizes deregulated miRNAs associated with carcinogenesis mechanisms in MPM.

For example, miR-145 was reported as downregulated in MPM tissue by three independent studies; importantly, its over-expression in MPM cell lines induces a reduction of proliferation and migration in two out of three transfected MPM cell lines (Tables 2 and 3).

Perhaps, greater effort should be devoted to elucidate which miRNAs are associated with neoplastic mechanisms along with the searching for candidate biomarkers.

Table 3. Deregulated miRNAs associated with carcinogenesis mechanisms in MPM.

miRNA with Deregulated Expression in MPM	Potential Function	Biological Effect of Experimental Manipulation of miRNA Expression	Other Effects	Reference
miR-16↓ (tissue)	Tumor suppressor	Restoring miR-16: growth inhibition, cell cycle arrest in G0/G1, increased apoptosis and reduced colony formation in MPM cell lines	Correlation with downregulation of Bcl2, CCND1. Administration of miR-16-containing minicells to xenografted mice inhibited tumor growth	[63]
miR-1↓ (tissue)	Tumor suppressor	Restoring miR-1: cell cycle arrest, increased apoptosis.	Correlation with upregulation of p53, BAX, p16/21; and downregulation of Bcl2 and survivin	[62]
miR-145↓ (tissue)	Tumor suppressor	Restoring miR-145: reduction of proliferation and migration of two out of three transfected MPM cell lines	Xenotransplant (transfected MPM cell line): inhibition of tumor growth in 6 of 8 treated mice compared to controls	[64]
miR-223↓ (tissue)	Tumor suppressor	Over-expression of miR-223: reduction of two MPM cell lines motility.	STMN1 levels were reduced and tubulin acetylation was induced	[68]

↑ = upregulated expression; ↓ = downregulated expression.

6. Conclusions and Future Perspectives

Accurate diagnosis of MPM is often difficult and complex. Difficulty in diagnosis has led to the search for new diagnostic tools that can be added to the resources currently used in the clinic. In this regard, accumulating evidence indicates that miRNAs are potential diagnostic biomarkers for several tumors, which prompted the study of microRNA expression levels as an important diagnostic and prognostic tool for MPM.

However, for this disease, limited availability of patient cohorts seemed to be an initial problem that had to be solved in order to perform research. Possibly, in order to advance in the field, it would be preferable to identify the most promising candidate miRNAs reported in the peer-reviewed literature and validate them in a multi-institutional/international coordinated effort using well-characterized biological samples from multiple research institutions in statistically well-powered prospective studies.

However, analysis of the published literature showed the heterogeneity of the data, samples, controls, and methods and the critical limitations and potential bias of several of the reviewed studies. Moreover, important information is often missing or unclear in various revised papers. Nevertheless, despite these limitations, several common candidate biomarker miRNAs were confirmed by various studies. Moreover, some of these miRNAs were associated with cellular mechanisms that are potentially involved in carcinogenesis in in vitro experiments. This result is telling of the true potential of miRNAs as diagnostic biomarkers for MPM.

This analysis allows some essential conclusions. (A) Larger and prospective studies are needed to confirm the true diagnostic value of all candidate miRNAs reported in the reviewed literature. (B) It is fundamental that research is reported clearly and transparently regarding study design, performance, and analysis. (C) It is necessary to critically evaluate the published data in order to identify deficiencies or bias and to overcome these issues in future or subjacent studies. (D) To date, none of the studies have successfully reached the final objective, which is the use of miRNAs as diagnostic biomarkers in the clinic.

Finally, miRNA-based biomarker tests could add relevant adjunct information to increase the probability of reaching the right diagnosis. Therefore, they may be used as complementary tests to gold standard immunohistochemical diagnostic tests, X-rays and clinical data.

Author Contributions: Vanessa Martínez-Rivera researched the content, wrote the original manuscript in Spanish and designed figures; María Cristina Negrete-García revised tables and figures; Federico Ávila-Moreno edited, revised and approved the final version of the review; Blanca Ortiz-Quintero researched and updated the content, wrote the manuscript version for a review paper, designed tables, edited original figures and revised and edited the final version of the review.

Conflicts of Interest: The authors declare no conflict of interest.

References

1. Kaufman, A.J.; Pass, H.I. Current concepts in malignant pleural mesothelioma. *Expert Rev. Anticancer Ther.* **2008**, *8*, 293–303. [CrossRef] [PubMed]
2. Robinson, B.W.; Lake, R.A. Advances in malignant mesothelioma. *N. Engl. J. Med.* **2005**, *353*, 1591–1603. [CrossRef] [PubMed]
3. Bianchi, C.; Bianchi, T. Malignant mesothelioma: Global incidence and relationship with asbestos. *Ind. Health* **2007**, *45*, 379–387. [CrossRef] [PubMed]
4. Roggli, V.L.; Sharma, A.; Butnor, K.J.; Sporn, T.; Vollmer, R.T. Malignant mesothelioma and occupational exposure to asbestos: A clinicopathological correlation of 1445 cases. *Ultrastruct. Pathol.* **2002**, *26*, 55–65. [CrossRef] [PubMed]
5. Ray, M.; Kindler, H.L. Malignant pleural mesothelioma: An update on biomarkers and treatment. *Chest* **2009**, *136*, 888–896. [CrossRef] [PubMed]
6. Delgermaa, V.; Takahashi, K.; Park, E.K.; Le, G.V.; Hara, T.; Sorahan, T. Global mesothelioma deaths reported to the world health organization between 1994 and 2008. *Bull. World Health Organ.* **2011**, *89*, 716–724. [CrossRef] [PubMed]
7. Park, E.K.; Takahashi, K.; Hoshuyama, T.; Cheng, T.J.; Delgermaa, V.; Le, G.V.; Sorahan, T. Global magnitude of reported and unreported mesothelioma. *Environ. Health Perspect.* **2011**, *119*, 514–518. [CrossRef] [PubMed]
8. Addis, B.; Roche, H. Problems in mesothelioma diagnosis. *Histopathology* **2009**, *54*, 55–68. [CrossRef] [PubMed]
9. Inai, K. Pathology of mesothelioma. *Environ. Health Prev. Med.* **2008**, *13*, 60–64. [CrossRef] [PubMed]
10. British Thoracic Society Standards of Care Committee. BTS statement on malignant mesothelioma in the UK, 2007. *Thorax* **2007**, *62* (Suppl. 2), ii1–ii19.
11. Evans, A.L.; Gleeson, F.V. Radiology in pleural disease: State of the art. *Respirology* **2004**, *9*, 300–312. [CrossRef] [PubMed]
12. Hallifax, R.J.; Talwar, A.; Wrightson, J.M.; Edey, A.; Gleeson, F.V. State-of-the-art: Radiological investigation of pleural disease. *Respir. Med.* **2017**, *124*, 88–99. [CrossRef] [PubMed]
13. Paintal, A.; Raparia, K.; Zakowski, M.F.; Nayar, R. The diagnosis of malignant mesothelioma in effusion cytology: A reappraisal and results of a multi-institution survey. *Cancer Cytopathol.* **2013**, *121*, 703–707. [CrossRef] [PubMed]
14. Whitaker, D. The cytology of malignant mesothelioma. *Cytopathology* **2000**, *11*, 139–151. [CrossRef] [PubMed]
15. Arif, Q.; Husain, A.N. Malignant mesothelioma diagnosis. *Arch. Pathol. Lab. Med.* **2015**, *139*, 978–980. [CrossRef] [PubMed]
16. Heffner, J.E.; Klein, J.S. Recent advances in the diagnosis and management of malignant pleural effusions. *Mayo Clin. Proc.* **2008**, *83*, 235–250. [CrossRef]
17. Husain, A.N.; Colby, T.V.; Ordonez, N.G.; Krausz, T.; Borczuk, A.; Cagle, P.T.; Chirieac, L.R.; Churg, A.; Galateau-Salle, F.; Gibbs, A.R.; et al. Guidelines for pathologic diagnosis of malignant mesothelioma: A consensus statement from the international mesothelioma interest group. *Arch. Pathol. Lab. Med.* **2009**, *133*, 1317–1331. [PubMed]
18. Carthew, R.W.; Sontheimer, E.J. Origins and mechanisms of miRNAs and siRNAs. *Cell* **2009**, *136*, 642–655. [CrossRef] [PubMed]
19. Sherrard, R.; Luehr, S.; Holzkamp, H.; McJunkin, K.; Memar, N.; Conradt, B. miRNAs cooperate in apoptosis regulation during c. Elegans development. *Genes Dev.* **2017**, *31*, 209–222. [CrossRef] [PubMed]
20. Shaw, W.R.; Armisen, J.; Lehrbach, N.J.; Miska, E.A. The conserved mir-51 microRNA family is redundantly required for embryonic development and pharynx attachment in caenorhabditis elegans. *Genetics* **2010**, *185*, 897–905. [CrossRef] [PubMed]
21. Koff, J.L.; Ramachandiran, S.; Bernal-Mizrachi, L. A time to kill: Targeting apoptosis in cancer. *Int. J. Mol. Sci.* **2015**, *16*, 2942–2955. [CrossRef] [PubMed]
22. Feitelson, M.A.; Arzumanyan, A.; Kulathinal, R.J.; Blain, S.W.; Holcombe, R.F.; Mahajna, J.; Marino, M.; Martinez-Chantar, M.L.; Nawroth, R.; Sanchez-Garcia, I.; et al. Sustained proliferation in cancer: Mechanisms and novel therapeutic targets. *Semin. Cancer Biol.* **2015**, *35*, S25–S54. [CrossRef] [PubMed]

23. Kanno, S.; Nosho, K.; Ishigami, K.; Yamamoto, I.; Koide, H.; Kurihara, H.; Mitsuhashi, K.; Shitani, M.; Motoya, M.; Sasaki, S.; et al. MicroRNA-196b is an independent prognostic biomarker in patients with pancreatic cancer. *Carcinogenesis* **2017**, *38*, 425–431. [CrossRef] [PubMed]

24. Lu, M.; Kong, X.; Wang, H.; Huang, G.; Ye, C.; He, Z. A novel microRNAs expression signature for hepatocellular carcinoma diagnosis and prognosis. *Oncotarget* **2017**, *8*, 8775–8784. [CrossRef] [PubMed]

25. Peng, Z.; Pan, L.; Niu, Z.; Li, W.; Dang, X.; Wan, L.; Zhang, R.; Yang, S. Identification of microRNAs as potential biomarkers for lung adenocarcinoma using integrating genomics analysis. *Oncotarget* **2017**, *8*, 64143–64156. [CrossRef] [PubMed]

26. Goto, A.; Tanaka, M.; Yoshida, M.; Umakoshi, M.; Nanjo, H.; Shiraishi, K.; Saito, M.; Kohno, T.; Kuriyama, S.; Konno, H.; et al. The low expression of miR-451 predicts a worse prognosis in non-small cell lung cancer cases. *PLoS ONE* **2017**, *12*, e0181270. [CrossRef] [PubMed]

27. Ahmadinejad, F.; Mowla, S.J.; Honardoost, M.A.; Arjenaki, M.G.; Moazeni-Bistgani, M.; Kheiri, S.; Teimori, H. Lower expression of miR-218 in human breast cancer is associated with lymph node metastases, higher grades, and poorer prognosis. *Tumour Biol. J. Int. Soc. Oncodev. Biol. Med.* **2017**, *39*. [CrossRef] [PubMed]

28. Rosignolo, F.; Memeo, L.; Monzani, F.; Colarossi, C.; Pecce, V.; Verrienti, A.; Durante, C.; Grani, G.; Lamartina, L.; Forte, S.; et al. MicroRNA-based molecular classification of papillary thyroid carcinoma. *Int. J. Oncol.* **2017**, *50*, 1767–1777. [CrossRef] [PubMed]

29. Santarelli, L.; Strafella, E.; Staffolani, S.; Amati, M.; Emanuelli, M.; Sartini, D.; Pozzi, V.; Carbonari, D.; Bracci, M.; Pignotti, E.; et al. Association of miR-126 with soluble mesothelin-related peptides, a marker for malignant mesothelioma. *PLoS ONE* **2011**, *6*, e18232. [CrossRef] [PubMed]

30. Deng, L.; Tang, J.; Yang, H.; Cheng, C.; Lu, S.; Jiang, R.; Sun, B. MTA1 modulated by miR-30e contributes to epithelial-to-mesenchymal transition in hepatocellular carcinoma through an ErbB2-dependent pathway. *Oncogene* **2017**, *36*, 3976–3985. [CrossRef] [PubMed]

31. Wang, Y.; Chen, T.; Huang, H.; Jiang, Y.; Yang, L.; Lin, Z.; He, H.; Liu, T.; Wu, B.; Chen, J.; et al. miR-363-3p inhibits tumor growth by targeting PCNA in lung adenocarcinoma. *Oncotarget* **2017**, *8*, 20133–20144. [CrossRef] [PubMed]

32. Gururajan, M.; Josson, S.; Chu, G.C.; Lu, C.L.; Lu, Y.T.; Haga, C.L.; Zhau, H.E.; Liu, C.; Lichterman, J.; Duan, P.; et al. miR-154* and miR-379 in the DLK1-DIO3 microRNA mega-cluster regulate epithelial to mesenchymal transition and bone metastasis of prostate cancer. *Clin. Cancer Res.* **2014**, *20*, 6559–6569. [CrossRef] [PubMed]

33. Asangani, I.A.; Rasheed, S.A.; Nikolova, D.A.; Leupold, J.H.; Colburn, N.H.; Post, S.; Allgayer, H. MicroRNA-21 (miR-21) post-transcriptionally downregulates tumor suppressor pdcd4 and stimulates invasion, intravasation and metastasis in colorectal cancer. *Oncogene* **2008**, *27*, 2128–2136. [CrossRef] [PubMed]

34. Schetter, A.J.; Leung, S.Y.; Sohn, J.J.; Zanetti, K.A.; Bowman, E.D.; Yanaihara, N.; Yuen, S.T.; Chan, T.L.; Kwong, D.L.; Au, G.K.; et al. MicroRNA expression profiles associated with prognosis and therapeutic outcome in colon adenocarcinoma. *JAMA* **2008**, *299*, 425–436. [CrossRef] [PubMed]

35. Patnaik, S.K.; Kannisto, E.; Knudsen, S.; Yendamuri, S. Evaluation of microRNA expression profiles that may predict recurrence of localized stage I non-small cell lung cancer after surgical resection. *Cancer Res.* **2010**, *70*, 36–45. [CrossRef] [PubMed]

36. Ortiz-Quintero, B. Cell-free microRNAs in blood and other body fluids, as cancer biomarkers. *Cell Prolif.* **2016**, *49*, 281–303. [CrossRef] [PubMed]

37. Mitchell, P.S.; Parkin, R.K.; Kroh, E.M.; Fritz, B.R.; Wyman, S.K.; Pogosova-Agadjanyan, E.L.; Peterson, A.; Noteboom, J.; O'Briant, K.C.; Allen, A.; et al. Circulating microRNAs as stable blood-based markers for cancer detection. *Proc. Natl. Acad. Sci. USA* **2008**, *105*, 10513–10518. [CrossRef] [PubMed]

38. Köberle, V.; Pleli, T.; Schmithals, C.; Augusto Alonso, E.; Haupenthal, J.; Bönig, H.; Peveling-Oberhag, J.; Biondi, R.M.; Zeuzem, S.; Kronenberger, B.; et al. Differential stability of cell-free circulating microRNAs: Implications for their utilization as biomarkers. *PLoS ONE* **2013**, *8*, 1–11. [CrossRef] [PubMed]

39. Lee, Y.; Jeon, K.; Lee, J.T.; Kim, S.; Kim, V.N. microRNA maturation: Stepwise processing and subcellular localization. *EMBO J.* **2002**, *21*, 4663–4670. [CrossRef] [PubMed]

40. Taylor, M. Circulating microRNAs as biomarkers and mediators of cell–cell communication in cancer. *Biomedicines* **2015**, *3*, 270–281. [CrossRef] [PubMed]

41. Zeng, Y.; Yi, R.; Cullen, B.R. microRNAs and small interfering RNAs can inhibit mRNA expression by similar mechanisms. *Proc. Natl. Acad. Sci. USA* **2003**, *100*, 9779–9784. [CrossRef] [PubMed]

42. Wang, B.; Yanez, A.; Novina, C.D. microRNA-repressed mRNAs contain 40 s but not 60 s components. *Proc. Natl. Acad. Sci. USA* **2008**, *105*, 5343–5348. [CrossRef] [PubMed]

43. Giraldez, A.J.; Mishima, Y.; Rihel, J.; Grocock, R.J.; Van Dongen, S.; Inoue, K.; Enright, A.J.; Schier, A.F. Zebrafish mir-430 promotes deadenylation and clearance of maternal mRNAs. *Science* **2006**, *312*, 75–79. [CrossRef] [PubMed]

44. Raiborg, C.; Stenmark, H. The escrt machinery in endosomal sorting of ubiquitylated membrane proteins. *Nature* **2009**, *458*, 445–452. [CrossRef] [PubMed]

45. Villarroya-Beltri, C.; Gutierrez-Vazquez, C.; Sanchez-Cabo, F.; Perez-Hernandez, D.; Vazquez, J.; Martin-Cofreces, N.; Martinez-Herrera, D.J.; Pascual-Montano, A.; Mittelbrunn, M.; Sanchez-Madrid, F. Sumoylated hnrnpa2b1 controls the sorting of miRNAs into exosomes through binding to specific motifs. *Nat. Commun.* **2013**, *4*, 2980. [CrossRef] [PubMed]

46. Kosaka, N.; Iguchi, H.; Hagiwara, K.; Yoshioka, Y.; Takeshita, F.; Ochiya, T. Neutral sphingomyelinase 2 (nsmase2)-dependent exosomal transfer of angiogenic microRNAs regulate cancer cell metastasis. *J. Biol. Chem.* **2013**, *288*, 10849–10859. [CrossRef] [PubMed]

47. Gibbings, D.J.; Ciaudo, C.; Erhardt, M.; Voinnet, O. Multivesicular bodies associate with components of miRNA effector complexes and modulate miRNA activity. *Nat. Cell Biol.* **2009**, *11*, 1143–1149. [CrossRef] [PubMed]

48. Pegtel, D.M.; Cosmopoulos, K.; Thorley-Lawson, D.A.; van Eijndhoven, M.A.; Hopmans, E.S.; Lindenberg, J.L.; de Gruijl, T.D.; Wurdinger, T.; Middeldorp, J.M. Functional delivery of viral miRNAs via exosomes. *Proc. Natl. Acad. Sci. USA* **2010**, *107*, 6328–6333. [CrossRef] [PubMed]

49. Iguchi, H.; Kosaka, N.; Ochiya, T. Secretory microRNAs as a versatile communication tool. *Commun. Integr. Biol.* **2010**, *3*, 478–481. [CrossRef] [PubMed]

50. Arroyo, J.D.; Chevillet, J.R.; Kroh, E.M.; Ruf, I.K.; Pritchard, C.C.; Gibson, D.F.; Mitchell, P.S.; Bennett, C.F.; Pogosova-Agadjanyan, E.L.; Stirewalt, D.L.; et al. Argonaute2 complexes carry a population of circulating microRNAs independent of vesicles in human plasma. *Proc. Natl. Acad. Sci. USA* **2011**, *108*, 5003–5008. [CrossRef] [PubMed]

51. Turchinovich, A.; Weiz, L.; Langheinz, A.; Burwinkel, B. Characterization of extracellular circulating microRNA. *Nucleic Acids Res.* **2011**, *39*, 7223–7233. [CrossRef] [PubMed]

52. Vickers, K.C.; Palmisano, B.T.; Shoucri, B.M.; Shamburek, R.D.; Remaley, A.T. microRNAs are transported in plasma and delivered to recipient cells by high-density lipoproteins. *Nat. Cell Biol.* **2011**, *13*, 423–433. [CrossRef] [PubMed]

53. Wagner, J.; Riwanto, M.; Besler, C.; Knau, A.; Fichtlscherer, S.; Roxe, T.; Zeiher, A.M.; Landmesser, U.; Dimmeler, S. Characterization of levels and cellular transfer of circulating lipoprotein-bound microRNAs. *Arterioscler. Thromb. Vasc. Biol.* **2013**, *33*, 1392–1400. [CrossRef] [PubMed]

54. Tabet, F.; Vickers, K.C.; Cuesta Torres, L.F.; Wiese, C.B.; Shoucri, B.M.; Lambert, G.; Catherinet, C.; Prado-Lourenco, L.; Levin, M.G.; Thacker, S.; et al. Hdl-transferred microRNA-223 regulates icam-1 expression in endothelial cells. *Nat. Commun.* **2014**, *5*, 3292. [CrossRef] [PubMed]

55. Wang, K.; Zhang, S.; Weber, J.; Baxter, D.; Galas, D.J. Export of microRNAs and microRNA-protective protein by mammalian cells. *Nucleic Acids Res.* **2010**, *38*, 7248–7259. [CrossRef] [PubMed]

56. Avila-Moreno, F.; Urrea, F.; Ortiz-Quintero, B. microRNAs in diagnosis and prognosis in lung cancer. *Rev. Investig. Clin. Organo Hosp. Enferm. Nutr.* **2011**, *63*, 516–535.

57. Biomarkers and surrogate endpoints: Preferred definitions and conceptual framework. *Clin. Pharmacol. Ther.* **2001**, *69*, 89–95.

58. Guled, M.; Lahti, L.; Lindholm, P.M.; Salmenkivi, K.; Bagwan, I.; Nicholson, A.G.; Knuutila, S. Cdkn2a, nf2, and jun are dysregulated among other genes by miRNAs in malignant mesothelioma -a miRNA microarray analysis. *Genes Chromosom. Cancer* **2009**, *48*, 615–623. [CrossRef] [PubMed]

59. Gee, G.V.; Koestler, D.C.; Christensen, B.C.; Sugarbaker, D.J.; Ugolini, D.; Ivaldi, G.P.; Resnick, M.B.; Houseman, E.A.; Kelsey, K.T.; Marsit, C.J. Downregulated microRNAs in the differential diagnosis of malignant pleural mesothelioma. *Int. J. Cancer* **2010**, *127*, 2859–2869. [CrossRef] [PubMed]

60. Benjamin, H.; Lebanony, D.; Rosenwald, S.; Cohen, L.; Gibori, H.; Barabash, N.; Ashkenazi, K.; Goren, E.; Meiri, E.; Morgenstern, S.; et al. A diagnostic assay based on microRNA expression accurately identifies malignant pleural mesothelioma. *J. Mol. Diagn.* **2010**, *12*, 771–779. [CrossRef] [PubMed]

61. Xi, Y.; Nakajima, G.; Gavin, E.; Morris, C.G.; Kudo, K.; Hayashi, K.; Ju, J. Systematic analysis of microRNA expression of RNA extracted from fresh frozen and formalin-fixed paraffin-embedded samples. *RNA* **2007**, *13*, 1668–1674. [CrossRef] [PubMed]

62. Xu, Y.; Zheng, M.; Merritt, R.E.; Shrager, J.B.; Wakelee, H.A.; Kratzke, R.A.; Hoang, C.D. Mir-1 induces growth arrest and apoptosis in malignant mesothelioma. *Chest* **2013**, *144*, 1632–1643. [CrossRef] [PubMed]

63. Reid, G.; Pel, M.E.; Kirschner, M.B.; Cheng, Y.Y.; Mugridge, N.; Weiss, J.; Williams, M.; Wright, C.; Edelman, J.J.; Vallely, M.P.; et al. Restoring expression of mir-16: A novel approach to therapy for malignant pleural mesothelioma. *Ann. Oncol.* **2013**, *24*, 3128–3135. [CrossRef] [PubMed]

64. Cioce, M.; Ganci, F.; Canu, V.; Sacconi, A.; Mori, F.; Canino, C.; Korita, E.; Casini, B.; Alessandrini, G.; Cambria, A.; et al. Protumorigenic effects of mir-145 loss in malignant pleural mesothelioma. *Oncogene* **2014**, *33*, 5319–5331. [CrossRef] [PubMed]

65. Ramirez-Salazar, E.G.; Salinas-Silva, L.C.; Vazquez-Manriquez, M.E.; Gayosso-Gomez, L.V.; Negrete-Garcia, M.C.; Ramirez-Rodriguez, S.L.; Chavez, R.; Zenteno, E.; Santillan, P.; Kelly-Garcia, J.; et al. Analysis of microRNA expression signatures in malignant pleural mesothelioma, pleural inflammation, and atypical mesothelial hyperplasia reveals common predictive tumorigenesis-related targets. *Exp. Mol. Pathol.* **2014**, *97*, 375–385. [CrossRef] [PubMed]

66. Andersen, M.; Grauslund, M.; Ravn, J.; Sorensen, J.B.; Andersen, C.B.; Santoni-Rugiu, E. Diagnostic potential of mir-126, mir-143, mir-145, and mir-652 in malignant pleural mesothelioma. *J. Mol. Diagn.* **2014**, *16*, 418–430. [CrossRef] [PubMed]

67. Ak, G.; Tomaszek, S.C.; Kosari, F.; Metintas, M.; Jett, J.R.; Metintas, S.; Yildirim, H.; Dundar, E.; Dong, J.; Aubry, M.C.; et al. microRNA and mRNA features of malignant pleural mesothelioma and benign asbestos-related pleural effusion. *Biomed. Res. Int.* **2015**, *2015*, 635748. [CrossRef] [PubMed]

68. Birnie, K.A.; Yip, Y.Y.; Ng, D.C.; Kirschner, M.B.; Reid, G.; Prele, C.M.; Musk, A.W.; Lee, Y.C.; Thompson, P.J.; Mutsaers, S.E.; et al. Loss of mir-223 and jnk signaling contribute to elevated stathmin in malignant pleural mesothelioma. *Mol. Cancer Res.* **2015**, *13*, 1106–1118. [CrossRef] [PubMed]

69. Cappellesso, R.; Nicole, L.; Caroccia, B.; Guzzardo, V.; Ventura, L.; Fassan, M.; Fassina, A. Young investigator challenge: MicroRNA-21/microRNA-126 profiling as a novel tool for the diagnosis of malignant mesothelioma in pleural effusion cytology. *Cancer Cytopathol.* **2016**, *124*, 28–37. [CrossRef] [PubMed]

70. Cappellesso, R.; Galasso, M.; Nicole, L.; Dabrilli, P.; Volinia, S.; Fassina, A. Mir-130a as a diagnostic marker to differentiate malignant mesothelioma from lung adenocarcinoma in pleural effusion cytology. *Cancer Cytopathol.* **2017**, *125*, 635–643. [CrossRef] [PubMed]

71. Weber, D.G.; Johnen, G.; Bryk, O.; Jockel, K.H.; Bruning, T. Identification of miRNA-103 in the cellular fraction of human peripheral blood as a potential biomarker for malignant mesothelioma—A pilot study. *PLoS ONE* **2012**, *7*, e30221. [CrossRef] [PubMed]

72. Weber, D.G.; Casjens, S.; Johnen, G.; Bryk, O.; Raiko, I.; Pesch, B.; Kollmeier, J.; Bauer, T.T.; Bruning, T. Combination of mir-103a-3p and mesothelin improves the biomarker performance of malignant mesothelioma diagnosis. *PLoS ONE* **2014**, *9*, e114483. [CrossRef] [PubMed]

73. Tomasetti, M.; Staffolani, S.; Nocchi, L.; Neuzil, J.; Strafella, E.; Manzella, N.; Mariotti, L.; Bracci, M.; Valentino, M.; Amati, M.; et al. Clinical significance of circulating mir-126 quantification in malignant mesothelioma patients. *Clin. Biochem.* **2012**, *45*, 575–581. [CrossRef] [PubMed]

74. Kirschner, M.B.; Cheng, Y.Y.; Badrian, B.; Kao, S.C.; Creaney, J.; Edelman, J.J.; Armstrong, N.J.; Vallely, M.P.; Musk, A.W.; Robinson, B.W.; et al. Increased circulating mir-625-3p: A potential biomarker for patients with malignant pleural mesothelioma. *J. Thorac. Oncol.* **2012**, *7*, 1184–1191. [CrossRef] [PubMed]

75. Valadi, H.; Ekstrom, K.; Bossios, A.; Sjostrand, M.; Lee, J.J.; Lotvall, J.O. Exosome-mediated transfer of mRNAs and microRNAs is a novel mechanism of genetic exchange between cells. *Nat. Cell Biol.* **2007**, *9*, 654–659. [CrossRef] [PubMed]

76. Pigati, L.; Yaddanapudi, S.C.; Iyengar, R.; Kim, D.J.; Hearn, S.A.; Danforth, D.; Hastings, M.L.; Duelli, D.M. Selective release of microRNA species from normal and malignant mammary epithelial cells. *PLoS ONE* **2010**, *5*, e13515. [CrossRef] [PubMed]

77. Jansen, F.; Schafer, L.; Wang, H.; Schmitz, T.; Flender, A.; Schueler, R.; Hammerstingl, C.; Nickenig, G.; Sinning, J.M.; Werner, N. Kinetics of circulating microRNAs in response to cardiac stress in patients with coronary artery disease. *J. Am. Heart Assoc.* **2017**, *6*. [CrossRef] [PubMed]

78. Schulte, C.; Karakas, M.; Zeller, T. microRNAs in cardiovascular disease—Clinical application. *Clin. Chem. Lab. Med.* **2017**, *55*, 687–704. [CrossRef] [PubMed]

79. Gayosso-Gómez, L.V.; Zárraga-Granados, G.; Paredes-Garcia, P.; Falfán-Valencia, R.; Vazquez-Manríquez, M.E.; Martinez-Barrera, L.M.; Castillo-Gonzalez, P.; Rumbo-Nava, U.; Guevara-Gutierrez, R.; Rivera-Bravo, B.; et al. Identification of circulating miRNA profiles that distinguish malignant pleural mesothelioma from lung adenocarcinoma. *EXCLI J.* **2014**, *13*, 740–750. [PubMed]

80. Lamberti, M.; Capasso, R.; Lombardi, A.; Di Domenico, M.; Fiorelli, A.; Feola, A.; Perna, A.F.; Santini, M.; Caraglia, M.; Ingrosso, D. Two different serum miRNA signatures correlate with the clinical outcome and histological subtype in pleural malignant mesothelioma patients. *PLoS ONE* **2015**, *10*, e0135331. [CrossRef] [PubMed]

81. Wong, L.L.; Wang, J.; Liew, O.W.; Richards, A.M.; Chen, Y.T. microRNA and heart failure. *Int. J. Mol. Sci.* **2016**, *17*, 502. [CrossRef] [PubMed]

82. Li, H.; Fan, J.; Yin, Z.; Wang, F.; Chen, C.; Wang, D.W. Identification of cardiac-related circulating microRNA profile in human chronic heart failure. *Oncotarget* **2016**, *7*, 33–45. [CrossRef] [PubMed]

83. Abd-El-Fattah, A.A.; Sadik, N.A.; Shaker, O.G.; Aboulftouh, M.L. Differential microRNAs expression in serum of patients with lung cancer, pulmonary tuberculosis, and pneumonia. *Cell Biochem. Biophys.* **2013**, *67*, 875–884. [CrossRef] [PubMed]

84. Zhang, H.; Sun, Z.; Wei, W.; Liu, Z.; Fleming, J.; Zhang, S.; Lin, N.; Wang, M.; Chen, M.; Xu, Y.; et al. Identification of serum microRNA biomarkers for tuberculosis using Rna-seq. *PLoS ONE* **2014**, *9*, e88909. [CrossRef] [PubMed]

85. Santarelli, L.; Staffolani, S.; Strafella, E.; Nocchi, L.; Manzella, N.; Grossi, P.; Bracci, M.; Pignotti, E.; Alleva, R.; Borghi, B.; et al. Combined circulating epigenetic markers to improve mesothelin performance in the diagnosis of malignant mesothelioma. *Lung Cancer* **2015**, *90*, 457–464. [CrossRef] [PubMed]

86. Bononi, I.; Comar, M.; Puozzo, A.; Stendardo, M.; Boschetto, P.; Orecchia, S.; Libener, R.; Guaschino, R.; Pietrobon, S.; Ferracin, M.; et al. Circulating microRNAs found dysregulated in ex-exposed asbestos workers and pleural mesothelioma patients as potential new biomarkers. *Oncotarget* **2016**, *7*, 82700–82711. [CrossRef] [PubMed]

87. Cavalleri, T.; Angelici, L.; Favero, C.; Dioni, L.; Mensi, C.; Bareggi, C.; Palleschi, A.; Rimessi, A.; Consonni, D.; Bordini, L.; et al. Plasmatic extracellular vesicle microRNAs in malignant pleural mesothelioma and asbestos-exposed subjects suggest a 2-miRNA signature as potential biomarker of disease. *PLoS ONE* **2017**, *12*, e0176680. [CrossRef] [PubMed]

88. Weber, D.G.; Gawrych, K.; Casjens, S.; Brik, A.; Lehnert, M.; Taeger, D.; Pesch, B.; Kollmeier, J.; Bauer, T.T.; Johnen, G.; et al. Circulating mir-132-3p as a candidate diagnostic biomarker for malignant mesothelioma. *Dis. Markers* **2017**, *2017*, 9280170. [CrossRef] [PubMed]

89. Han, H.S.; Yun, J.; Lim, S.N.; Han, J.H.; Lee, K.H.; Kim, S.T.; Kang, M.H.; Son, S.M.; Lee, Y.M.; Choi, S.Y.; et al. Downregulation of cell-free mir-198 as a diagnostic biomarker for lung adenocarcinoma-associated malignant pleural effusion. *Int. J. Cancer* **2013**, *133*, 645–652. [CrossRef] [PubMed]

90. Micolucci, L.; Akhtar, M.M.; Olivieri, F.; Rippo, M.R.; Procopio, A.D. Diagnostic value of microRNAs in asbestos exposure and malignant mesothelioma: Systematic review and qualitative meta-analysis. *Oncotarget* **2016**, *7*, 58606–58637. [CrossRef] [PubMed]

91. Kemp, C.D.; Rao, M.; Xi, S.; Inchauste, S.; Mani, H.; Fetsch, P.; Filie, A.; Zhang, M.; Hong, J.A.; Walker, R.L.; et al. Polycomb repressor complex-2 is a novel target for mesothelioma therapy. *Clin. Cancer Res.* **2012**, *18*, 77–90. [CrossRef] [PubMed]

92. Peltier, H.J.; Latham, G.J. Normalization of microRNA expression levels in quantitative rt-pcr assays: Identification of suitable reference RNA targets in normal and cancerous human solid tissues. *RNA* **2008**, *14*, 844–852. [CrossRef] [PubMed]

93. Vandesompele, J.; De Preter, K.; Pattyn, F.; Poppe, B.; Van Roy, N.; De Paepe, A.; Speleman, F. Accurate normalization of real-time quantitative rt-pcr data by geometric averaging of multiple internal control genes. *Genome Biol.* **2002**, *3*, research0034.1–research0034.11. [CrossRef] [PubMed]

94. Bustin, S.A.; Benes, V.; Garson, J.A.; Hellemans, J.; Huggett, J.; Kubista, M.; Mueller, R.; Nolan, T.; Pfaffl, M.W.; Shipley, G.L.; et al. The miqe guidelines: Minimum information for publication of quantitative real-time pcr experiments. *Clin. Chem.* **2009**, *55*, 611–622. [CrossRef] [PubMed]

95. Rutjes, A.W.; Reitsma, J.B.; Di Nisio, M.; Smidt, N.; van Rijn, J.C.; Bossuyt, P.M. Evidence of bias and variation in diagnostic accuracy studies. *CMAJ* **2006**, *174*, 469–476. [CrossRef] [PubMed]

96. Lijmer, J.G.; Mol, B.W.; Heisterkamp, S.; Bonsel, G.J.; Prins, M.H.; van der Meulen, J.H.; Bossuyt, P.M. Empirical evidence of design-related bias in studies of diagnostic tests. *JAMA* **1999**, *282*, 1061–1066. [CrossRef] [PubMed]

97. Bossuyt, P.M.; Reitsma, J.B.; Bruns, D.E.; Gatsonis, C.A.; Glasziou, P.P.; Irwig, L.M.; Lijmer, J.G.; Moher, D.; Rennie, D.; de Vet, H.C. Towards complete and accurate reporting of studies of diagnostic accuracy: The stard initiative. *BMJ* **2003**, *326*, 41–44. [CrossRef] [PubMed]

98. Altman, D.G.; McShane, L.M.; Sauerbrei, W.; Taube, S.E. Reporting recommendations for tumor marker prognostic studies (remark): Explanation and elaboration. *BMC Med.* **2012**, *10*, 51. [CrossRef] [PubMed]

99. Von Elm, E.; Altman, D.G.; Egger, M.; Pocock, S.J.; Gotzsche, P.C.; Vandenbroucke, J.P. The strengthening the reporting of observational studies in epidemiology (strobe) statement: Guidelines for reporting observational studies. *PLoS Med.* **2007**, *4*, e296. [CrossRef] [PubMed]

100. Esquela-Kerscher, A.; Slack, F.J. Oncomirs—microRNAs with a role in cancer. *Nat. Rev. Cancer* **2006**, *6*, 259–269. [CrossRef] [PubMed]

International Journal of
Molecular Sciences

MDPI

Communication

Genomic Deletion of *BAP1* and *CDKN2A* Are Useful Markers for Quality Control of Malignant Pleural Mesothelioma (MPM) Primary Cultures

Kadir Harun Sarun [1], Kenneth Lee [1,2,3], Marissa Williams [1,3], Casey Maree Wright [1], Candice Julie Clarke [2], Ngan Ching Cheng [4], Ken Takahashi [1] and Yuen Yee Cheng [1,3,*]

[1] Asbestos Diseases Research Institute, University of Sydney, Sydney, NSW 2139, Australia; kadir.sarun@sydney.edu.au (K.H.S.); kenneth.Lee@health.nsw.gov.au (K.L.); marissa.williams@sydney.edu.au (M.W.); cmdodds84@gmail.com (C.M.W.); ken.takahashi@sydney.edu.au (K.T.)
[2] Anatomical Pathology Department, Concord Repatriation General Hospital, Sydney, NSW 2139, Australia; Candice.Clarke@health.nsw.gov.au
[3] School of Medicine, University of Sydney, Sydney, NSW 2006, Australia
[4] Liver Injury and Cancer Program, Centenary Institute, Sydney, NSW 2050, Australia; ngan.cheng@gmail.com
* Correspondence: yycheng@sydney.edu.au; Tel.: +61-2-9767-9800; Fax: +61-2-9767-9860

Received: 7 September 2018; Accepted: 30 September 2018; Published: 7 October 2018

Abstract: Malignant pleural mesothelioma (MPM) is a deadly cancer that is caused by asbestos exposure and that has limited treatment options. The current standard of MPM diagnosis requires the testing of multiple immunohistochemical (IHC) markers on formalin-fixed paraffin-embedded tissue to differentiate MPM from other lung malignancies. To date, no single biomarker exists for definitive diagnosis of MPM due to the lack of specificity and sensitivity; therefore, there is ongoing research and development in order to identify alternative biomarkers for this purpose. In this study, we utilized primary MPM cell lines and tested the expression of clinically used biomarker panels, including CK8/18, Calretinin, CK 5/6, CD141, HBME-1, WT-1, D2-40, EMA, CEA, TAG72, BG8, CD15, TTF-1, BAP1, and Ber-Ep4. The genomic alteration of *CDNK2A* and *BAP1* is common in MPM and has potential diagnostic value. Changes in *CDKN2A* and *BAP1* genomic expression were confirmed in MPM samples in the current study using Fluorescence In situ Hybridization (FISH) analysis or copy number variation (CNV) analysis with digital droplet PCR (ddPCR). To determine whether MPM tissue and cell lines were comparable in terms of molecular alterations, IHC marker expression was analyzed in both sample types. The percentage of MPM biomarker levels showed variation between original tissue and matched cells established in culture. Genomic deletions of *BAP1* and *CDKN2A*, however, showed consistent levels between the two. The data from this study suggest that genomic deletion analysis may provide more accurate biomarker options for MPM diagnosis.

Keywords: mesothelioma; biomarker; FISH; genomic deletion; copy number variation; ddPCR

1. Introduction

Malignant pleural mesothelioma (MPM) is a tumor originating from the mesothelium, the membrane lining the thoracic and peritoneal cavities [1]. MPM is strongly linked to previous asbestos exposure [2] and asbestiform minerals such as erionite and fluoroedenite [3]. Australia has one of the world's highest incidences of MPM due to the heavy industrial utilization of asbestos in the past [4]. MPM is a deadly cancer with poor prognosis [1,5,6], and treatment options are mainly palliative [7]. Most MPM patients are diagnosed at a late stage of the disease where limited treatment options are available; this is due to a lack of symptoms at early stages and the long latency period

Int. J. Mol. Sci. **2018**, *19*, 3056

between asbestos exposure and the development of MPM. Mesothelioma is especially difficult to diagnose, as symptoms closely resemble those of lung cancer. Delays or errors in diagnosis hinder treatment intervention that can subsequently adversely affect the patients' survival and quality-of-life (QoL); therefore, accurate diagnosis is essential for prognostic and therapeutic purposes [8].

Immunohistochemistry (IHC) is the standard method for biomarker detection of MPM, and multiple mesothelial markers have been identified to enable the distinction between epithelioid MPM and adenocarcinomas in routine practice. The three predominant subtypes differentiated by their MPM histomorphology are epithelioid, biphasic, and sarcomatoid. The proteins assessed using IHC vary in different laboratories, but the use of antibodies for the identification of calretinin and CEA is prominent [8]. To date, it is generally accepted that no single biomarker is absolutely sensitive or specific for MPM, and multipanel immunohistochemical tests are essential for diagnosis [9,10]. Therefore, further molecular characterization of the tumor is required to potentially identify more specific markers to aid in the diagnosis of MPM.

As well as intertumor heterogeneity, MPM tumors also exhibit intratumor heterogeneity. This tumour complexity limits the ability to delegate suitable treatment options due to the existence of several tumor clones and subclones within a single patient [11]. Intertumor heterogeneity is inclusive of the variable molecular phenotype of MPM. While genomic loss and gain are evident in MPM tumors, they exhibit low levels of drivers and recurrent mutations in comparison to other cancers [12]. The most commonly reported mutations are identified in genes such as *NF2*, *BAP1*, *TP53*, *NRAS*, and *EGFR* [13,14]. Asbestos fibers have been demonstrated to induce chromosome instability resulting in dysfunctional DNA damage response [15]. The most frequently reported chromosomal losses are those affecting chromosomal arms 3p, 9p, and 22q. Genes located in these regions include *BAP1*, *CDKN2A*, and *NF2*, respectively [16,17]. Among the three, *CDKN2A* represents the highest number of homozygous deletions in MPM-patient tumors [18]. To date, the mechanisms underlying the poor response of MPM patients to a wide range of therapeutic interventions are largely undetermined. Molecular intertumor heterogeneity, including a diversity of mutation, epigenetic, expression, and microscopic (phenotypic) changes may cause inefficacy of the treatment regimens. In contrast to nonsmall cell lung cancer, many mutations, such as in *EGFR* or *TP53*, are uncommon in the majority of MPM cases [9,19].

Due to the lack of single, accurate biomarkers for MPM, recent studies have focused on the analysis of biomarker combinations or panels, as well as the development of new diagnostic methods separate from IHC. For example, the determination of *p16* (*CDKN2A*) homozygous deletion using Fluorescence In Situ Hybridization (FISH) and identification of BRCA1-associated protein 1 (BAP1) loss by IHC are particularly useful to differentiate mesothelial hyperplasia (MH) from MPM. Currently, these two markers are not widely used in the clinic, potentially due to the low sensitivity of the existing detection method [20–23]. Considering this, new markers with increased sensitivity and specificity are thus required. Cell-line models derived from MPM tumors are useful for the discovery of biomarkers and testing their efficacy. Cell culture is limitlessly renewable and can be manipulated to study cell function and gene signatures. Cultured cells derived from tumors have been shown to maintain many of the hallmarks of cancer apart from tumor-specific angiogenesis [24,25]. The development of a primary MPM cell culture provides an inexpensive and more homogeneous MPM cell population for genomic marker identification. Primary MPM cell lines have provided a medium to better understand the genomic alterations that exist in mesothelioma [26]. Further, in tumor samples, the inevitable infiltration of stromal and inflammatory cell populations can influence the molecular phenotype. The implementation of cell-culture models allows the exclusion of such populations, and allows exclusive testing of the tumor cells and accurate estimation of gene copy number.

This study aimed to utilize the MPM cell lines that were established between 2013–2017 [27] to study different types of biomarkers, including protein markers using IHC and genomic markers using qualitative FISH analysis coupled with absolute quantification analysis with droplet digital PCR (ddPCR).

2. Results

2.1. Immunohistochemistry Analysis Demonstrates Variable Marker Expression between MPM Tissue and Derivative Cell-Line Samples

A total of 15 biomarkers used in clinical practices for differential diagnosis of MPM were assessed in all samples, including MPM tumor samples together with derivative primary cell-line samples from the corresponding parent tumor tissue and/or plural effusion. Short Tandem Repeat (STR) profiling was employed to provide genomic signatures for confirmation of cell-line identity (Supplementary Table S1). Figure 1 shows the IHC staining of a MPM tumor and the expression of 15 protein markers used in the clinic. MPM primary cells were extracted from tumor tissue and/or pleural effusion samples and grown in cell culture until they reached passage 15 to eliminate normal cell contamination, after which they were cultured into 3D. Two-dimensional and three-dimensional MPM cell blocks were used for protein marker analysis, and our results indicated that the 3D model more closely represents the tumor architecture (Supplementary Figure S1). Protein levels were compared between MPM tumor samples and subsequent derivative cell lines from the same tissue. The detailed percentage scoring of protein marker expression in MPM tumor samples and cell lines is listed in Table 1 (each MM ID represents samples from one patient). It was found that the majority of IHC protein markers are not correlative between tumor tissue and their derived cells (Table 1). Due to the observed variability of protein markers between tissue and cell lines, other molecular biomarker strategies were considered.

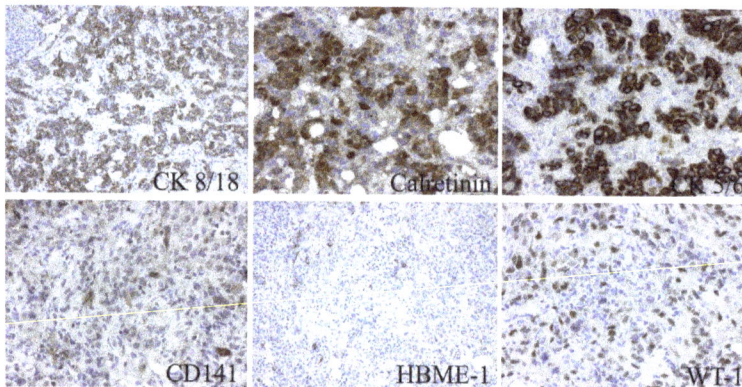

Figure 1. *Cont.*

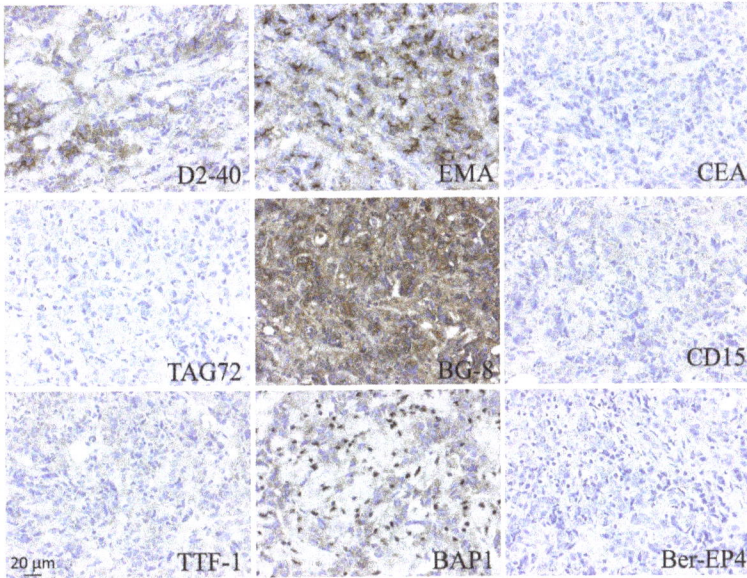

Figure 1. Representative immunohistochemical (IHC) staining of a malignant pleural mesothelioma (MPM) (sample ID 1157) patient tumor sample with the 15 biomarkers currently used for clinical diagnosis. *BAP1* is not expressed in sample MM ID 1157, therefore *BAP1* staining of sample MM ID 1518 is included as an example of positive *BAP1* expression in this figure. All pictures are taken at same magnification with scale bar indicated left bottom corner of TFF-1 staining.

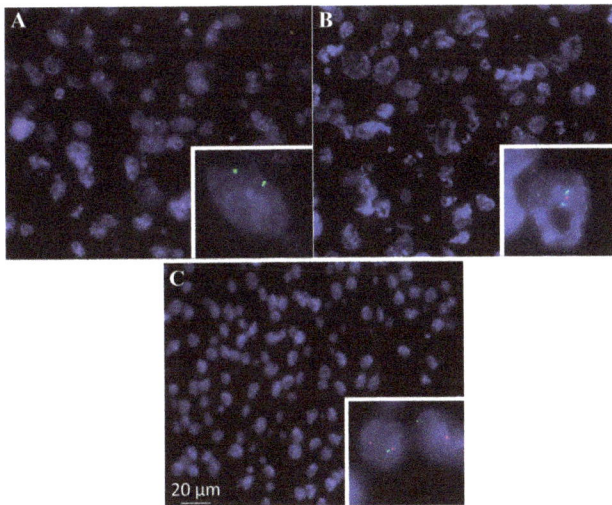

Figure 2. Representative example of Fluorescence In situ Hybridization (FISH) to visualise method of *CDKN2A* deletion in MPM samples. Images depict (**A**) homozygous, (**B**) heterozygous and (**C**) no loss of *CDKN2A* in MPM. The bottom-right corner of each image shows *CDNK2A* (red) as well as control *CEP9* (green) signals. Images were taken using ZEISS Axio Imager M2. All pictures taken at same magnification with scale bar indicated at panel C.

Table 1. Mesothelioma biomarker scoring.

MM ID	Sample Type	CK 8/18	Calretinin	CK 5/6	CD141	HBME-1	WT-1	D2-40	EMA	CEA	TAG72	BG8	CD15	TTF-1	BAP1	Ber-EP4
1137	Tissue	(+++) 90%	-	(+++) <5%	(++) 10%	-	(++) 10%	(+) <5%	-	-	-	(++) 70%	-	-	(++) 90–100%	-
	3D cells	(+++) 100%	-	-	-	-	-	-	-	-	-	(++/+++) 90–100%	-	-	(+++) 30%	-
1137 T	2D cells	(+++) 90%	(++) 5%	(+) 10%	(+) 20%	(+++) 10%	(+) 10%	-	(+) <5%	(++) 10%	-	(+) 60%	-	(+++) <5%	-	(++/+++) 60%
1157	tissue	(+++) 90–100%	(+++) 80%	(+++) 90–100%	(+) 50%	(+++) 10%	(++) 90%	(++) 30%	(+++) 40%	-	-	(+++) 90–100%	-	-	-	-
	3D cells	(+++) 100%	(+) 40%	(+++) 70%	-	-	(+++) 90%	(+) <10%	(+++) 40%	-	-	(+) 40%	-	-	-	-
1157 T	2D cells	(+++) 90–100%	-	(++/+++) 90%	(+) <5%	-	-	-	(+) <5%	-	-	-	-	-	-	(+) 50%
1180	Tissue	(+++++) 90–100%	(+++) 30%	(+++) <5%	(+++) 30%	(+++) <5%	-	(+++) 10%	-	-	-	(++) 80%	-	-	(+++) 80%	-
	3D cells	(+++) 100%	-	-	-	-	(++) 30%	-	(++) 40%	-	-	(+) 30%	-	(+) 10%	(+++) 95%	-
1180 T	2D cells	(+++) 90–100%	-	-	-	-	(+++) <10%	-	-	-	-	-	-	(++) <10%	(+++) 95%	-
1187	tissue	(+++) 90–100%	(+++) 90–100%	(+++) 90–100%	(+++) 90–100%	(+++) 90–100%	(+++) 90–100%	(+++) 40%	-	-	(+++) 90–100%	-	-	(++) 90–100%	-	(++) 90–100%
	3D cells	(++) 10%	-	-	-	-	(+++) 90%	(++) 40%	-	-	-	-	-	-	(++) 80%	-
1187 T	2D cells	(+++) 100%	-	-	(++) 40%	-	(++) 80%	(+) 10%	-	-	-	(+) 70%	-	-	(+++) 95%	(+) 80%
1505	Tissue	(+++) 90–100%	(+++) 40%	(+++) 10%	(++) 30%	-	(++) <5%	(+++) <5%	-	-	-	(+++) 90–100%	-	(+++) 40%	(+++) 90–100%	-
1505 T	3D cells	(+++) 100%	(+) 20%	(+) <5%	(+) <5%	(+) <5%	(+) <5%	-	(++) 60%	(++) 60%	(+) 20%	(+++) 100%	-	(+++) 80%	(+++) 100%	(+++) 50%
	2D cells	(+++) 100%	(++) 80%	(+) 80%	(+) 70%	(+) 20%	-	(+) 30%	-	(+) <5%	(++) 60%	(+++) 100%	(+) 80%	(+++) 80%	(+++) 100%	(+++) 100%

Table 1. *Cont.*

MM ID	Sample Type	CK 8/18	Calretinin	CK 5/6	CD141	HBME-1	WT-1	D2-40	EMA	CEA	TAG72	BG8	CD15	TTF-1	BAP1	Ber-EP4
1506	Tissue	(+++) 90%	(+++) 80%	-	(++) 70%	(+++) 100%	(++) 60%	(+) 10%	-	-	-	(++) 40%	-	-	(+++) 90%	(+) 10%
1506 T	3D cells	-	-	(+) <5%	(+++) 40%	-	-	-	(++) 80%	-	-	(+++) 90%	-	-	(+++) 100%	(+++) 80%
1506 T	2D cells	(+) 10%	-	(+) 40%	(+) 10%	(+) <5%	-	-	(++) 40%	(+) 20%	(++) 70%	(+++) 100%	(+) 90%	(+) <5%	(+++) 80%	(+++) 100%
1518	Tissue	(+++) 40%	-	(+++) 30%	(+) 20%	-	(++) 10%	(++) 30%	(++) 10%	-	-	(+) 40%	-	-	(+++) 20%	-
1518 P	3D cells	(+++) 100%	(+) 10%	(+) <5%	-	(+) 5%	(+++) 40%	(+) <5%	(+) <5%	-	-	(+++) 100%	-	-	(+++) 90%	(+++) 90%
1518 P	2D cells	(+++) 100%	(+) 5%	-	(+) 10%	-	(+) 10%	(+) <5%	(+) <5%	(+) 10%	-	(+++) 100%	(+) 80%	-	(++) 80%	(+++) 100%
1518 T	3D cells	(++/+++) 70%	(++) 10%	-	(+) <5%	-	-	-	-	-	-	-	-	-	-	(+) 10%
1518 T	2D cells	(++) 70%	(++) 4%	-	(+) <5%	-	(+++) 40%	(++) 30%	-	(++/+++) 10%	-	-	(+) <5%	(+++) 20%	-	(+++) 40%
1170 T	3D cells	-	-	-	(+) 40%	(++) <5%	(+++) 20%	-	(+++) 30%	-	(+) 10%	(+) 10%	-	(+++) 90-100%	(+++) 95%	(++/+++) 40%
1170 T	2D cells	-	-	-	-	(++) 70%	(++/+++) 80%	-	-	-	-	-	-	(+++) 80%	(+++) 95%	(++) 20%
1843	Tissue	(++) 40%	(+) <10%	(+++) 40%	(+++) 30%	(+++) 30%	(+) 5%	-	(+++) 20%	-	-	(+) 20%	-	(+) <5%	-	-
1843 T	3D cells	(+++) 50%	-	-	-	-	(+++) 100%	(+) 20%	(++/+++) 40%	-	-	-	-	(+++) 90%	(+++) 90%	-
1843 T	2D cells	(++) 80%	-	-	-	-	(++) 70%	-	-	-	-	(+) 50%	-	(++) 80%	(++) 80%	(+) 50%
2164	Tissue	(+++) 40%	(+++) 40%	-	(++) 40%	-	(+) <5%	(+++) 70%	(+++) 60%	-	-	(+) 10%	-	-	-	-
2164 P	3D cells	-	-	-	-	-	(++) 90%	-	-	-	-	-	-	-	-	-
2164 P	2D cells	-	-	-	-	-	(++) 90-100%	(++) 30%	(+) 10%	-	-	(+) 40%	-	-	-	-

Int. J. Mol. Sci. **2018**, *19*, 3056

Table 1. *Cont.*

MM ID	Sample Type	CK 8/18	Calretinin	CK 5/6	CD141	HBME-1	WT-1	D2-40	EMA	CEA	TAG72	BG8	CD15	TTF-1	BAP1	Ber-EP4
2170	Tissue	(+++) 90%	(+++) 60%	(+) 5%	(++) 40%	(+++) 90%	(+++) 40%	(+++) 80%	(+++) 80%	-	-	(++) 10%	-	-	-	(+++) 5%
2170 T	3D cells	-	(+) <5%	-	-	-	-	-	-	(+) 15%	-	(+++) 100%	(++) 5%	-	(+++) 100%	-
	2D cells	(+++) 100%	-	-	(++) 10%	(+) 5%	-	(+) <5%	(+) 10%	-	-	(+++) 100%	(+) 10%	-	(+++) 100%	(+) 80%
2174 P	3D	(++) <5%	(++) <5%	-	(+++) 50%	(+++) 100%	(+++) 80%	-	(++) 40%	-	-	(+++) 100%	-	-	(+++) 100%	-
	2D	(+++) 80%	(++) 80%	-	(+++) 90%	(+) 20%	-	(+++) 90%	(+++) 80%	(+) 10%	-	(+++) 100%	(+) 90%	-	(+++) 70%	(+) 90%
2175	Tissue	(+++) 100%	(+++) 60%	(+++80%)	-	(+) 10%	(+) 20%	(+) 50%	(+++) 70%	-	-	-	-	-	(+++) 80%	-
2175 P	3D cells	-	-	-	(+++) 60%	-	-	-	(++/+++) 80%	-	-	-	-	-	(+++) 95%	-
	2D cells	-	-	-	(++) <5%	-	-	-	(+) <5%	-	-	-	-	-	(+++) 95%	-
2280	Tissue	(+++) 100%	(+) <5%	-	(+++) 90%	-	-	-	(+) <5%	-	-	-	-	-	(+) 100%	-
2280 T	3D cells	-	-	-	(+++) 60%	-	-	-	(+++) 80%	-	-	(+++) 80%	-	-	(+++) 100%	-
	2D cells	(+++) <5%	-	-	(+) 5%	(+) 10%	-	(+) <5%	(+) 10%	-	-	(+++) 100%	(+) 20%	-	(+++) 100%	(+) 90%

T = cell lines established from MPM tumor tissue; P = cell lines established from MPM pleural effusion. +, ++, +++ = 1 positive, 2, positive, 3, positive of IHC intensity.

2.2. Genomic Deletion of CDKN2A Was Identified in MPM Samples Using FISH

In this study, we have established FISH analysis to identify the genomic deletion of *CDKN2A* using specific probes in MPM tissue samples. FISH staining identified heterozygous or homozygous loss of the *CDNK2A* region in MPM tumor samples (Figure 2); alternatively, normal cells retained expression of both alleles. Figure 2 demonstrates the homozygous loss in the majority of tumor cells (Figure 2A), whereas a smaller portion of samples displayed heterozygous loss (Figure 2B) or no loss (Figure 2C). Of the 12 MPM tumor samples analyzed, 66% (8/12, cut-off 15%) showed homozygous and 16% (2/12, cut-off 40%) showed heterozygous loss of *CDKN2A*. These data indicate that FISH provides a qualitative presentation of genomic deletion in MPM. Although FISH is a well-established technique for the identification of genomic deletion, it is not largely accessible in every laboratory and difficult to provide quantitative assessment; therefore, a more accessible approach to identify genomic changes would be beneficial. To be able to quantitatively analyze genomic deletion, we have assessed the absolute quantification of genomic expression using ddPCR.

2.3. Copy Number Variation Contributes to Loss of BAP1 and CDKN2A Expression in MPM

Our initial attempt to study the DNA content of MPM cells was carried out using metaphase spread and flow-cytometry analysis of DNA content (Supplementary Figure S2). Results indicated cell lines 1180, 1843, 2164, 1157, and 1518 showed tetraploidy. However, these data did not provide conclusive information in regard to specific genes containing copy number variation (CNV). Prior to this study, we reviewed that about 50% MPM cases tested show loss of BAP1 protein expression. Many studies have reported that loss of BAP1 protein expression is due to genetic mutation or DNA methylation in the genomic region. To better understand the mechanism causing *BAP1* loss in MPM, we performed genomic Sanger sequencing of the genetic regions spanning exon 6 and 7, where the majority of mutations reside, as reported in the literature [28,29]. In addition, DNA methylation status was determined by methylation-specific PCR (MSP) analysis to study the involvement of DNA methylation in *BAP1* loss. Results from DNA sequencing and MSP studies (Supplementary Figure S3) indicated no evidence of genomic mutation near exon 6 and 7, and there was no promoter hypermethylation in the nine samples tested. We therefore performed CNV to assess loss of heterozygosity (LOH) of *BAP1* using ddPCR. Results obtained from ddPCR analysis confirm *BAP1* deletion (Figure 3) in MPM samples that correlate with the loss of *BAP1* protein expression observed using IHC analysis (Figure 1). Loss of *CDKN2A* is a common event in mesothelioma [30]. We tested its genomic alteration using FISH and, similar to previous studies, we showed that either heterozygous or homozygous loss of *CDKN2A* is prevalent in our MPM cohort. To assess the potential of detecting copy number loss of *CDKN2A* in MPM samples using ddPCR, we performed CNV ddPCR analysis in MPM tissues and their matched primary cell lines. The percentage of *CDKN2A* loss assessed using ddPCR correlated to results observed using FISH ananlysis (Figure 2). Our results also showed the stability and consistency of CNV detection across formalin-fixed paraffin-embedded (FFPE) samples and established cell lines. Using a normal mesothelial cell line and healthy individual buffy coat (BCN7) samples as normal controls for the presence of both *CDKN2A* alleles and MPM cell lines (H2052 and H28) as controls for gene deletion. Seven percent (one out of 14) of samples showed the retention of both alleles, and the majority of cases (93%: 13 out of 14) showed deletion of the *CDKN2A* genomic region.

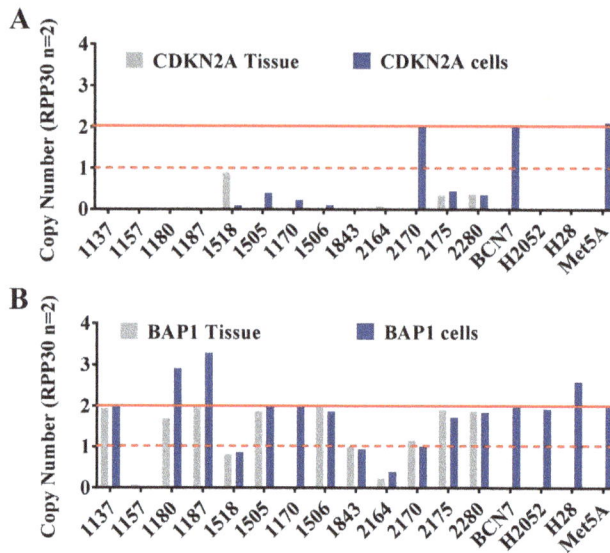

Figure 3. *CDKN2A* and BAP1 protein loss is due to copy number. Copy number (Blue = results from cell line, grey = results from tumor tissue, red line = two copies, red dotter line = one copy) of (**A**) *CDKN2A* and (**B**) *BAP1* in MPM tumor and its derivative cells normalized to the ribonuclease P protein subunit p30 (*RPP30*) assessed by droplet digital PCR (ddPCR).

2.4. Concordance of BAP1 Protein Expression and Genomic Deletion

Table 2 demonstrates BAP1 protein expression and the corresponding ddPCR CNV analysis in matched cell line and tissue samples. In our sample cohort, 54% of MPM FFPE samples showed high levels of BAP1 protein expression, and 46% showed low or no expression of the BAP1 protein. When examining cell lines established from tissue collection (matched samples), 64% showed a high level of expression, while 36% showed low to no expression of BAP1. Strong correlation (92%, 24 out of 26) was demonstrated between the BAP1 protein and genomic expression in FFPE and cell lines samples, as measured by IHC and ddPCR, respectively. Our results indicated low correlation between BAP1 protein and genomic expression, as determined by IHC and CNV. respectively (Supplementary Figure S4). Both FFPE and matched cell-line samples exhibited 50% of *BAP1* genomic retention (13 out of 26) and 70% BAP1 protein expression (19 out of 27), whereas 42% (11 out of 26) of samples that displayed low BAP1 protein expression also showed genomic deletion in the *BAP1* region. SPSS software was utilized for the measurement of sensitivity and specificity of biomarkers for detection of MPM. *CDKN2A* genomic loss had 96.4% of sensitivity and 100% specificity for MPM detection. BAP1 protein expression (using IHC) and genomic deletion (using ddPCR) results were analyzed for their concordance by measuring their combined sensitivity and specificity. Results from SPSS indicated there was 93.3% sensitivity and 63.6% specificity when combining the protein expression and genomic deletion of *BAP1* as a marker for MPM. When analyzing genomic deletion (ddPCR results) alone, *BAP1* genomic loss had a sensitivity of 42.9% and specificity of 100% in identifying MPM compared to normal mesothelial cells and healthy donor buffy-coat controls. In comparison, BAP1 protein expression identifies MPM with lower sensitivity and specificity levels of 67.9% and 28.6%, respectively. These results indicate that detection of *BAP1* and *CDKN2A* by genomic analysis (CNV using ddPCR) is a more distinctive method to identify MPM.

Table 2. Concordance of BAP1 IHC and ddPCR analysis.

MM ID	Sample Type	BAP1 IHC	BAP1 ddPCR (Reference to RPP30)
1137	Tissue	(+++) 90–100%	1.94
	cells	(+++) 30%	2
1157	Tissue	ND	0.028
	cells	ND	0.07
1180	Tissue	(+++) 80%	2.91
	cells	(+++) 95%	1.69
1187	Tissue	ND	3.28
	cells	(+++) 80%	2
1505	Tissue	(+++) 90–100%	1.87
	cells	(+++) 100%	2.01
1506	Tissue	(+++) 90%	1.96
	cells	(+++) 100%	1.87
1518	Tissue	(+++) 20%	0.82
	cells	(+++) 90%	0.86
1170	Tissue	(+++) 95%	No tissue availible
	Cells	(+++) 95%	1.98
1843	Tissue	ND	1
	Cells	ND	0.94
2164	Tissue	ND	0.24
	Cells	ND	0.39
2170	Tissue	ND	1.15
	Cells	(+++) 100%	1
2174	Tissue	No tissue availible	No tissue availible
	Cells	(+++) 70%	1.327814
2175	Tissue	(+++) 80%	1.91
	Cells	(+++) 95%	1.92
2280	Tissue	(+) 100%	1.88
	Cells	(+++) 100%	1.85

ND = not detected. +, ++, +++ = 1 positive, 2, positive, 3 positive of IHC intensity.

3. Discussion

MPM tumors are histologically heterogeneous and have distinct morphological subtypes that range from epithelioid, sarcomatoid, and biphasic. The biphasic histological subtype consists of epithelioid and sarcomatoid components, with each contributing to at least 10% of the tumor. Further, histomorphological features such as mitotic numbers and nuclear atypia have been included to conclude a total score [31]. This heterogeneity within morphological subgroups further adds to the complexity of the definitive identification of MPM [32]. It is therefore important to establish cultured cells from mesothelioma biopsies and pleural effusions to be utilized for studying the cellular, molecular, and genetic levels of this tumor. Over a period of four years (2013–2017), we established fifteen cell lines (defined as successful subculture) from a primary culture [27] from fourteen human tumor and/or plural effusion samples. In the current study, we utilized these established mesothelioma cell lines to determine biomarker expression in a system that closely parallels the tumor.

Currently, no single immunohistochemical marker offers high specificity and sensitivity or definitive negative predictive value for the diagnosis of mesothelioma. The most useful mesothelial and epithelial markers proposed for the diagnosis of mesothelioma are a combination of markers, often including calretinin (a vitamin D-dependent calcium-binding protein involved in calcium signalling) [33], HBME-1, thrombomodulin, WT-1, mesothelin, and podoplanin as mesothelial markers, and pCEA, Ber-EP4, TTF-1, and TAG72 as epithelial markers [34]. The cytokeratin-19/CEA ratio is a useful marker for mesothelioma diagnosis due to the high level of cytokeratin (18 and 19) expression in mesothelial cells. Additionally, cytokeratin-19 was previously found in two hereditary cases of mesothelioma [35]. The IHC panel included a total of 15 markers, and these were tested in the matched tumor and primary cell-line sample set. Unexpectedly, we did not observe a complimentary pattern in expression between tumor samples and derivative cell lines; often, the cell-line marker expression scores deviated from what was observed in the tumor samples. This finding suggests that in terms of MPM identification, IHC marker subtyping is not an ideal method for use in cell lines. Cell-to-cell communication in tumors creates distinct protein expression phenotypes that differ from those in cell culture [36] and this could explain the observed difference in protein marker expression. Analysis of the genomic phenotype could, instead, provide an alternative method of tumor identification.

Homozygous deletion (HD) of *CDKN2A* is one of the most common gene alterations associated with MPM [30]. Detection of *CDKN2A* HD using FISH can be used to differentiate between MPM and RMH (43% to 93% sensitivity; 100% specificity) [37]. However, the *CDKN2A* FISH assay only provides qualitative measurement of genomic expression and is inaccessible to many laboratories due to high costs and a highly specialised workflow. Furthermore, the complex structural chromosomal instability in mesothelioma cell lines can lead to the complication of karyotype differentiation and problems in chromosomal abnormality detection that may not be revealed by routine G-banding or FISH techniques [38]. ddPCR offers an alternative method for genomic analysis that provides absolute quantification, thereby enabling CNV analysis. Additionally, it is relatively cost-effective for routine use [39], thus enabling high-throughput application in the clinic. Hida et al. previously carried out BAP1 IHC analysis and *CDKN2A*-specific FISH in 40 MPM and 20 reactive mesothelial hyperplasia (RMH) samples [40]. Results indicated that BAP1 expression loss and *CDKN2A* homozygous deletion were present in 27 (67.5%) and 17 (42.5%) MPM cases, respectively. Three MPM cases (7.5%) and all 20 RMH cases had neither *BAP1* loss nor *CDKN2A* homozygous deletion. The combination of two markers produced higher sensitivity (92.5%, 37/40) and estimated probability than BAP1 IHC and *CDKN2A* FISH used alone. In our study, the combination of the two markers produced 96.4% of sensitivity and 100% specificity. Our results indicate that CNV analysis of tumor and matched cell-line samples were concurrent and both indicated *CDKN2A* deletion. *BAP1* loss in MPM is attributed to multiple mechanisms including mutation, DNA methylation, or copy number loss [41]. These data show that BAP1 protein loss was due to genomic deletion and *BAP1* CNV at the tumor level was also found to be reiterated in the matched MPM cell lines.

The discovery of alternative molecular markers for MPM is required to facilitate effective diagnosis to improve the dire prognosis of the disease. This study suggests that the CNV of *CDKN2A* is identifiable in MPM tumor samples and derivative cell lines alike using ddPCR. Additionally, *BAP1* CNV was demonstrated using ddPCR and was correlated between tumor and cell-line samples. This highlights the stability of *CDKN2A* and *BAP1* genomic deletion in MPM tumors and suggests identification of CNV could offer a potential alternative in MPM diagnostic testing.

4. Materials and Methods

4.1. Patient Tissue-Sample Collection and MPM Cell-Line Establishment

MPM Cell-Line Establishment

Formalin-fixed paraffin-embedded (FFPE) tumor tissues and fresh tumor samples were collected from MPM patients. All patients gave informed written consent and the project was approved by the Human Research Ethics Committees at Concord Repatriation General Hospital (HREC/11/CRGH/75 approved since 2011). Patient demographics are listed in Table 3.

Table 3. Patient demographics.

MM ID	Aga at Diagnose	Gender	Histological Subtype	Surgery Procedure	Survival (months)	Asbestos Exposure
1137	83	Male	Desmoplastic	Biopsy, Decortication, Pleurodesis	0.7	Yes
1157	51	Male	Epitheliod	Biopsy, Decortication, Pleurodesis	20.6	Yes
1170	74	Male	Biphasic	Biopsy, Decortication, Pleurodesis	0.8	Yes
1180	72	Male	Biphasic	Biopsy, Decortication, Pleurodesis	7.4	Yes
1187	64	Male	Biphasic	Extrapleural pneumonectomy	33.5	Yes
1505	57	Male	Epitheliod	Biopsy, Surgical exploration	7.2	Yes
1506	72	Male	Epitheliod	Biopsy, Pleurodesis	24.0	Yes
1518	80	Male	Biphasic	Biopsy, Decortication, Pleurodesis	22.7	Yes
1843	60	Male	Biphasic	Decortication, Pleurodesis	15.6	Yes
2164	73	Male	Epitheliod	Biopsy, Decortication, Pleurodesis	*	Yes
2170	75	Male	Epitheliod	Biopsy, Decortication, Pleurodesis	*	ND
2174	78	Male	Epitheliod	Biopsy, Decortication, Pleurodesis	3.0	Yes
2175	66	Male	Epitheliod	Biopsy, Decortication, Pleurodesis	12.5	Yes
2280	69	Male	Epitheliod	Extrapleural pneumonectomy	*	Yes

* = still alive; ND = no data provided.

4.2. Immunohistochemical Analysis of MPM Tissue Sections and Established Cell-Line Blocks

MPM cell lines were cultured into a 3D model and embedded into cell blocks that were further processed into paraffin blocks. MPM tissue blocks and cell blocks were sectioned at 0.4 μm thickness, deparaffinised, and rehydrated in graded concentrations of xylene and ethanol. Antigen retrieval and immunohistochemical staining were performed on an automated Leica Bond III (Leica Microsystems, Melbourne, Australia) using a Bond Polymer Refine Detection Kit (Leica Biosystems, Milton Keynes, UK). Either enzyme 1 (Leica Biosystems, UK) or Heat-Induced Epitope Retrieval (HIER) was performed on all slides in either Bond Epitope Retrieval Solution (Leica Biosystems, UK) 1 (pH6) or 2 (pH9) for 20 min. Primary antibody was applied and incubated for 20 min at room temperature (Table 4). Slides were then immersed in H_2O_2 for 5 min to quench endogenous peroxidases. Slides were processed for postprimary detection for 15 min, followed by a polymer for 15 min. 3,3′-diaminobenzidine (DAB with enhancer) chromogenic detection and haematoxylin counterstaining were used. Diagnostic clinical procedures related to diagnosis of the cases were performed in a NATA-approved laboratory using external quality-assurance program (QAP)-validated tests. The method of scoring for each antibody in each case was as per usual clinical diagnostic practice. A negative staining pattern was defined as no staining. Positive staining cells were defined as 1+ (weak), 2+ (moderate), or 3+ (strong) staining intensity in the cells, and the number of cells showing the relevant positive intensity were scored as a percentage over the total number of cells present.

4.3. FISH Analysis of CDKN2A Genomic Analysis

FISH dual-color analysis was performed with a *CEP9* Spectrum Green-labelled probe and a Spectrum Orange-labelled, locus-specific *CDKN2A* (p16) probe (Cat. 05J51-001, Abbott Molecular, Sydney, Australia). Briefly, paraffin sections were deparaffinised, dehydrated in ethanol, and washed 3 times with H_2O. Sections were digested with protease K (0.5 mg/mL) at 37 °C for 20 min. The slides were washed in SSC twice, dehydrated with ethanol, and air-dried. The probes were denatured for 5 min at 95 °C before hybridization. Slides were hybridized overnight at 37 °C and washed in 0.2 × SSC/NP40 71 °C for 2 min. Nuclei were counterstained with DAPI/antifade (Vysis, Abbott Molecular, Sydney, Australia). Each FISH assay included normal lung-tissue sections as a negative control, and sections of mesothelioma previously identified as carrying p16 deletion as a positive control. Analyses were performed with a fluorescence microscope (Axio M2, ZEISS, Oberkochen, Germany) equipped with filter sets with single- and dual-band exciters for Spectrum Green, Spectrum Orange, and DAPI (UV 360 nm). The histologic areas previously selected on the hematoxylin-eosin-stained sections were identified on the FISH-treated slides. Overlapping cells were excluded and individual, and well-defined cells were analyzed and scored. At least 100 cells were scored for each mesothelioma case. Homozygous deletion was defined as the absence of both red *CDKN2A* signals (9p21) in the presence of at least one green chromosome 9 signal (*CEP9*). A heterozygous deletion was defined as the presence of one 9p21 signal, and two *CEP9* signals. The cut-off values were established by methods previously described [42]. The cut-off value was the mean percentage plus three SD using normal mesothelial cell nuclei. We established a cut-off value of 15% for homozygous deletion and 40% for heterozygous deletion.

Table 4. Antibodies used in this study.

Antibody	Clone	Manufacturer	Product Code	Species	Dilution
CK8/18	EP17/30	Dako	M3652	Rabbit	1:100
Calretinin	Polyclonal	Biocare	CP092C	Rabbit	1:100
CK5/6	D5 & 16B4	Cell Marque	358M-16	Mouse	1:150
CD141	15CB	Novocastra	NCL-CD141	Mouse	1:50
HBME1	HBME-1	Dako	M3505	Mouse	1:50
WT1	WT49	Novocastra	NCL-L-WT1-562	Mouse	1:50
D2-40	D2-40	Biocare	CM266C	Mouse	1:100
EMA	GP1.4	Novocastra	NCL-L-EMA	Mouse	1:350
CEA	11-7	Dako	M7072	Mouse	1:200
TAG72	B72-3	Cell Marque	337M-84	Mouse	1:2000
BG8	F3	Covance	SIG-3317-1000	Mouse	1:100
CD15	Carb-3	Dako	M3631	Mouse	1:100
TTF-1	SPT24	Novocastra	NCL-L-TTF-1	Mouse	1:100
BAP1	C-4	Santa Cruz	SC-28383	Mouse	1:200
HEA	Ber-EP4	Dako	M0804	Mouse	1:100

4.4. CDKN2A and BAP1 Genomic Loss Were Suggested by CNV Using ddPCR

Primers for the amplification of the genomic region of *BAP1*, *CDKN2A*, and *RPP30* were optimized using ddPCR EvaGreen (Bio-Rad, California, CA, USA) according to the manufacturer's recommendations. Eighty nanograms of total genomic DNA was isolated from MPM tissues and MPM cell lines for use as a template for ddPCR. ddPCR reaction mixtures were assembled using 2× EvaGreen ddPCR Supermix (Bio-Rad) and primers at a final concentration of 0.2 μM in a total reaction volume of 20 μL. Reactions were dispensed into each well of droplet generator DG8 cartridge (Bio-Rad). Seventy microliters of Evagreen specific droplet generation oil (Bio-Rad) was used to generate approximately 15,000 to 20,000 droplets using the droplet generator (Bio-Rad). A 40 μL droplet emulsion was then loaded onto a 96-well PCR plate (Bio-Rad). The plate was then heat-sealed with a pierceable foil in the PX1 PCR Plate Sealer (Bio-Rad), and placed in the thermocycler (Bio-Rad T1000). Optimal thermal-cycling conditions were used: 95 °C for 5 min; 35 cycles of 95 °C for 30 s, 60 °C for 30 s, 72 °C for 1 min; and a final step at 72 °C for 1 min. The reaction mixtures were then held at 4 °C until needed. The cycled droplets were read individually with the QX200 droplet reader (Bio-Rad), and analyzed with QuantaSoft droplet-reader software, version 1.7 (Bio-Rad). The error reported for a single well was the Poisson 95% confidence interval. No template controls (NTC) were used to monitor contaminations and primer–dimer formation and determination of the cut-off threshold (Supplementary Figure S5), copy number for each genomic region was calculated by normalization to the included reference gene *RPP30* (retains two copies per cell). Homozygous deletion was considered in cases where no detection of the target genomic region was determined, but a distinctive *RPP30* population was apparent. MPM tissue samples and matched cell-line positive populations were used to calculate the positive expression values, and results were plotted as copy number detected per sample.

4.5. Statistics

All statistical analyses were carried out using IBM SPSS software version 25. The sensitivity and specificity of *BAP1* and *CDKN2A* deletion were performed using the crosstabs function of in the descriptive statistics of SPSS.

5. Conclusions

Loss of *BAP1* and *CDKN2A* are important diagnostic biomarkers in MPM. This study demonstrated the feasibility of genomic deletion as an appropriate biomarker for MPM detection that is consistent in both MPM tumor tissue and matched MPM cell lines.

Supplementary Materials: Supplementary materials can be found at http://www.mdpi.com/1422-0067/19/10/3056/s1.

Author Contributions: K.H.S. carried out the majority of the experiments in this study, including establishment and maintenance of the cell lines, preparation of cell blocks, sectioning, DNA isolation, and drafting the manuscript. K.L. is an experienced pathologist that provided professional scoring of MPM biomarkers using IHC staining and *CDKN2A* FISH analysis. M.W. provided conceptual input and edited the manuscript, C.J.C. carried out IHC and *CDKN2A* staining, C.M.W. and N.C.C. established the MPM cell lines, K.T. was involved in establishing the concept of the study and editing the manuscript. Y.Y.C. conceived the project, established the MPM cell lines, performed sectioning, prepared cell blocks, performed ddPCR analysis of copy number variation, and drafted the manuscript. All authors approved the submitted version.

Conflicts of Interest: The authors declare no conflicts of interest.

References

1. Robinson, B.W.; Lake, R.A. Advances in malignant mesothelioma. *N. Engl. J. Med.* **2005**, *353*, 1591–1603. [CrossRef] [PubMed]
2. LaDou, J.; Castleman, B.; Frank, A.; Gochfeld, M.; Greenberg, M.; Huff, J.; Joshi, T.K.; Landrigan, P.J.; Lemen, R.; Myers, J.; et al. The case for a global ban on asbestos. *Environ. Health Perspect.* **2010**, *118*, 897–901. [CrossRef] [PubMed]
3. Carbone, M.; Kanodia, S.; Chao, A.; Miller, A.; Wali, A.; Weissman, D.; Adjei, A.; Baumann, F.; Boffetta, P.; Buck, B.; et al. Consensus Report of the 2015 Weinman International Conference on Mesothelioma. *J. Thorac. Oncol.* **2016**, *11*, 1246–1262. [CrossRef] [PubMed]
4. Kao, S.C.; Reid, G.; Lee, K.; Vardy, J.; Clarke, S.; van Zandwijk, N. Malignant mesothelioma. *Intern. Med. J.* **2010**, *40*, 742–750. [CrossRef] [PubMed]
5. Tsao, A.S.; Wistuba, I.; Roth, J.A.; Kindler, H.L. Malignant pleural mesothelioma. *J. Clin. Oncol.* **2009**, *27*, 2081–2090. [CrossRef] [PubMed]
6. Johnson, D.H.; Schiller, J.H.; Bunn, P.A., Jr. Recent clinical advances in lung cancer management. *J. Clin. Oncol.* **2014**, *32*, 973–982. [CrossRef] [PubMed]
7. Scherpereel, A.; Astoul, P.; Baas, P.; Berghmans, T.; Clayson, H.; de Vuyst, P.; Dienemann, H.; Galateau-Salle, F.; Hennequin, C.; Hillerdal, G.; et al. Guidelines of the European Respiratory Society and the European Society of Thoracic Surgeons for the management of malignant pleural mesothelioma. *Eur. Respir. J.* **2010**, *35*, 479–495. [CrossRef] [PubMed]
8. Linton, A.; Kao, S.; Vardy, J.; Clarke, S.; van Zandwijk, N.; Klebe, S. Immunohistochemistry in the diagnosis of malignant pleural mesothelioma: Trends in Australia and a literature review. *Asia Pac. J. Clin. Oncol.* **2013**, *9*, 273–279. [CrossRef] [PubMed]
9. Husain, A.N.; Colby, T.; Ordonez, N.; Krausz, T.; Attanoos, R.; Beasley, M.B.; Borczuk, A.C.; Butnor, K.; Cagle, P.T.; Chirieac, L.R.; et al. Guidelines for pathologic diagnosis of malignant mesothelioma: 2012 Update of the consensus statement from the International Mesothelioma Interest Group. *Arch. Pathol. Lab. Med.* **2013**, *137*, 647–667. [CrossRef] [PubMed]
10. Tischoff, I.; Neid, M.; Neumann, V.; Tannapfel, A. Pathohistological diagnosis and differential diagnosis. *Recent Results Cancer Res.* **2011**, *189*, 57–78. [PubMed]
11. Oehl, K.; Vrugt, B.; Opitz, I.; Meerang, M. Heterogeneity in Malignant Pleural Mesothelioma. *Int. J. Mol. Sci.* **2018**, *19*, 1603. [CrossRef] [PubMed]

12. Guo, G.; Chmielecki, J.; Goparaju, C.; Heguy, A.; Dolgalev, I.; Carbone, M.; Seepo, S.; Meyerson, M.; Pass, H.I. Whole-exome sequencing reveals frequent genetic alterations in BAP1, NF2, CDKN2A, and CUL1 in malignant pleural mesothelioma. *Cancer Res.* **2015**, *75*, 264–269. [CrossRef] [PubMed]

13. Hylebos, M.; Van Camp, G.; van Meerbeeck, J.P.; Op de Beeck, K. The Genetic Landscape of Malignant Pleural Mesothelioma: Results from Massively Parallel Sequencing. *J. Thorac. Oncol.* **2016**, *11*, 1615–1626. [CrossRef] [PubMed]

14. Kim, J.E.; Kim, D.; Hong, Y.S.; Kim, K.P.; Yoon, Y.K.; Lee, D.H.; Kim, S.W.; Chun, S.M.; Jang, S.J.; Kim, T.W. Mutational Profiling of Malignant Mesothelioma Revealed Potential Therapeutic Targets in EGFR and NRAS. *Transl. Oncol.* **2018**, *11*, 268–274. [CrossRef] [PubMed]

15. Jaurand, M.C.; Fleury-Feith, J. Pathogenesis of malignant pleural mesothelioma. *Respirology* **2005**, *10*, 2–8. [CrossRef] [PubMed]

16. Borczuk, A.C.; Pei, J.; Taub, R.N.; Levy, B.; Nahum, O.; Chen, J.; Chen, K.; Testa, J.R. Genome-wide analysis of abdominal and pleural malignant mesothelioma with DNA arrays reveals both common and distinct regions of copy number alteration. *Cancer Biol. Ther.* **2016**, *17*, 328–335. [CrossRef] [PubMed]

17. Ivanova, A.V.; Goparaju, C.M.; Ivanov, S.V.; Nonaka, D.; Cruz, C.; Beck, A.; Lonardo, F.; Wali, A.; Pass, H.I. Protumorigenic role of HAPLN1 and its IgV domain in malignant pleural mesothelioma. *Clin. Cancer Res.* **2009**, *15*, 2602–2611. [CrossRef] [PubMed]

18. Illei, P.B.; Rusch, V.W.; Zakowski, M.F.; Ladanyi, M. Homozygous deletion of CDKN2A and codeletion of the methylthioadenosine phosphorylase gene in the majority of pleural mesotheliomas. *Clin. Cancer Res.* **2003**, *9*, 2108–2113. [PubMed]

19. Dahabreh, I.J.; Linardou, H.; Siannis, F.; Kosmidis, P.; Bafaloukos, D.; Murray, S. Somatic EGFR mutation and gene copy gain as predictive biomarkers for response to tyrosine kinase inhibitors in non-small cell lung cancer. *Clin Cancer Res.* **2010**, *16*, 291–303. [CrossRef] [PubMed]

20. Nasu, M.; Emi, M.; Pastorino, S.; Tanji, M.; Powers, A.; Luk, H.; Baumann, F.; Zhang, Y.A.; Gazdar, A.; Kanodia, S.; et al. High Incidence of Somatic BAP1 alterations in sporadic malignant mesothelioma. *J. Thorac. Oncol.* **2015**, *10*, 565–576. [CrossRef] [PubMed]

21. Cigognetti, M.; Lonardi, S.; Fisogni, S.; Balzarini, P.; Pellegrini, V.; Tironi, A.; Bercich, L.; Bugatti, M.; Rossi, G.; Murer, B.; et al. BAP1 (BRCA1-associated protein 1) is a highly specific marker for differentiating mesothelioma from reactive mesothelial proliferations. *Mod. Pathol.* **2015**, *28*, 1043–1057. [CrossRef] [PubMed]

22. Bott, M.; Brevet, M.; Taylor, B.S.; Shimizu, S.; Ito, T.; Wang, L.; Creaney, J.; Lake, R.A.; Zakowski, M.F.; Reva, B.; et al. The nuclear deubiquitinase BAP1 is commonly inactivated by somatic mutations and 3p21.1 losses in malignant pleural mesothelioma. *Nat. Genet.* **2011**, *43*, 668–672. [CrossRef] [PubMed]

23. Sheffield, B.S.; Hwang, H.C.; Lee, A.F.; Thompson, K.; Rodriguez, S.; Tse, C.H.; Gown, A.M.; Churg, A. BAP1 immunohistochemistry and p16 FISH to separate benign from malignant mesothelial proliferations. *Am. J. Surg. Pathol.* **2015**, *39*, 977–982. [CrossRef] [PubMed]

24. Sato, G. Tissue culture: The unrealized potential. *Cytotechnology* **2008**, *57*, 111–114. [CrossRef] [PubMed]

25. Sato, G.H.; Sato, J.D.; Okamoto, T.; McKeehan, W.L.; Barnes, D.W. Tissue culture: The unlimited potential. *In Vitro Cell. Dev. Biol. Anim.* **2010**, *46*, 590–594. [CrossRef] [PubMed]

26. Ivanov, S.V.; Ivanova, A.V.; Goparaju, C.M.; Chen, Y.; Beck, A.; Pass, H.I. Tumorigenic properties of alternative osteopontin isoforms in mesothelioma. *Biochem. Biophys. Res. Commun.* **2009**, *382*, 514–518. [CrossRef] [PubMed]

27. Cheng, Y.Y.S.K.; Lee, K.; Clarke, C.J.; Cheng, N.C.; van Zandwijk, N.; Klebe, S.; Reid, G. Establishing malignant pleural mesothelioma cell lines using the spheroid method produces a model with better 3D architecture. *J. Thorac. Oncol.* **2017**, *12*, S2265–S2266. [CrossRef]

28. Yoshikawa, Y.; Sato, A.; Tsujimura, T.; Emi, M.; Morinaga, T.; Fukuoka, K.; Yamada, S.; Murakami, A.; Kondo, N.; Matsumoto, S.; et al. Frequent inactivation of the BAP1 gene in epithelioid-type malignant mesothelioma. *Cancer Sci.* **2012**, *103*, 868–874. [CrossRef] [PubMed]

29. Cheung, M.; Testa, J.R. BAP1, a tumor suppressor gene driving malignant mesothelioma. *Transl. Lung Cancer Res.* **2017**, *6*, 270–278. [CrossRef] [PubMed]

30. Port, J.; Murphy, D.J. Mesothelioma: Identical Routes to Malignancy from Asbestos and Carbon Nanotubes. *Curr. Biol.* **2017**, *27*, R1173–R1176. [CrossRef] [PubMed]

31. Rosen, L.E.; Karrison, T.; Ananthanarayanan, V.; Gallan, A.J.; Adusumilli, P.S.; Alchami, F.S.; Attanoos, R.; Brcic, L.; Butnor, K.J.; Galateau-Sallé, F.; et al. Nuclear grade and necrosis predict prognosis in malignant epithelioid pleural mesothelioma: A multi-institutional study. *Mod. Pathol.* **2018**, *31*, 598–606. [CrossRef] [PubMed]

32. Marusyk, A.; Almendro, V.; Polyak, K. Intra-tumour heterogeneity: A looking glass for cancer? *Nat. Rev. Cancer* **2012**, *12*, 323–334. [CrossRef] [PubMed]

33. Kao, S.C.; Klebe, S.; Henderson, D.W.; Reid, G.; Chatfield, M.; Armstrong, N.J.; Yan, T.D.; Vardy, J.; Clarke, S.; van Zandwijk, N.; et al. Low calretinin expression and high neutrophil-to-lymphocyte ratio are poor prognostic factors in patients with malignant mesothelioma undergoing extrapleural pneumonectomy. *J. Thorac. Oncol.* **2011**, *6*, 1923–1929. [CrossRef] [PubMed]

34. Marchevsky, A.M. Application of immunohistochemistry to the diagnosis of malignant mesothelioma. *Arch. Pathol. Lab. Med.* **2008**, *132*, 397–401. [PubMed]

35. Relan, V.; Morrison, L.; Parsonson, K.; Clarke, B.E.; Duhig, E.E.; Windsor, M.N.; Matar, K.S.; Naidoo, R.; Passmore, L.; McCaul, E.; et al. Phenotypes and karyotypes of human malignant mesothelioma cell lines. *PLoS ONE* **2013**, *8*, e58132. [CrossRef] [PubMed]

36. Ravi, M.; Mohan, D.K.; Sahu, B. Protein Expression Differences of 2-Dimensional and Progressive 3-Dimensional Cell Cultures of Non-Small-Cell-Lung-Cancer Cell Line H460. *J. Cell. Biochem.* **2017**, *118*, 1648–1652. [CrossRef] [PubMed]

37. Nabeshima, K.; Matsumoto, S.; Hamasaki, M.; Hida, T.; Kamei, T.; Hiroshima, K.; Tsujimura, T.; Kawahara, K. Use of p16 FISH for differential diagnosis of mesothelioma in smear preparations. *Diagn. Cytopathol.* **2016**, *44*, 774–780. [CrossRef] [PubMed]

38. Riegel, M. Human molecular cytogenetics: From cells to nucleotides. *Genet. Mol. Biol.* **2014**, *37*, 194–209. [CrossRef] [PubMed]

39. Cheng, Y.Y.; Mok, E.; Tan, S.; Leygo, C.; McLaughlin, C.; George, A.M.; Reid, G. SFRP Tumour Suppressor Genes Are Potential Plasma-Based Epigenetic Biomarkers for Malignant Pleural Mesothelioma. *Dis. Mark.* **2017**. [CrossRef] [PubMed]

40. Hida, T.; Hamasaki, M.; Matsumoto, S.; Sato, A.; Tsujimura, T.; Kawahara, K.; Iwasaki, A.; Okamoto, T.; Oda, Y.; Honda, H.; et al. BAP1 immunohistochemistry and p16 FISH results in combination provide higher confidence in malignant pleural mesothelioma diagnosis: ROC analysis of the two tests. *Pathol. Int.* **2016**, *66*, 563–570. [CrossRef] [PubMed]

41. Bueno, R.; Stawiski, E.W.; Goldstein, L.D.; Durinck, S.; De Rienzo, A.; Modrusan, Z.; Gnad, F.; Nguyen, T.T.; Jaiswal, B.S.; Chirieac, L.R.; et al. Comprehensive genomic analysis of malignant pleural mesothelioma identifies recurrent mutations, gene fusions and splicing alterations. *Nat. Genet.* **2016**, *48*, 407–416. [CrossRef] [PubMed]

42. Chung, C.T.; Santos Gda, C.; Hwang, D.M.; Ludkovski, O.; Pintilie, M.; Squire, J.A.; Tsao, M.-S. FISH assay development for the detection of p16/CDKN2A deletion in malignant pleural mesothelioma. *J. Clin. Pathol.* **2010**, *63*, 630–634. [CrossRef] [PubMed]

International Journal of
Molecular Sciences

MDPI

Review

In Silico and In Vitro Analyses of LncRNAs as Potential Regulators in the Transition from the Epithelioid to Sarcomatoid Histotype of Malignant Pleural Mesothelioma (MPM)

Anand S. Singh [1,2], Richard Heery [1,2] and Steven G. Gray [1,3,4,5,]*

[1] Thoracic Oncology Research Group, Trinity Translational Medical Institute, St. James's Hospital, Dublin D08 W9RT, Ireland; rheery@tcd.ie (A.S.S.); anandsimarsingh@gmail.com (R.H.)
[2] MSc in Translational Oncology Program, Trinity College Dublin, Dublin 2, Ireland
[3] HOPE Directorate, St. James's Hospital, Dublin 8, Ireland
[4] Department of Clinical Medicine, Trinity College Dublin, Dublin 8, Ireland
[5] Labmed Directorate, St. James's Hospital, Dublin 8, Ireland
[*] Correspondence: sgray@stjames.ie; Tel.: +353-1-4284945

Received: 27 February 2018; Accepted: 24 April 2018; Published: 26 April 2018

Abstract: Malignant pleural mesothelioma (MPM) is a rare malignancy, with extremely poor survival rates. At present, treatment options are limited, with no second line chemotherapy for those who fail first line therapy. Extensive efforts are ongoing in a bid to characterise the underlying molecular mechanisms of mesothelioma. Recent research has determined that between 70–90% of our genome is transcribed. As only 2% of our genome is protein coding, the roles of the remaining proportion of non-coding RNA in biological processes has many applications, including roles in carcinogenesis and epithelial–mesenchymal transition (EMT), a process thought to play important roles in MPM pathogenesis. Non-coding RNAs can be separated loosely into two subtypes, short non-coding RNAs (<200 nucleotides) or long (>200 nucleotides). A significant body of evidence has emerged for the roles of short non-coding RNAs in MPM. Less is known about the roles of long non-coding RNAs (lncRNAs) in this disease setting. LncRNAs have been shown to play diverse roles in EMT, and it has been suggested that EMT may play a role in the aggressiveness of MPM histological subsets. In this report, using both in vitro analyses on mesothelioma patient material and in silico analyses of existing RNA datasets, we posit that various lncRNAs may play important roles in EMT within MPM, and we review the current literature regarding these lncRNAs with respect to both EMT and MPM.

Keywords: malignant pleural mesothelioma; long non-coding RNAs (lncRNAs); epithelial-mesenchymal transition

1. Introduction

Malignant pleural mesothelioma (MPM) is a rare, but aggressive form of cancer, predominantly associated with prior exposure to asbestos [1]. Whilst many countries have banned the use of asbestos [2], it is still used in developing countries. A recent report based on extrapolations for asbestos use estimated global mesothelioma deaths at 38,400 per annum [3], and while there have been some recent advances in this disease, particularly with respect to immune-oncology [4,5], the current standard of care (a combination of pemetrexed/raltitrexed and cisplatin chemotherapy) [6,7] is non-curative, and results in a response rate of approximately 40% [8].

Epithelial–mesenchymal transition (EMT) is a process by which epithelial cells shed many of their epithelial traits and acquire various features observed in mesenchymal cells. During this transition, epithelial cells lose their polarity and many of their intercellular contacts, such as desmosomes, adherens junctions, and tight junctions, resulting in their disassociation from epithelial sheets. At the end of this process, cells undergoing EMT assume a variety of mesenchymal-like properties: enhanced migratory capacity, invasiveness, heightened resistance to apoptosis, and greatly increased production of extracellular matrix components [9].

Most MPMs have three main histologic subtypes, divided into epithelioid, sarcomatoid, or mixed (biphasic) [10,11]. However, multiple morphological patterns have also been described within these subtypes, and similarities in clinical presentation and histological appearance of MPM, primary lung carcinoma, pleural metastases, reactive pleural diseases, and rare pleural malignancies can pose challenges to MPM diagnosis [12]. Indeed, "The current gold standard of MPM diagnosis is a combination of two positive and two negative immune-histochemical markers in the epithelioid and biphasic type, but sarcomatous type do not have specific markers, making diagnosis more difficult." [12]. Because MPM has a partial fibroblastic phenotype in the context of EMT, it has been postulated that this may, in part, explain the aggressiveness of this cancer conferring both high invasiveness and chemoresistance [13], and in this regard, it may be applied to the epithelioid versus sarcomatoid histotype of MPM [13]. In this regard, the epithelioid and sarcomatoid histologic variants of MPM can be considered as E- and M-parts of the EMT axis, with the biphasic histotype considered an intermediate [14]. In support of this, hierarchical clustering of transcriptomic data from MPM separates this cancer into two distinct molecular subgroups, and one subgroup (C2) with an associated EMT molecular signature has worse overall survival (OS) [15].

A significant body of work has examined the roles of other forms of non-coding RNA such as microRNAs in both EMT [16,17] and MPM [18,19], and there is some evidence that miRNAs and lncRNAs interact or cross-talk to orchestrate EMT [20]. Despite the known roles of lncRNAs in the establishment of EMT in cancer, a topic recently reviewed in detail by us and others [21,22], very few studies have specifically examined the functional roles of lncRNAs in MPM [23–26].

With the advent of high-throughput sequencing technology, transcriptomic data for MPM is emerging. Using unsupervised consensus clustering of RNA-seq-derived expression data from 211 MPM samples, Bueno et al. [27] identified four major clusters: sarcomatoid, epithelioid, biphasic-epithelioid (biphasic-E), and biphasic-sarcomatoid (biphasic-S). Of these, differential expression analysis of the sarcomatoid and epithelioid consensus clusters identified a significant number of lncRNAs which could distinguish between these, as shown in Table 1.

A discussion of the putative roles for these and other lncRNAs in EMT will be presented in subsequent sections.

Table 1. Differentially expressed long non-coding RNAs (lncRNAs) between sarcomatoid versus epithelioid samples as identified by Bueno et al. [27], and discussed in this article.

Name	log2 Fold Change	Unadjusted *p*-Value	Comments
PCAT1	−1.227580845	0.000168412	
HOTAIR	4.342211972	1.09×10^{-10}	Associates with chromatin remodelling complexes to regulate EMT [21]
MALAT1	−0.902533139	2.72×10^{-7}	
NEAT1	−0.534058107	0.012990525	Identified as an lncRNA with altered (−2.8 fold) expression in MPM [26]
GAS5	0.053707959	0.785538121	GAS5 shown to have altered expression in MPM
HULC	−0.724711448	0.03946186	Known roles in EMT in other cancers [28–30]
H19	2.155715056	1.09×10^{-9}	Promotes EMT in NSCLC [31], and various other cancers [21]
ZFAS1	−0.443662478	0.018761094	Known regulator of EMT in other cancer settings [32–36]
PVT1	−0.64835701	7.75×10^{-5}	Previously identified as an lncRNA with altered expression in MPM [24] Overexpression shown to inhibit EMT in lung adenocarcinoma [37].
CASC2	−1.434979397	5.64×10^{-12}	Associated with Epithelioid and Biphasic samples and high expression associated with better OS in The Cancer Genome Atlas (TCGA) dataset

In this manuscript, we examined the expression of a novel series of lncRNAs (Epidermal Growth Factor Receptor- antisense RNA 1 EGFR-AS1, prostate cancer associated transcript 6 PCAT6 and zinc finger E-box binding homeobox 2 antisense RNA 1 ZEB2-AS1) for altered expression in MPM. We show that all three of these lncRNAs are overexpressed in MPM, and that one of them, PCAT6, is significantly altered across all of the histological subtypes.

Subsequently, using in silico meta-analysis of existing The Cancer Genome Atlas (TCGA) and other datasets (www.cbioportal.org; http://watson.compbio.iupui.edu/chirayu/proggene/database/?url=proggene; www.oncomine.org), we review the known lncRNAs previously described by us and others in MPM (PVT1, NEAT1, PAX8-AS1, and GAS5). Finally, using in silico analyses, combined with a review of the current literature, we examine additional lncRNAs with known roles in EMT for the dysregulated expression in MPM, and show that for many of these, this dysregulated expression is often associated with the biphasic histological subtype. These results suggest that many lncRNAs may be a factor in the transition from the epithelioid to the more aggressive sarcomatoid histotype of malignant pleural mesothelioma.

2. Results

2.1. Novel LncRNAs with Altered Expression in MPM

Several lncRNAs have recently been identified by our unit as having potentially significant roles in MPM. In the following sections we describe their expression and putative roles in both EMT and MPM.

2.1.1. EGFR-AS1

High expression of EGFR is associated with MPM [38]. However, clinical trials of EGFR tyrosine kinase inhibitors (TKIs) as single agents in MPM failed [39–41]. However, more recently, expression of EGFR on MPM has been used for the targeted delivery of microRNA mimics delivered by targeted bacterial minicells (TargomiRs) in a recent clinical trial in MPM [42], while most recently a patient harbouring mutations in EGFR (G719C and S768I) was successfully treated with Afatinib an EGFR TKI [43]; an lncRNA associated with EGFR called EGFR-AS1 has been identified. This lncRNA was shown to regulate EGFR expression in liver cancer [44], and most recently, expression of this lncRNA has been shown to be associated with sensitivity to EGFR TKIs in patients with head and neck SCC (HNSCC) [45]. Strikingly, knockdown of EGFR-AS1 in vitro and in vivo lead to increased sensitivity, whereas overexpression is sufficient to induce resistance to EGFR TKIs [45]. In this regard, preliminary data from our group has shown that EGFR-AS1 is significantly overexpressed in MPM (Figure 1); this may explain in part why EGFR TKIs failed as single agents in clinical trials of MPM. The role of EGFR-AS1 in EMT has as yet to be determined. However, the known role of EMT in bypassing EGFR dependence [46] suggests that this lncRNA may indeed play a role in orchestrating EMT transitions in MPM.

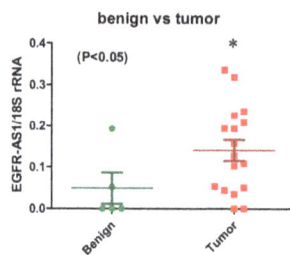

Figure 1. Overexpression of EGFR-AS1 in primary malignant pleural mesothelioma (MPM). EGFR-AS1 lncRNA expression was examined by RT-PCR in a series of primary MPM ($n = 17$) versus benign pleura ($n = 5$). Semi-quantitative densitometric analysis of the results determined that EGFR-AS1 lncRNA was significantly elevated in the tumours compared to benign pleura. Statistical significance was assessed using a 1-tailed unpaired Students *t*-test (* $p = 0.0445$).

2.1.2. PCAT6

PCAT6 is a lncRNA linked to KDM5B (also known as JARID1B). This lysine demethylase has been shown to induce EMT in various cancers, including lung cancer [47–49]. Expression of PCAT6 has also been shown to be altered in NSCLC [50,51], and circulating levels of this lncRNA in patient blood has potential as both a diagnostic and prognostic biomarker in NSCLC [51].

Preliminary data from our group indicates that expression of KDM5B is significantly upregulated in primary MPM (Figure 2A), remaining significant across all histological subtypes (Figure 2B). Similar significant overexpression of KDM5B is also observed in MPM samples in the Gordon et al. [52] mesothelioma dataset (Figure 2C). Across the TCGA dataset, KDM5B appears to have significant alterations in about 14% of MPM cases, including amplification of its genomic region, overexpression, or indeed downregulation of its mRNA (Figure 2D), all of which are found in either the epithelioid or biphasic subtypes (Figure 2E).

Moreover, we have also shown that PCAT6 itself is upregulated in MPM (Figure 2F). However, when examined across histological subtypes, the upregulation observed was significant only in the biphasic subset (Figure 2G). In the TCGA dataset, expression of this lncRNA does not appear to be upregulated, although amplification of its genomic location occurs in 3% of MPM specimens (Figure 2H), again, similar to KDM5B, these are spread over the epithelioid and biphasic subsets. (Figure 2I).The functional role of this lncRNA in EMT is as yet unknown, but knockdown of this lncRNA in lung cancer is associated with inhibited cellular proliferation and metastasis [50].

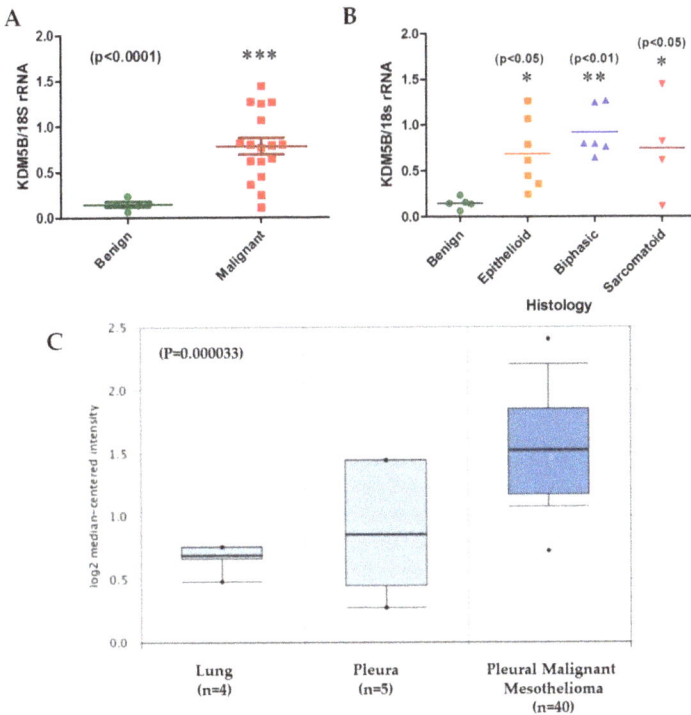

Figure 2. *Cont.*

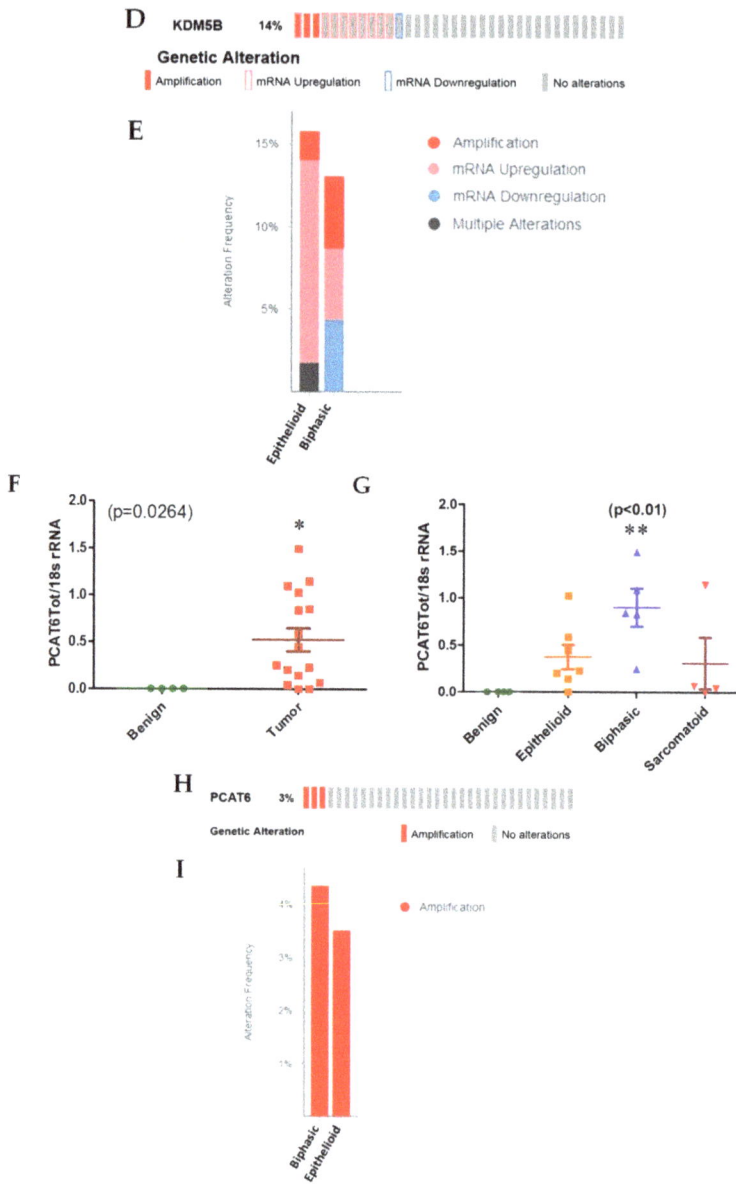

Figure 2. An examination of KDM5B and PCAT6 expression/alterations in MPM. (**A**) KDM5B mRNA is significantly elevated in tumours (*n* = 17) compared to benign pleura (*n* = 5), (**B**) the same samples stratified by histological subtype, (**C**) Oncomine analysis of the Gordon mesothelioma dataset confirming significant overexpression of KDM5B, (**D**) in silico examination using cBioPortal reveals that 14% of samples had alterations to KDM5B, (**E**) when stratified by histotype, these alterations were restricted to epithelioid or biphasic subtypes, (**F**) total PCAT6 lncRNA is significantly elevated in tumours (*n* = 16(red) compared to benign pleura (*n* = 4—green), (**G**) when stratified by histological subtype (Benign = green; Epithelioid = yellow; Biphasic = blue; Sarcomatoid = red), elevated expression of total PCAT6 is significant only in the biphasic subset. Statistical significance was assessed using a Mann–Whitney *t*-test (* *p* < 0.05), or by an ANOVA using Dunnett's Multiple Comparison Test (* *p* < 0.05; ** *p* < 0.01; *** *p* < 0.001), (**H**) in silico examination using cBioPortal reveals that 3% of samples had amplification of PCAT6, (**I**) when stratified by histotype, these alterations were restricted to biphasic or epithelioid subtypes.

2.1.3. ZEB2-AS1

ZEB2 is a known regulator of EMT [21]. Originally called ZEB2NAT, but now more often described as ZEB2-AS1, this natural anti-sense lncRNA of ZEB2 was shown to regulate ZEB2 during the process of EMT [53]. This lncRNA has been found to be upregulated in both urinary bladder cancer [54] and hepatocellular carcinoma [55], and in bladder cancer cells is partly responsible for activation of ZEB2 during EMT induction by Transforming growth factor beta (TGF-β) [54]. Furthermore, knockdown of this lncRNA in Hepatocellular Carcinoma (HCC) cells results in reduced vimentin and N-caherin expression with restoration of E-cadherin expression [55], further supporting a role for this lncRNA in the regulation of EMT.

ZEB2 was found to be a significantly altered gene between the sarcomatoid vs. epithelioid clusters (unadjusted *p*-value: $p < 2.03 \times 10^{-26}$) in the analysis by Bueno et al. [27], but this has not been supported by earlier analysis in the Gordon dataset [52]. There is some suggestion that in the larger dataset by Lopez-Rios that higher expression of ZEB2 is associated with the sarcomatoid subtype ($p = 0.065$) [56]. Very little is known about the expression of ZEB2-AS1 in MPM. Our preliminary analysis suggests that expression of this lncRNA is potentially dysregulated in MPM (Figure 3), but further studies will be required to validate these observations.

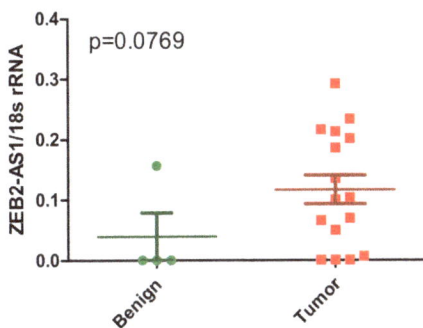

Figure 3. Altered expression of ZEB2-AS1 in primary MPM. ZEB2-AS1 lncRNA expression was examined by RT-PCR in a series of primary MPM (*n* = 16—red) vs. benign pleura (*n* = 4—green). Semi-quantitative densitometric analysis of the results suggests that ZEB2-AS1 lncRNA was elevated in the tumours compared to benign pleura. Statistical significance was assessed using a 1-tailed unpaired Students *t*-test ($p = 0.0769$).

2.2. Previously Published lncRNAs with Known Links to MPM

A significant body of research has shown that many short non-coding RNAs, such as microRNAs (miRNAs), have extensive alterations and diverse roles in MPM, and have been discussed by us in depth in a previous review [18]. The evidence for altered expression or roles of lncRNAs in MPM has not as yet been exhaustively analysed in MPM. In the following sections we review the current knowledge of the known lncRNAs associated with MPM, and whether or not these lncRNAs can be linked to EMT processes.

2.2.1. PVT1 and c-Myc

PVT1 is an lncRNA which has been shown to be associated with poor prognosis in many cancers [57]. Its expression has also been linked to EMT in various cancers. For example, in breast cancer, PVT1 is significantly upregulated, and directly interacts with SOX2 to drive EMT [58]. In pancreatic cancer, PVT1 has been found to promote EMT by downregulation of the cyclin-dependent kinase p21 [59]. The other ways PVT1 has been shown to elicit responses include by acting as a competitive endogenous RNA (ceRNA) for various miRNAs [60–67], or by interactions with EZH2 to epigenetically regulate genes associated with EMT [68–75].

Both PVT1 and c-Myc are located at the same chromosomal location (8q24.21) and an increase in PVT1 expression is required for high MYC protein levels in 8q24-amplified human cancer cells [76]. In this regard, frequent coamplification and cooperation between c-MYC and PVT1 oncogenes have been observed to promote malignant pleural mesothelioma [24]. Next Generation Sequencing (NGS) demonstrated a downregulation of PVT1 in a sarcomatoid subset compared to epithelioid (Table 1) [27].

In silico analysis of the TCGA provisional dataset demonstrated that amplification occurred only in epithelioid samples (Figure 4A,B), which is somewhat in agreement with the observations made by Riquelme et al., where copy number gains were seen in the biphasic (6 of 26, 23%) and epithelioid (5 of 37, 13%) histotypes but not in the sarcomatoid cases [24]. In samples where PVT1 overexpression is observed it is either associated with the epithelioid or biphasic histology (Figure 4B).

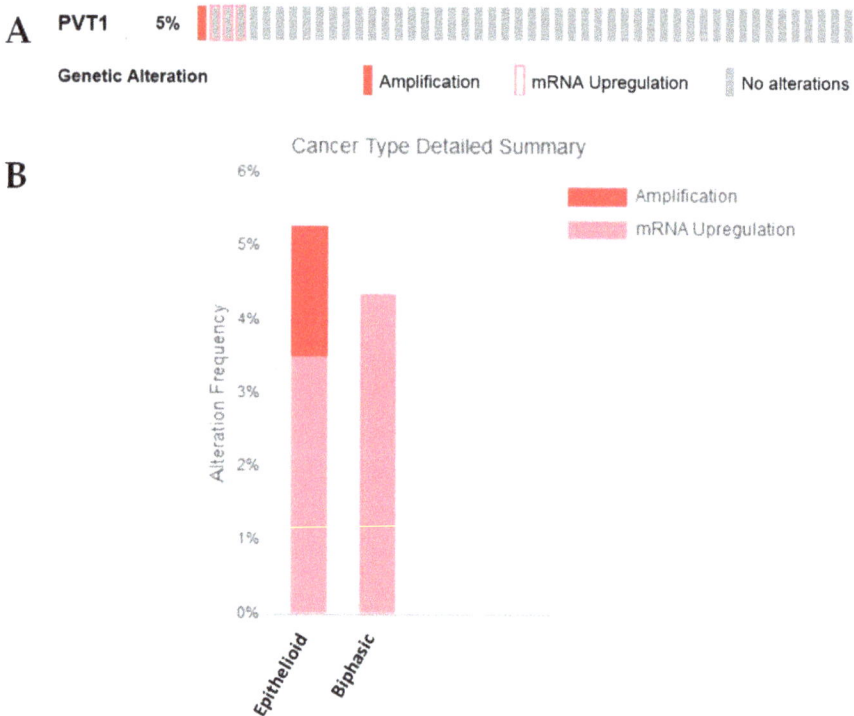

Figure 4. An examination of PVT1 expression/alterations in the TCGA dataset. (**A**) In silico examination using cBioPortal reveals that 5% of samples had overexpression of PVT1 RNA; (**B**) when stratified by histotype, only the epithelioid subtype had amplification of PVT1, whereas some patients with epithelioid and biphasic but not sarcomatoid subtypes had overexpression of this lncRNA.

2.2.2. NEAT1

Neat1 was identified by our group as an lncRNA altered in MPM [26]. It is now well-established that this lncRNA promotes EMT [21,77], and one of the means by which it affects EMT is through regulation of EZH2 [78,79]. Most recently, the expression of NEAT1 has been shown to be BAP1 dependent [80]. Given that it is estimated that approximately 65% of mesotheliomas harbour mutations inactivating BAP1 [81], this may have implications with respect to the role of this lncRNA in MPM pathogenesis. Both our data, and that of Bueno et al. (Table 1) [27], showed an overall downregulation of this lncRNA in MPM. Further analysis of the TCGA dataset shows that a proportion of samples have upregulation of this lncRNA (Figure 5A), which when stratified by histology, is found mostly in the Epithelioid subset, with a smaller proportion in the Biphasic subset also showing elevated expression (Figure 5B).

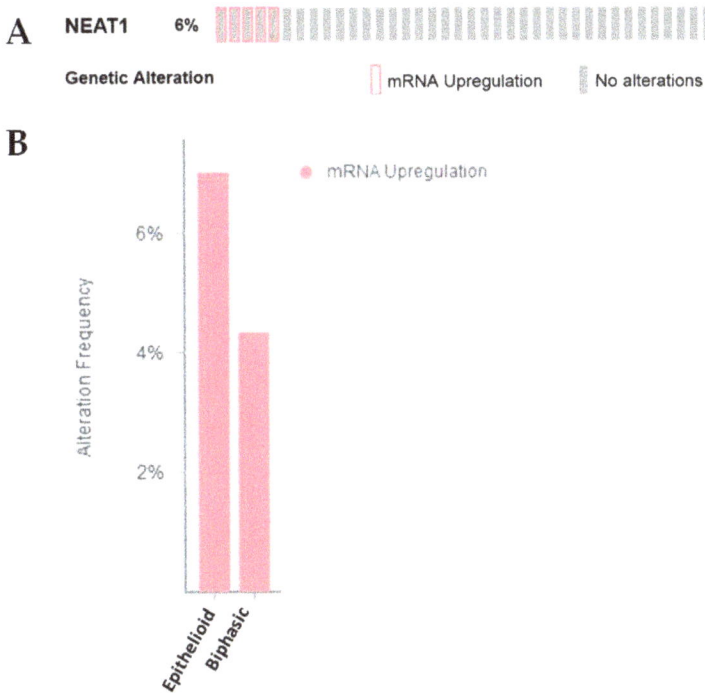

Figure 5. An examination of Neat1 expression/alterations in the TCGA dataset. (**A**) In silico examination using cBioPortal reveals that 6% of samples had overexpression of Neat1 lncRNA. (**B**) When stratified by histotype, the majority of samples with elevated Neat1 are found in the epithelioid subset, followed by a proportion in the biphasic subset.

2.2.3. PAX8-AS1

This lncRNA was also identified [26] as being significantly altered in MPM. The gene associated with this lncRNA, *PAX8*, has been shown to play important roles in the development of ovarian cancer [82], and may do this through upregulation of markers of EMT [83]; although conflicting results have emerged [84]. Interestingly, PAX8 expression is observed in peritoneal mesotheliomas [85,86], but not in pleural mesotheliomas [86]. In MPM, no significant changes in expression of PAX8 were seen in the Gordon dataset [52], whereas high expression of PAX8 was observed in 4 of 87 MPM samples (5%) in the TCGA dataset (data not shown).

Whilst our previous publication found that PAX8-AS1 was significantly altered in MPM [26], analysis of the TCGA dataset using cBioPortal found no alterations in this lncRNAs expression, suggesting that perhaps this lncRNA may not play a direct role in the regulation of EMT and/or the pathogenesis of MPM.

2.2.4. GAS5

A link between GAS5 and EGFR TKI sensitivity has also been identified. Levels of GAS5 were downregulated in the EGFR TKI resistant lung adenocarcinoma cell line A549 compared to sensitive cell lines. Moreover, restoration of GAS5 expression could greatly sensitise these cells to gefitinib treatment in xenograft mouse models [87]. In a separate study relating to prostate cancer, increased expression of GAS5 was associated with decreased Akt signalling [88]. Therefore, it could be suggested that lncRNA mediated regulation of Akt signalling seems to highly important in determining the sensitivity of NSCLC cells to EGFR TKI, such as gefitinib.

In MPM, Felley-Bosco and colleagues have shown that this lncRNA is overexpressed in malignant tumours compared to non-tumoural tissue, (* $p < 0.0001$ expression, Mann–Whitney test) [25]. While loss of this lncRNA is associated with a shortening of the cell-cycle in MPM cell lines, the role of this lncRNA in regulating EMT in MPM is unknown, however, studies in other cancers, such as osteosarcoma, have shown that expression of this lncRNA decreased in tumours compared to adjacent normal tissue. Furthermore, overexpression of GAS5 suppressed cellular proliferation, migration, and EMT in osteosarcoma cell lines [89].

2.3. Previously Published lncRNAs with Known Links to EMT

It is now well established that various lncRNAs play essential roles in the regulation of EMT, a subject we recently reviewed in depth [21]. Despite this, several of these key lncRNAs have not been studied in depth in MPM. In the following sections we discuss the known roles of several of these key lncRNAs, and using in silico analyses to describe the current evidence for their altered expression in mesothelioma histological subtypes.

2.3.1. HOTAIR

HOTAIR is a lncRNA transcribed from the HOXC gene cluster that promotes epigenetic silencing of target genes, including the HOXD gene cluster, through the recruitment of the PRC2 and LSD1/CoREST/REST chromatin remodelling complexes [90,91]. It is well established that HOTAIR is overexpressed in a wide variety of solid malignancies, and moreover, that this overexpression is associated with metastasis and tumour recurrence [21]. Critically, HOTAIR has been linked extensively to the promotion of EMT in solid tumours [21]. In this regard, HOTAIR has been found to regulate EMT through recruitment of PRC2 to the CDH1 promoter [92]. HOTAIR also forms a tripartite complex with Snail and EZH2, facilitating the recruitment of EZH2 to Snail binding sites at the promoters of the epithelial genes E-cadherin, Hepatocyte nuclear factor (HNF), HNF1α, and HNF4α, resulting in their epigenetic silencing [93]. HOTAIR also positively regulates the expression of JMJD3 and Snail to regulate EMT [94]. In addition, this lncRNA plays roles in the silencing of many anti-EMT regulators, such as the miRNAs miR-7, miR-34a, and miR-568 [95–97].

In MPM, overexpression of HOTAIR was found in the sarcomatoid subset of the Bueno NGS dataset (Table 1) [27], suggesting that HOTAIR is a lncRNA associated with the progression of MPM from the epithelioid to the sarcomatoid subtype. In silico analysis of an existing TCGA dataset shows that for those samples showing overexpression of this lncRNA the majority were biphasic (Figure 6A,B), and further analysis reveals that higher expression of HOTAIR in mesothelioma is associated with an poorer overall survival (Figure 6C).

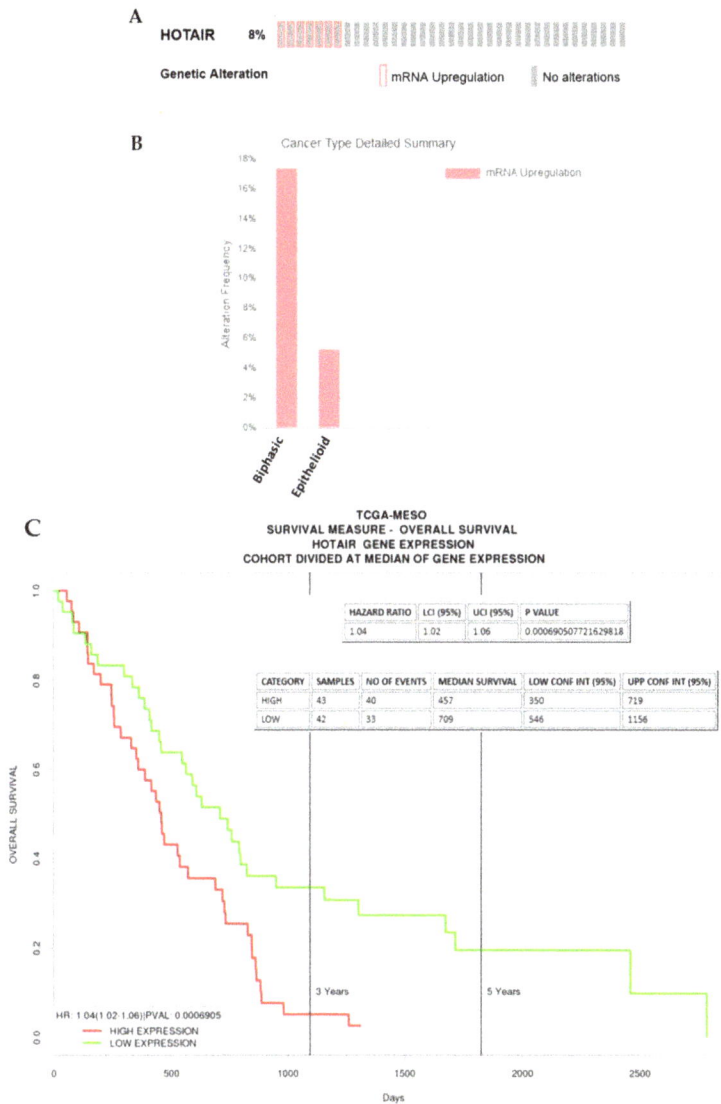

Figure 6. An examination of HOTAIR expression/alterations in the TCGA dataset. (**A**) In silico examination using cBioPortal reveals that 8% of samples had overexpression of HOTAIR RNA; (**B**) when stratified, the majority of these samples were associated with the Biphasic subtype; (**C**) when examined using ProGeneV2 (http://watson.compbio.iupui.edu/chirayu/proggene/database/?url=proggene), higher expression of HOTAIR was associated with a worse overall survival.

2.3.2. MALAT1

MALAT-1 (metastasis-associated lung adenocarcinoma transcript 1 also called NEAT2 or nuclear enriched abundant transcript 2) was first identified in NSCLC as a predictive marker associated with metastatic disease and shorter survival in early stage lung adenocarcinoma [98]. Since its initial discovery, MALAT-1 has been shown to be overexpressed and linked to the promotion of EMT in many cancers [21,99,100]. However, there are conflicting results which suggest that this lncRNA can either promote or inhibit EMT [77,101,102]. This may be in part because MALAT-1 can regulate EMT and other processes in various ways. For example, MALAT-1 can act as a competing endogenous RNA (ceRNA) for various miRNAs including miR-1, miR-200c, miRNA-204, and miR-205 resulting in the subsequent promotion of EMT [103–106]. Another mechanism by which MALAT-1 induces EMT is via the recruitment of the PRC2 components Suz12 and EZH2 to regulate E-Cadherin [105,107] and β-catenin [108,109].

MALAT-1 has been shown to activate EMT through either MAPK/ERK or PI3K/Akt signalling. MALAT-1 knockdown significantly reduced MAPK/ERK signalling in gallbladder cancer cells [110], and in glioma, MALAT-1 acts as a tumour suppressor by attenuating ERK/MAPK mediated signalling [111]. In osteosarcoma cells, downregulation of MALAT-1 inhibits PI3K/Akt signalling [112], whereas in breast and ovarian cancer cells, knockdown of MALAT-1 knockdown results in increased PI3K/Akt signalling and induction of EMT [102,113]. In this regard, in amodel of silica induced pulmonary fibrosis, MALAT-1 acts as a ceRNA for miR-503, one of whose targets is PI3K p85. By "sponging" this miRNA, MALAT-1 allows stimulation of EMT through a MALAT-1-miR-503-PI3K/Akt/mTOR/Snail pathway [114].

MALAT-1 is induced by TGF-β and plays a critical role during the promotion of EMT by TGF-β in bladder cancer cells [107]. TGF-β often elicits its effect through the Wnt signalling pathway, and significant evidence now suggests that lncRNAs play a major role in this process [115]. For example, MALAT-1 induces EMT in various cancers via the Wnt/β-catenin signalling pathway [116–118], while loss of WIF1 enhances the migratory potential of glioblastoma cells through WNT5A activation mediated by MALAT1 [119]. Intriguingly, MALAT1 expression was found to be overexpressed in the sarcomatoid subset of the Bueno NGS dataset (Table 1) [27]. In silico analysis of an existing TCGA dataset also shows that for MPM samples with overexpression of this lncRNA, the majority were epithelioid with some in the biphasic category (Figure 7A,B).

In renal cell carcinoma, a link between MALAT-1 and c-MYC, a downstream effector of Wnt/β-Catenin signalling, was found to be an element in the regulation of β-catenin and transcription factor c-Myc [116]; other lncRNAs have now been shown to play additional roles in regulating EMT via either c-Myc or n-Myc.

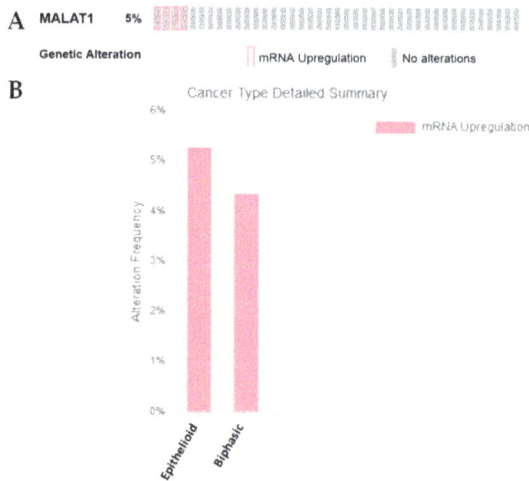

Figure 7. An examination of MALAT1 expression/alterations in the TCGA dataset. (**A**) In silico examination using cBioPortal reveals that 5% of samples had overexpression of MALAT1 RNA; (**B**) when stratified by histotype, the majority of these samples were associated with the epithelioid subtype.

2.3.3. MYCNOS and N-MYC

N-Myc (MYCN) belongs to the MYC family and was originally identified as being amplified in 20–30% of neuroblastoma tumours, but it is now well established that dysregulation of this transcription factor is common in many non-neuronal tumours [120]. N-Myc has also been shown to play roles in driving EMT in cancer [121]. In this regard, an lncRNA called MYCNOS has been shown to regulate the expression of N-Myc [122–124].

While a role for this lncRNA has not yet been identified in MPM, MYCNOS is upregulated in a proportion of MPM (5%—Figure 8A), and is mostly upregulated in the biphasic subset—Figure 8B. N-Myc also shows overexpression in a subset of MPM samples, but the majority of the samples do not fall into a defined histotype (Figure 8C). In these samples only two patients have co-overexpression of both MYCNOS and MYCN.

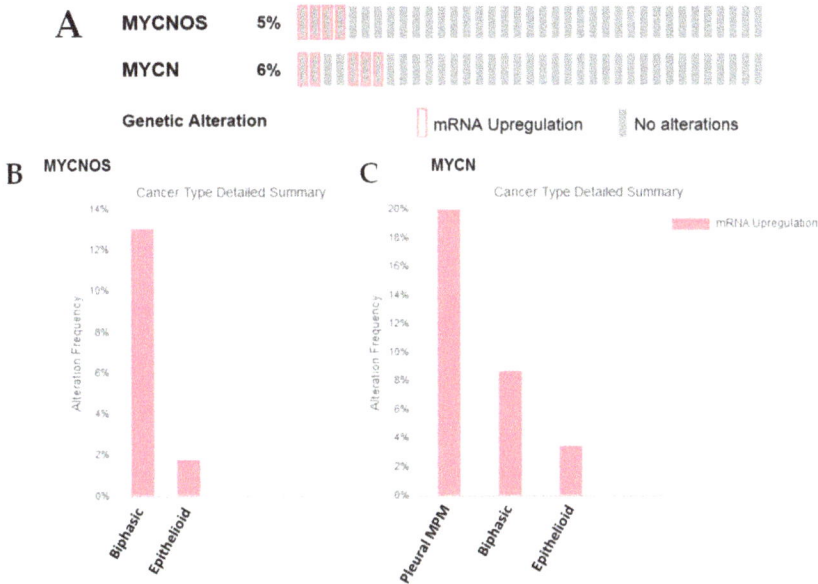

Figure 8. An examination of MYCNOS and N-Myc expression/alterations in the TCGA dataset. (**A**) In silico examination using cBioPortal reveals that 5% of samples had overexpression of MYCNOS RNA, while 6% had overexpression of N-Myc, (**B**) when stratified by histotype, the majority of samples with elevated MYCNOS were found in the biphasic subset, (**C**) N-Myc stratification does not fall into any defined histotype.

2.3.4. H19

H19 is an imprinted lncRNA, and has long been identified as an aberrantly expressed non-coding RNA in a great number of cancers, and has been shown to play multi-faceted roles during the tumourigenic process [125]; and is considered to be a critical element in EMT [126]. Indeed, overexpression of this lncRNA is associated with the activation of EMT in numerous cancers, including pancreatic cancer, CRC, nasopharyngeal carcinoma, bladder cancer, gallbladder cancer, and oesophageal cancer4 [21], where it has been shown to silence E-cadherin through recruitment of EZH2 to its promoter, or functions as a ceRNA for several pro-EMT miRNAs [21].

Upregulation of this lncRNA is found in the differential analysis between the epithelioid versus sarcomatoid clusters in the analysis by Bueno et al. (Table 1) [27]. In silico analysis of the TCGA dataset suggests that a small number of samples have higher expression of H19 (Figure 9A), which are distributed between the epithelioid (*n* = 1) and biphasic (*n* = 2) (Figure 9B). However, higher median expression of H19 is associated with a worse overall survival in this dataset (Figure 9C).

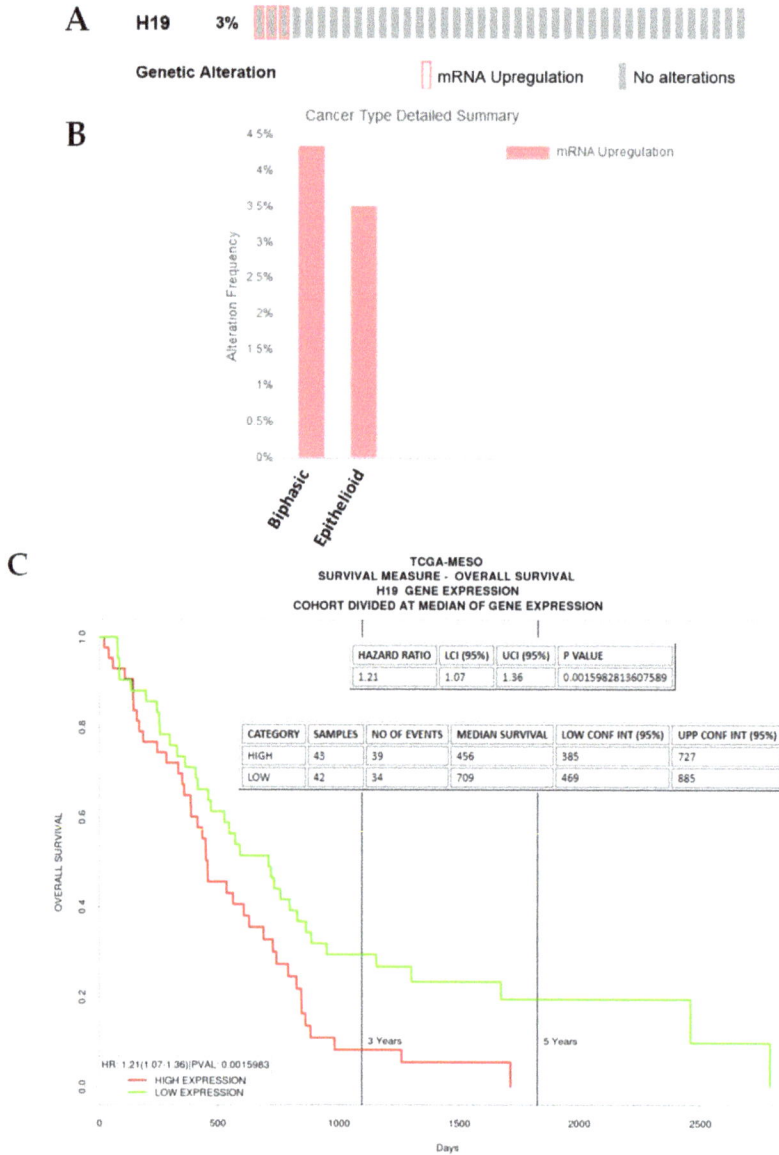

Figure 9. An examination of H19 in MPM. (**A**) H19 lncRNA is altered/overexpressed in a small proportion of MPM patients, as assessed using cBioPortal; (**B**) when separated by histology these samples fall into either the biphasic or epithelioid subsets; (**C**) when overall survival is assessed in this dataset, high median expression is associated with a significantly worse OS (*p* = 0.0016).

2.3.5. HULC

The lncRNA Highly Upregulated in Liver Cancer (HULC) was originally identified as one of the most upregulated genes in hepatocellular carcinoma (HCC) [127]; this lncRNA has now been shown to be aberrantly upregulated in several cancers [128]. Some evidence has also been reported suggesting that HULC can also act to inhibit c-Myc expression and PI3K/Akt signalling [129,130], and HULC has also been shown to cooperate with MALAT1 to promote liver cancer stem cell growth/aggressiveness [131]. Moreover, HULC has been shown to affect transcription through interaction with EZH2 [132].

A role for HULC in the regulation of EMT has been observed in HCC where it functions as a ceRNA for miRNAs (miR-122, miR-200a-3p, miR-372, and miR-488) [29,30,133,134] to mediate EMT via upregulation of Snail [135], ZEB1 [29], or ADAM9 [30], and this lncRNA has also been reported to induce EMT in gastric cancer [28].

A functional role for this lncRNA in MPM has not yet been identified. However, it was observed to be significantly downregulated in the sarcomatoid compared to the epithelioid subgroup (Table 1) [27]. cBioPortal analysis of the current TCGA mesothelioma dataset finds that 7% of samples have either amplifications or deletions in HULC, or overexpress this lncRNA (Figure 10A). When separated according to histology, the majority of alterations observed were found to be of the biphasic subtype (Figure 10B).

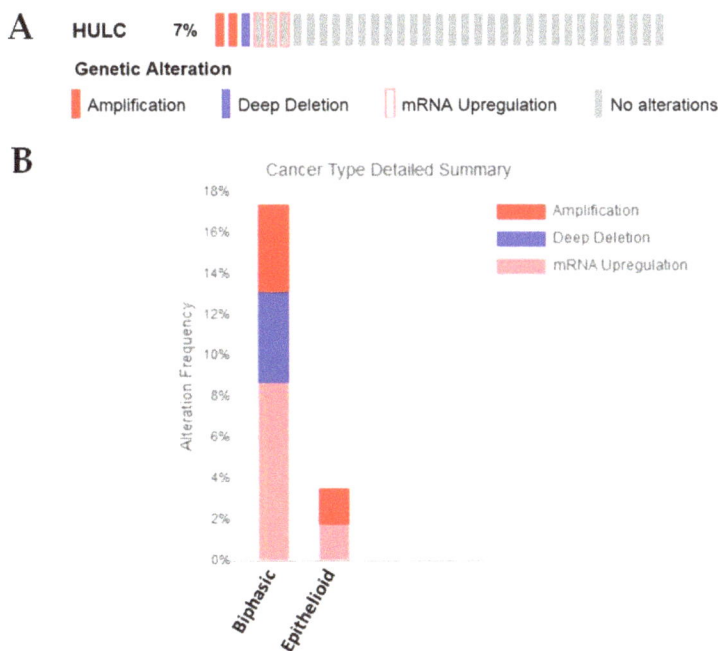

Figure 10. An examination of HULC in MPM. (**A**) HULC is either amplified/deleted or overexpressed in a small proportion (7%) of MPM patients, as assessed using cBioPortal, (**B**) when separated by histology the majority of these samples fall into the biphasic subgroup.

2.3.6. CASC2

In a study of NSCLC, expression of this lncRNA in the adenocarcinoma subtype was associated with inhibition of EMT through regulation of SOX4 [37]. A similar role for this lncRNA in regulating EMT in HCC has been identified, where this lncRNA has been shown to act as a ceRNA for miR-367 via a CASC2/miR-367/FBXW7 axis [136]. Furthermore, CASC2 has been shown to inhibit HCC by acting as a ceRNA for miR-362-5p, which resulted in the inhibition of the Nuclear Factor Kappa Beta (NF-κB) pathway [137].

A functional role for this lncRNA in MPM has not yet been defined, but decreased expression of this lncRNA is significantly associated with the sarcomatoid subtype in the Bueno NGS samples [27] (Table 1). Moreover, analysis of the TCGA dataset in cBioPortal reveals that those samples showing high expression of this lncRNA are associated with more with epithelial and biphasic subtypes, with the majority of the overexpression being observed in the epithelioid subset, while amplifications/deletions of this lncRNA were observed in biphasic samples (Figure 11A,B). When expression of this lncRNA was examined for Overall Survival (OS) benefit using ProGeneV2, high median expression was associated with better overall survival (Figure 11C).

As CASC2 is downregulated in human HCC samples, it may therefore be of interest to examine the levels of this lncRNA in MPM to see if loss of CASC is associated with a more aggressive histological phenotype as observed in Table 1.

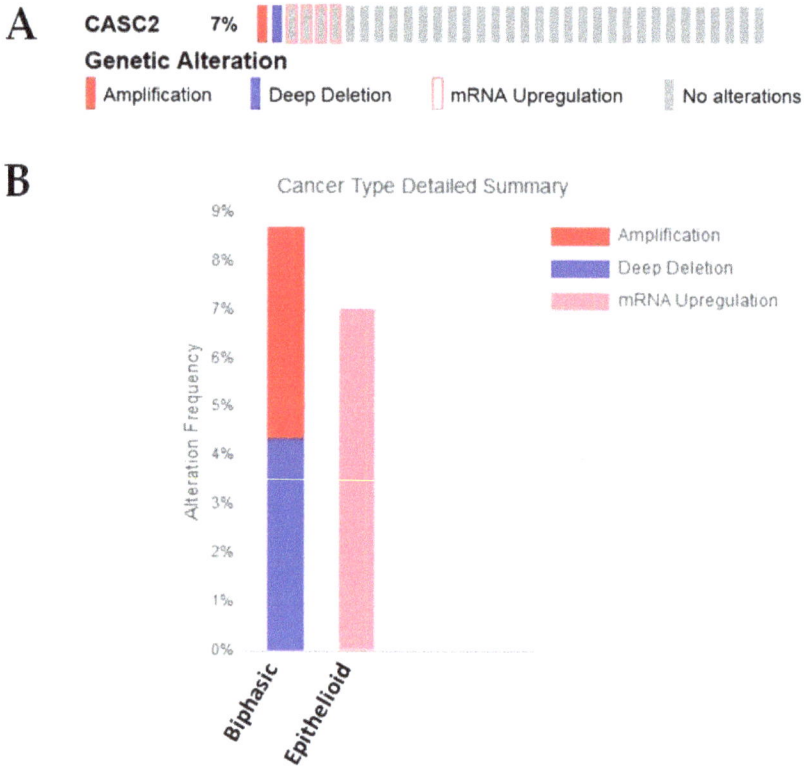

Figure 11. *Cont.*

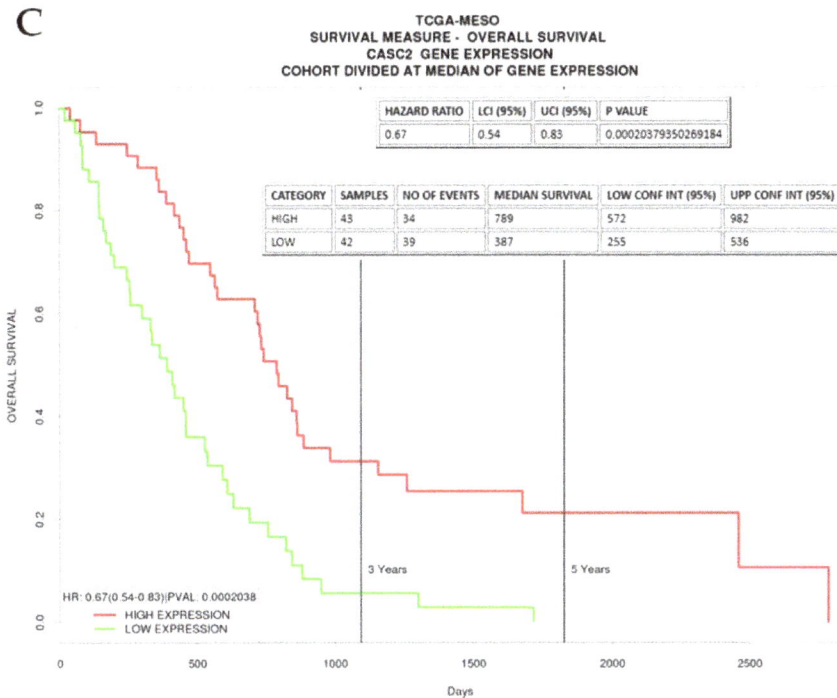

Figure 11. An examination of CASC alterations and expression in MPM. (**A**) CASC2 mRNA is altered/overexpressed in a small proportion of MPM patients as assessed using cBioPortal, (**B**) when separated by histology these samples fall into either the biphasic or epithelioid subsets, (**C**) when overall survival is assessed in this dataset, high median expression is associated with a significantly better OS (p = 0.000203).

2.3.7. ZFAS1

ZFAS1 is a lncRNA transcribed antisense to the ZNFX1 protein-coding gene, first identified as an lncRNA involved in mammary development and subsequently found to have altered expression in breast cancer [138]. Since this initial finding, ZFAS1 has been shown to be pro-tumourigenic and promote EMT in a number of other cancers, including colon cancer, gastric carcinoma, and glioma [33–36,139–148].

The role of this lncRNA has not yet been identified in MPM, but this lncRNA was found to be significantly altered between epithelioid versus sarcomatoid samples (Table 1) [27]. In the TCGA dataset, ZFAS1 shows overexpression in 5% of the samples; this was associated in samples with epithelioid or biphasic histologies (Figure 12A,B).

3. Materials and Methods

3.1. Primary Tumor Samples

Surgical specimens were obtained as discarded tumour samples from patients who had undergone an extended pleuropneumonectomy at Glenfield Hospital, Leicester, UK. Benign specimens were acquired from patients never diagnosed with MPM. Informed consent was obtained from each patient, and the study was conducted after formal approval from the relevant Hospital Ethics Committee (Leicestershire Research Ethics Committee (REC) references 6742 and 6948). Samples consisted of 5 benign lesions and 17 MPM samples (epithelioid: $n = 7$; sarcomatoid: $n = 4$; biphasic: $n = 6$), details of which are provided in Table 2.

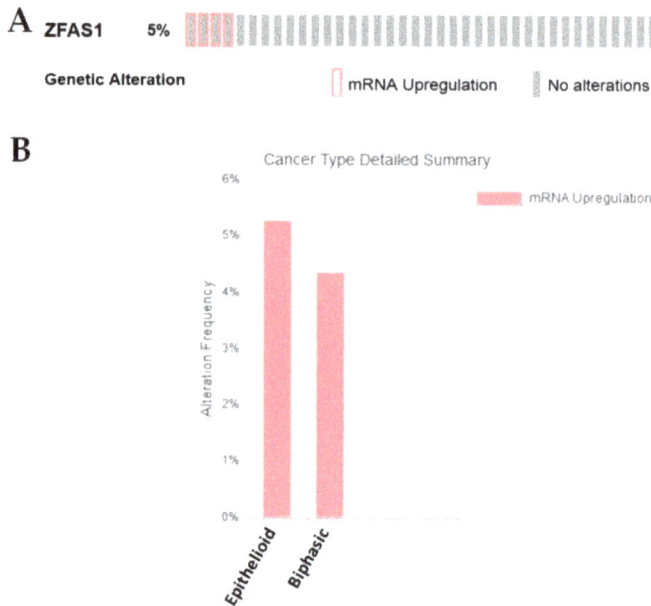

Figure 12. ZFAS1 is altered in a subset of MPM. (**A**) ZFAS1 is overexpressed in a small proportion (5%) of MPM patients as assessed using cBioPortal; (**B**) when separated by histology the majority of these samples fall into epithelioid or biphasic subgroups.

Table 2. Details of pleura/mesothelioma samples used in this study.

Sample	Pathology (Benign, Epithelial, Biphasic, Sarcomatoid)	Age	Gender
JE29	Benign—pleural plaque	55	Male
JE30	Benign—pleural plaque	55	Male
JE32	Benign—pneumothorax	30	Male
JE41	Benign—empyema	68	Male
JE48	Benign—pleural plaque	55	Male
JE31	Epithelioid	62	Male
JE139	Epithelioid	73	Male
JE149	Epithelioid	66	Male
JE155	Epithelioid	56	Female
JE157	Epithelioid	52	Male
JE162	Epithelioid	56	Male
JE173	Epithelioid	54	Male
JE86	Biphasic	54	Male
JE89	Biphasic	54	Female
JE136	Biphasic	41	Male
JE150	Biphasic	58	Male
JE151	Biphasic	N/A	Male
JE160	Biphasic	60	Female
JE106	Sarcomatoid	74	Male
JE125	Sarcomatoid	64	Male
JE133	Sarcomatoid	59	Male
JE145	Sarcomatoid (desmoplastic)	64	Male

N/A—not available.

3.2. Ethics Statement

Investigations were conducted in accordance with the relevant ethical standards, the Declaration of Helsinki, national, and international guidelines, and were approved by the relevant institutional review board (041018/8804, 13 October 2004, St James's Hospital/The Adelaide and Meath incorporating the National Childrens Hospital (SJH/AMNCH) REC).

Ethics Approval and Consent to Participate

All subjects gave their informed consent for inclusion before they participated in the study. The study was conducted in accordance with the Declaration of Helsinki.

Fresh Frozen Samples: The study was conducted after formal approval from the relevant Hospital Ethics Committee (Leicestershire REC references 6742 and 6948).

3.3. RNA Isolation and RT-PCR Amplification

Total RNA was extracted from fresh frozen patient samples using TRI reagent® (Cincinnati, OH, USA) according to manufacturer's instructions. Prior to first strand cDNA synthesis, 200 ng of total RNA was pre-treated by digestion with amplification grade DNase (Sigma-Aldrich, St. Louis, MO, USA) according to the manufacturer's instructions. cDNA was then generated using an all-in-one cDNA Synthesis Supermix (Bimake, Houston, TX, USA) according to the manufacturer's instructions. Patient samples were examined for the expression of various lncRNAs and 18S rRNA at the end point of PCR, using primers and annealing temperatures as outlined in Table 3. Each analysis was carried out once.

Table 3. Primers and associated annealing temperatures.

Gene/lncRNA	Primer Sequence	Temp	Source
EGFR-AS1	F: 5′-CTTTGCGATCTGCACACACC-3′ R: 5′-GAAGCCTACGTGATGGCCAG-3′	62	This study
PCAT6	F: 5′-CCCTAGATACACCCGCCTGGT-3′ R: 5′-ACATTCCAGGGCACCGAGAG-3′	64	This study
ZEB2-AS1	F: 5′-GAGAGAGACGAGAGACCCTGAA-3′ R: 5′-AAATTCATCATGCACACACCC-3′	60	This Study
KDM5B (JARID1B)	F: 5′-GCTACCCCCTCCAGCTACTCAGA-3′ R: 5′-TCCTCCTCGACTTCCTCCTCATC-3′	62	This study
18S rRNA	F: 5′-GATGGGCGGCGGAAAATAG-3′ R: 5′-GCGTGGATTCTGCATAATGGT-3′	60	[149]

PCR cycling conditions were 1 min at 95 °C, 1 min at the appropriate annealing temperature as per Table 2, 1 min at 72 °C for 35 cycles, with a final extension of 72 °C for 10 min. RT-PCR products for each experimental gene and appropriate housekeeping genes (18S rRNA) were run on 2% agarose gels. Following image capture, product quantification was performed using TINA 2.09c (Raytest, Isotopenmeßgeräte GmbH, Straubenhardt, Germany) densitometry software. The mRNA expression was normalised to loading controls, and was expressed as a ratio of target mRNA expression: loading control expression.

3.4. Statistical Analysis

All data are expressed as mean ± SEM unless stated otherwise. Statistical analysis was performed with Prism 5.01 (GraphPad, La Jolla, CA, USA) using either *t*-tests or one-way analysis of variance (ANOVA) where groups in the experiment were three or more. Following ANOVA, post-test analyses utilised the Dunnett's Multiple comparison test.

3.5. In Silico Analysis

In silico analysis was conducted on three additional mesothelioma datasets as follows:

(a) The dataset previously published by Gordon et al. [52], which was interrogated using Oncomine, (b) and an existing TCGA data set (TCGA Mesothelioma; raw data at the NCI; the dataset consists of *n* = 87 samples: epithelioid (57), biphasic (23), sarcomatoid (2), other mesothelioma (5).

Data-mining of the available mesothelioma datasets was conducted using Oncomine [150,151] cBioportal [152–154], or PROGgeneV2 [155,156], using their respective default settings.

4. Conclusions

Despite intensive efforts, the range of treatment options available to clinicians for the treatment of patients with MPM remains low. The current mainstay of treatment is a combination of cisplatin and pemetrexed (or alternatively raltitrexed), and only approximately 40% of patients will respond to this regimen. At present, no second-line strategy has been approved to date, except rechallenging the patients with long-lasting tumour control after first-line treatment with pemetrexed-based chemotherapy [157].

Across the histological subtypes of MPM, patients who have an epithelioid histology generally have the best OS. Because MPM has a partial fibroblastic phenotype in the context of EMT, it has been postulated that this may, in part, explain the aggressiveness of this cancer by conferring both its high invasiveness and chemoresistance [13]; in particular, with regard to the epithelioid rather than sarcomatoid histotype of MPM [13]. In this regard, the epithelioid and sarcomatoid histologic variants of MPM can be considered as E- and M-parts of the EMT axis, with the biphasic histotype considered an intermediate [14].

In this report, we have shown that many lncRNAs associated with EMT have predominantly altered expression, associated for the most part with the sarcomatoid histologies. Therefore, a greater understanding of the molecular mechanisms governing EMT remains imperative for the development of novel therapies that can slow or prevent metastasis, the current great unmet need of cancer therapy.

In a previous review, we discussed the role of many lncRNAs as elements associated with resistance mechanisms to cisplatin [21], and many of the lncRNAs discussed in this article such as HOTAIR or MALAT1 have well defined roles in cisplatin resistance [21].

If these lncRNAs are both associated with driving MPM from the epithelioid subtype to the more aggressive forms with poorer OS (biphasic and sarcomatoid) with resistance to cisplatin, then potentially targeting these may have therapeutic applicability. Methodologies to restore ncRNAs in MPM, such as the recently completed Phase I MesomiR 1 clinical trial [42], suggest that this technology could also be utilised or adapted to specifically target lncRNAs in MPM.

In conclusion, a large body of evidence suggests that lncRNAs associated with EMT are dysregulated in MPM, and their alteration may be associated with the more aggressive histological subtypes. More work remains to delineate how we may be able to take advantage of this clinically.

Author Contributions: A.S.S., R.H. and S.G.G. designed the experiments; A.S.S., R.H., and S.G.G. performed the experiments; S.G.G. wrote the paper. All authors have read and approved the final manuscript for publication.

Funding: The research in this publication was in part supported by funding for consumables from the Masters in Translational Oncology program (TCD) for Anand S. Singh and Richard Heery.

Acknowledgments: The authors wish to acknowledge all those whose research has been used in this manuscript and apologise to those whose work we may have missed.

Conflicts of Interest: The authors declare no conflicts of interest.

Int. J. Mol. Sci. **2018**, *19*, 1297

Abbreviations

ceRNA	competitive endogenous RNA
EMT	Epithelial Mesenchymal Transition
lncRNA	long non-coding RNA
NSCLC	Non-Small Cell Lung Cancer
miRNA	microRNA
MPM	Malignant Pleural Mesothelioma
NGS	Next Generation Sequencing
OS	Overall Survival
TCGA	The Cancer Genome Atlas

References

1. Wagner, J.C.; Sleggs, C.A.; Marchand, P. Diffuse pleural mesothelioma and asbestos exposure in the North Western Cape Province. *Br. J. Ind. Med.* **1960**, *17*, 260–271. [CrossRef] [PubMed]
2. Remon, J.; Reguart, N.; Corral, J.; Lianes, P. Malignant pleural mesothelioma: New hope in the horizon with novel therapeutic strategies. *Cancer Treat. Rev.* **2015**, *41*, 27–34. [CrossRef] [PubMed]
3. Odgerel, C.O.; Takahashi, K.; Sorahan, T.; Driscoll, T.; Fitzmaurice, C.; Yoko, O.M.; Sawanyawisuth, K.; Furuya, S.; Tanaka, F.; Horie, S.; et al. Estimation of the global burden of mesothelioma deaths from incomplete national mortality data. *Occup. Environ. Med.* **2017**, *74*, 851–858. [CrossRef] [PubMed]
4. Alley, E.W.; Lopez, J.; Santoro, A.; Morosky, A.; Saraf, S.; Piperdi, B.; van Brummelen, E. Clinical safety and activity of pembrolizumab in patients with malignant pleural mesothelioma (KEYNOTE-028): Preliminary results from a non-randomised, open-label, phase 1b trial. *Lancet Oncol.* **2017**, *18*, 623–630. [CrossRef]
5. Quispel-Janssen, J.; Zago, G.; Schouten, R.; Buikhuisen, W.; Monkhorst, K.; Thunissen, E.; Baas, P. OA13.01 A Phase II Study of Nivolumab in Malignant Pleural Mesothelioma (NivoMes): With Translational Research (TR) Biopies. *J. Thorac. Oncol.* **2016**, *12*, S292–S293. [CrossRef]
6. Baas, P.; Fennell, D.; Kerr, K.M.; Van Schil, P.E.; Haas, R.L.; Peters, S. Malignant pleural mesothelioma: ESMO Clinical Practice Guidelines for diagnosis, treatment and follow-up. *Ann. Oncol.* **2015**, *26* (Suppl. 5), v31–v39. [CrossRef] [PubMed]
7. Kindler, H.L.; Ismaila, N.; Armato, S.G., 3rd; Bueno, R.; Hesdorffer, M.; Jahan, T.; Jones, C.M.; Miettinen, M.; Pass, H.; Rimner, A.; et al. Treatment of Malignant Pleural Mesothelioma: American Society of Clinical Oncology Clinical Practice Guideline. *J. Clin. Oncol.* **2018**. [CrossRef] [PubMed]
8. Vogelzang, N.J.; Rusthoven, J.J.; Symanowski, J.; Denham, C.; Kaukel, E.; Ruffie, P.; Gatzemeier, U.; Boyer, M.; Emri, S.; Manegold, C.; et al. Phase III study of pemetrexed in combination with cisplatin versus cisplatin alone in patients with malignant pleural mesothelioma. *J. Clin. Oncol.* **2003**, *21*, 2636–2644. [CrossRef] [PubMed]
9. Brabletz, T.; Kalluri, R.; Nieto, M.A.; Weinberg, R.A. EMT in cancer. *Nat. Rev. Cancer* **2018**, *18*, 128–134. [CrossRef] [PubMed]
10. Husain, A.N.; Colby, T.V.; Ordonez, N.G.; Allen, T.C.; Attanoos, R.L.; Beasley, M.B.; Butnor, K.J.; Chirieac, L.R.; Churg, A.M.; Dacic, S.; et al. Guidelines for Pathologic Diagnosis of Malignant Mesothelioma 2017 Update of the Consensus Statement From the International Mesothelioma Interest Group. *Arch. Pathol. Lab. Med.* **2018**, *142*, 89–108. [CrossRef] [PubMed]
11. Sun, H.H.; Vaynblat, A.; Pass, H.I. Diagnosis and prognosis-review of biomarkers for mesothelioma. *Ann. Transl. Med.* **2017**, *5*, 244. [CrossRef] [PubMed]
12. Panou, V.; Vyberg, M.; Weinreich, U.M.; Meristoudis, C.; Falkmer, U.G.; Roe, O.D. The established and future biomarkers of malignant pleural mesothelioma. *Cancer Treat. Rev.* **2015**, *41*, 486–495. [CrossRef] [PubMed]
13. Schramm, A.; Opitz, I.; Thies, S.; Seifert, B.; Moch, H.; Weder, W.; Soltermann, A. Prognostic significance of epithelial-mesenchymal transition in malignant pleural mesothelioma. *Eur. J. Cardiothorac. Surg.* **2010**, *37*, 566–572. [CrossRef] [PubMed]
14. Thies, S.; Friess, M.; Frischknecht, L.; Korol, D.; Felley-Bosco, E.; Stahel, R.; Vrugt, B.; Weder, W.; Opitz, I.; Soltermann, A. Expression of the Stem Cell Factor Nestin in Malignant Pleural Mesothelioma Is Associated with Poor Prognosis. *PLoS ONE* **2015**, *10*, e0139312. [CrossRef] [PubMed]

15. De Reynies, A.; Jaurand, M.C.; Renier, A.; Couchy, G.; Hysi, I.; Elarouci, N.; Galateau-Salle, F.; Copin, M.C.; Hofman, P.; Cazes, A.; et al. Molecular classification of malignant pleural mesothelioma: Identification of a poor prognosis subgroup linked to the epithelial-to-mesenchymal transition. *Clin. Cancer Res.* **2014**, *20*, 1323–1334. [CrossRef] [PubMed]
16. Legras, A.; Pecuchet, N.; Imbeaud, S.; Pallier, K.; Didelot, A.; Roussel, H.; Gibault, L.; Fabre, E.; Le Pimpec-Barthes, F.; Laurent-Puig, P.; et al. Epithelial-to-Mesenchymal Transition and MicroRNAs in Lung Cancer. *Cancers* **2017**, *9*, 101. [CrossRef] [PubMed]
17. Jafri, M.A.; Al-Qahtani, M.H.; Shay, J.W. Role of miRNAs in human cancer metastasis: Implications for therapeutic intervention. *Semin. Cancer Biol.* **2017**, *44*, 117–131. [CrossRef] [PubMed]
18. Quinn, L.; Finn, S.P.; Cuffe, S.; Gray, S.G. Non-coding RNA repertoires in malignant pleural mesothelioma. *Lung Cancer* **2015**, *90*, 417–426. [CrossRef] [PubMed]
19. Martinez-Rivera, V.; Negrete-Garcia, M.C.; Avila-Moreno, F.; Ortiz-Quintero, B. Secreted and Tissue miRNAs as Diagnosis Biomarkers of Malignant Pleural Mesothelioma. *Int. J. Mol. Sci.* **2018**, *19*, 595. [CrossRef] [PubMed]
20. Cao, M.X.; Jiang, Y.P.; Tang, Y.L.; Liang, X.H. The crosstalk between lncRNA and microRNA in cancer metastasis: Orchestrating the epithelial-mesenchymal plasticity. *Oncotarget* **2017**, *8*, 12472–12483. [CrossRef] [PubMed]
21. Heery, R.; Finn, S.P.; Cuffe, S.; Gray, S.G. Long Non-Coding RNAs: Key Regulators of Epithelial-Mesenchymal Transition, Tumour Drug Resistance and Cancer Stem Cells. *Cancers* **2017**, *9*, 38. [CrossRef] [PubMed]
22. Wang, L.; Yang, F.; Jia, L.T.; Yang, A.G. Missing Links in Epithelial-Mesenchymal Transition: Long Non-Coding RNAs Enter the Arena. *Cell. Physiol. Biochem.* **2017**, *44*, 1665–1680. [CrossRef] [PubMed]
23. Oehl, K.; Kresoja-Rakic, J.; Opitz, I.; Vrugt, B.; Weder, W.; Stahel, R.; Wild, P.; Felley-Bosco, E. Live-Cell Mesothelioma Biobank to Explore Mechanisms of Tumor Progression. *Front. Oncol.* **2018**, *8*, 40. [CrossRef] [PubMed]
24. Riquelme, E.; Suraokar, M.B.; Rodriguez, J.; Mino, B.; Lin, H.Y.; Rice, D.C.; Tsao, A.; Wistuba, I.I. Frequent coamplification and cooperation between C-MYC and PVT1 oncogenes promote malignant pleural mesothelioma. *J. Thorac. Oncol.* **2014**, *9*, 998–1007. [CrossRef] [PubMed]
25. Renganathan, A.; Kresoja-Rakic, J.; Echeverry, N.; Ziltener, G.; Vrugt, B.; Opitz, I.; Stahel, R.A.; Felley-Bosco, E. GAS5 long non-coding RNA in malignant pleural mesothelioma. *Mol. Cancer* **2014**, *13*, 119. [CrossRef] [PubMed]
26. Wright, C.M.; Kirschner, M.B.; Cheng, Y.Y.; O'Byrne, K.J.; Gray, S.G.; Schelch, K.; Hoda, M.A.; Klebe, S.; McCaughan, B.; van Zandwijk, N.; et al. Long non coding RNAs (lncRNAs) are dysregulated in Malignant Pleural Mesothelioma (MPM). *PLoS ONE* **2013**, *8*, e70940. [CrossRef] [PubMed]
27. Bueno, R.; Stawiski, E.W.; Goldstein, L.D.; Durinck, S.; De Rienzo, A.; Modrusan, Z.; Gnad, F.; Nguyen, T.T.; Jaiswal, B.S.; Chirieac, L.R.; et al. Comprehensive genomic analysis of malignant pleural mesothelioma identifies recurrent mutations, gene fusions and splicing alterations. *Nat. Genet.* **2016**, *48*, 407–416. [CrossRef] [PubMed]
28. Zhao, Y.; Guo, Q.; Chen, J.; Hu, J.; Wang, S.; Sun, Y. Role of long non-coding RNA HULC in cell proliferation, apoptosis and tumor metastasis of gastric cancer: A clinical and in vitro investigation. *Oncol. Rep.* **2014**, *31*, 358–364. [CrossRef] [PubMed]
29. Li, S.P.; Xu, H.X.; Yu, Y.; He, J.D.; Wang, Z.; Xu, Y.J.; Wang, C.Y.; Zhang, H.M.; Zhang, R.X.; Zhang, J.J.; et al. LncRNA HULC enhances epithelial-mesenchymal transition to promote tumorigenesis and metastasis of hepatocellular carcinoma via the miR-200a-3p/ZEB1 signaling pathway. *Oncotarget* **2016**, *7*, 42431–42446. [CrossRef] [PubMed]
30. Hu, D.; Shen, D.; Zhang, M.; Jiang, N.; Sun, F.; Yuan, S.; Wan, K. MiR-488 suppresses cell proliferation and invasion by targeting ADAM9 and lncRNA HULC in hepatocellular carcinoma. *Am. J. Cancer Res.* **2017**, *7*, 2070–2080. [PubMed]
31. Zhang, Q.; Li, X.; Chen, Z. LncRNA H19 promotes epithelial-mesenchymal transition (EMT) by targeting miR-484 in human lung cancer cells. *J. Cell. Biochem.* **2017**. [CrossRef] [PubMed]
32. Xu, W.; He, L.; Li, Y.; Tan, Y.; Zhang, F.; Xu, H. Silencing of lncRNA ZFAS1 inhibits malignancies by blocking Wnt/beta-catenin signaling in gastric cancer cells. *Biosci. Biotechnol. Biochem.* **2018**, *82*, 456–465. [CrossRef] [PubMed]

33. Zhou, H.; Wang, F.; Chen, H.; Tan, Q.; Qiu, S.; Chen, S.; Jing, W.; Yu, M.; Liang, C.; Ye, S.; et al. Increased expression of long-noncoding RNA ZFAS1 is associated with epithelial-mesenchymal transition of gastric cancer. *Aging* **2016**, *8*, 2023–2038. [CrossRef] [PubMed]

34. Fang, C.; Zan, J.; Yue, B.; Liu, C.; He, C.; Yan, D. Long non-coding ribonucleic acid zinc finger antisense 1 promotes the progression of colonic cancer by modulating ZEB1 expression. *J. Gastroenterol. Hepatol.* **2017**, *32*, 1204–1211. [CrossRef] [PubMed]

35. Guo, H.; Wu, L.; Zhao, P.; Feng, A. Overexpression of long non-coding RNA zinc finger antisense 1 in acute myeloid leukemia cell lines influences cell growth and apoptosis. *Exp. Ther. Med.* **2017**, *14*, 647–651. [CrossRef] [PubMed]

36. Pan, L.; Liang, W.; Fu, M.; Huang, Z.H.; Li, X.; Zhang, W.; Zhang, P.; Qian, H.; Jiang, P.C.; Xu, W.R.; et al. Exosomes-mediated transfer of long noncoding RNA ZFAS1 promotes gastric cancer progression. *J. Cancer Res. Clin. Oncol.* **2017**, *143*, 991–1004. [CrossRef] [PubMed]

37. Wang, D.; Gao, Z.M.; Han, L.G.; Xu, F.; Liu, K.; Shen, Y. Long noncoding RNA CASC2 inhibits metastasis and epithelial to mesenchymal transition of lung adenocarcinoma via suppressing SOX4. *Eur. Rev. Med. Pharmacol. Sci.* **2017**, *21*, 4584–4590. [PubMed]

38. Edwards, J.G.; Swinson, D.E.; Jones, J.L.; Waller, D.A.; O'Byrne, K.J. EGFR expression: Associations with outcome and clinicopathological variables in malignant pleural mesothelioma. *Lung Cancer* **2006**, *54*, 399–407. [CrossRef] [PubMed]

39. Garland, L.L.; Rankin, C.; Gandara, D.R.; Rivkin, S.E.; Scott, K.M.; Nagle, R.B.; Klein-Szanto, A.J.; Testa, J.R.; Altomare, D.A.; Borden, E.C. Phase II study of erlotinib in patients with malignant pleural mesothelioma: A Southwest Oncology Group Study. *J. Clin. Oncol.* **2007**, *25*, 2406–2413. [CrossRef] [PubMed]

40. Govindan, R.; Kratzke, R.A.; Herndon, J.E., 2nd; Niehans, G.A.; Vollmer, R.; Watson, D.; Green, M.R.; Kindler, H.L. Gefitinib in patients with malignant mesothelioma: A phase II study by the Cancer and Leukemia Group B. *Clin. Cancer Res.* **2005**, *11*, 2300–2304. [CrossRef] [PubMed]

41. Bononi, A.; Napolitano, A.; Pass, H.I.; Yang, H.; Carbone, M. Latest developments in our understanding of the pathogenesis of mesothelioma and the design of targeted therapies. *Expert Rev. Respir. Med.* **2015**, *9*, 633–654. [CrossRef] [PubMed]

42. Van Zandwijk, N.; Pavlakis, N.; Kao, S.C.; Linton, A.; Boyer, M.J.; Clarke, S.; Huynh, Y.; Chrzanowska, A.; Fulham, M.J.; Bailey, D.L.; et al. Safety and activity of microRNA-loaded minicells in patients with recurrent malignant pleural mesothelioma: A first-in-man, phase 1, open-label, dose-escalation study. *Lancet Oncol.* **2017**, *18*, 1386–1396. [CrossRef]

43. Agatsuma, N.; Yasuda, Y.; Ozasa, H. Malignant Pleural Mesothelioma Harboring Both G719C and S768I Mutations of EGFR Successfully Treated with Afatinib. *J. Thorac. Oncol.* **2017**, *12*, e141–e143. [CrossRef] [PubMed]

44. Qi, H.L.; Li, C.S.; Qian, C.W.; Xiao, Y.S.; Yuan, Y.F.; Liu, Q.Y.; Liu, Z.S. The long noncoding RNA, EGFR-AS1, a target of GHR, increases the expression of EGFR in hepatocellular carcinoma. *Tumour Biol.* **2016**, *37*, 1079–1089. [CrossRef] [PubMed]

45. Tan, D.S.W.; Chong, F.T.; Leong, H.S.; Toh, S.Y.; Lau, D.P.; Kwang, X.L.; Zhang, X.; Sundaram, G.M.; Tan, G.S.; Chang, M.M.; et al. Long noncoding RNA EGFR-AS1 mediates epidermal growth factor receptor addiction and modulates treatment response in squamous cell carcinoma. *Nat. Med.* **2017**, *23*, 1167–1175. [CrossRef] [PubMed]

46. Barr, S.; Thomson, S.; Buck, E.; Russo, S.; Petti, F.; Sujka-Kwok, I.; Eyzaguirre, A.; Rosenfeld-Franklin, M.; Gibson, N.W.; Miglarese, M.; et al. Bypassing cellular EGF receptor dependence through epithelial-to-mesenchymal-like transitions. *Clin. Exp. Metast.* **2008**, *25*, 685–693. [CrossRef] [PubMed]

47. Tang, B.; Qi, G.; Tang, F.; Yuan, S.; Wang, Z.; Liang, X.; Li, B.; Yu, S.; Liu, J.; Huang, Q.; et al. JARID1B promotes metastasis and epithelial-mesenchymal transition via PTEN/AKT signaling in hepatocellular carcinoma cells. *Oncotarget* **2015**, *6*, 12723–12739. [CrossRef] [PubMed]

48. Haley, J.A.; Haughney, E.; Ullman, E.; Bean, J.; Haley, J.D.; Fink, M.Y. Altered Transcriptional Control Networks with Trans-Differentiation of Isogenic Mutant-KRas NSCLC Models. *Front. Oncol.* **2014**, *4*, 344. [CrossRef] [PubMed]

49. Enkhbaatar, Z.; Terashima, M.; Oktyabri, D.; Tange, S.; Ishimura, A.; Yano, S.; Suzuki, T. KDM5B histone demethylase controls epithelial-mesenchymal transition of cancer cells by regulating the expression of the microRNA-200 family. *Cell Cycle* **2013**, *12*, 2100–2112. [CrossRef] [PubMed]

50. Wan, L.; Zhang, L.; Fan, K.; Cheng, Z.X.; Sun, Q.C.; Wang, J.J. Knockdown of Long Noncoding RNA PCAT6 Inhibits Proliferation and Invasion in Lung Cancer Cells. *Oncol. Res.* **2016**, *24*, 161–170. [CrossRef] [PubMed]

51. Wan, L.; Zhang, L.; Fan, K.; Wang, J.J. Diagnostic significance of circulating long noncoding RNA PCAT6 in patients with non-small cell lung cancer. *Onco Targets Ther.* **2017**, *10*, 5695–5702. [CrossRef] [PubMed]

52. Gordon, G.J.; Rockwell, G.N.; Jensen, R.V.; Rheinwald, J.G.; Glickman, J.N.; Aronson, J.P.; Pottorf, B.J.; Nitz, M.D.; Richards, W.G.; Sugarbaker, D.J.; et al. Identification of novel candidate oncogenes and tumor suppressors in malignant pleural mesothelioma using large-scale transcriptional profiling. *Am. J. Pathol.* **2005**, *166*, 1827–1840. [CrossRef]

53. Beltran, M.; Puig, I.; Pena, C.; Garcia, J.M.; Alvarez, A.B.; Pena, R.; Bonilla, F.; de Herreros, A.G. A natural antisense transcript regulates ZEB2/Sip1 gene expression during Snail1-induced epithelial-mesenchymal transition. *Genes Dev.* **2008**, *22*, 756–769. [CrossRef] [PubMed]

54. Zhuang, J.; Lu, Q.; Shen, B.; Huang, X.; Shen, L.; Zheng, X.; Huang, R.; Yan, J.; Guo, H. TGFbeta1 secreted by cancer-associated fibroblasts induces epithelial-mesenchymal transition of bladder cancer cells through lncRNA-ZEB2NAT. *Sci. Rep.* **2015**, *5*, 11924. [CrossRef] [PubMed]

55. Lan, T.; Chang, L.; Wu, L.; Yuan, Y. Downregulation of ZEB2-AS1 decreased tumor growth and metastasis in hepatocellular carcinoma. *Mol. Med. Rep.* **2016**, *14*, 4606–4612. [CrossRef] [PubMed]

56. Lopez-Rios, F.; Chuai, S.; Flores, R.; Shimizu, S.; Ohno, T.; Wakahara, K.; Illei, P.B.; Hussain, S.; Krug, L.; Zakowski, M.F.; et al. Global gene expression profiling of pleural mesotheliomas: Overexpression of aurora kinases and P16/CDKN2A deletion as prognostic factors and critical evaluation of microarray-based prognostic prediction. *Cancer Res.* **2006**, *66*, 2970–2979. [CrossRef] [PubMed]

57. Zhu, S.; Shuai, P.; Yang, C.; Zhang, Y.; Zhong, S.; Liu, X.; Chen, K.; Ran, Q.; Yang, H.; Zhou, Y. Prognostic value of long non-coding RNA PVT1 as a novel biomarker in various cancers: A meta-analysis. *Oncotarget* **2017**, *8*, 113174–113184. [CrossRef] [PubMed]

58. Wang, Y.; Zhou, J.; Wang, Z.; Wang, P.; Li, S. Upregulation of SOX2 activated LncRNA PVT1 expression promotes breast cancer cell growth and invasion. *Biochem. Biophys. Res. Commun.* **2017**, *493*, 429–436. [CrossRef] [PubMed]

59. Wu, B.Q.; Jiang, Y.; Zhu, F.; Sun, D.L.; He, X.Z. Long Noncoding RNA PVT1 Promotes EMT and Cell Proliferation and Migration through Downregulating p21 in Pancreatic Cancer Cells. *Technol. Cancer Res. Treat.* **2017**. [CrossRef] [PubMed]

60. Yang, T.; Zhou, H.; Liu, P.; Yan, L.; Yao, W.; Chen, K.; Zeng, J.; Li, H.; Hu, J.; Xu, H.; et al. lncRNA PVT1 and its splicing variant function as competing endogenous RNA to regulate clear cell renal cell carcinoma progression. *Oncotarget* **2017**, *8*, 85353–85367. [CrossRef] [PubMed]

61. Li, H.; Chen, S.; Liu, J.; Guo, X.; Xiang, X.; Dong, T.; Ran, P.; Li, Q.; Zhu, B.; Zhang, X.; et al. Long non-coding RNA PVT1–5 promotes cell proliferation by regulating miR-126/SLC7A5 axis in lung cancer. *Biochem. Biophys. Res. Commun.* **2018**, *495*, 2350–2355. [CrossRef] [PubMed]

62. Wang, C.; Han, C.; Zhang, Y.; Liu, F. LncRNA PVT1 regulate expression of HIF1alpha via functioning as ceRNA for miR199a5p in nonsmall cell lung cancer under hypoxia. *Mol. Med. Rep.* **2018**, *17*, 1105–1110. [PubMed]

63. Gao, Y.L.; Zhao, Z.S.; Zhang, M.Y.; Han, L.J.; Dong, Y.J.; Xu, B. Long Noncoding RNA PVT1 Facilitates Cervical Cancer Progression via Negative Regulating of miR-424. *Oncol. Res.* **2017**, *25*, 1391–1398. [CrossRef] [PubMed]

64. Conte, F.; Fiscon, G.; Chiara, M.; Colombo, T.; Farina, L.; Paci, P. Role of the long non-coding RNA PVT1 in the dysregulation of the ceRNA-ceRNA network in human breast cancer. *PLoS ONE* **2017**, *12*, e0171661. [CrossRef] [PubMed]

65. Yang, S.; Ning, Q.; Zhang, G.; Sun, H.; Wang, Z.; Li, Y. Construction of differential mRNA-lncRNA crosstalk networks based on ceRNA hypothesis uncover key roles of lncRNAs implicated in esophageal squamous cell carcinoma. *Oncotarget* **2016**, *7*, 85728–85740. [CrossRef] [PubMed]

66. Lan, T.; Yan, X.; Li, Z.; Xu, X.; Mao, Q.; Ma, W.; Hong, Z.; Chen, X.; Yuan, Y. Long non-coding RNA PVT1 serves as a competing endogenous RNA for miR-186–5p to promote the tumorigenesis and metastasis of hepatocellular carcinoma. *Tumour Biol.* **2017**, *39*. [CrossRef] [PubMed]

67. Chang, Z.; Cui, J.; Song, Y. Long noncoding RNA PVT1 promotes EMT via mediating microRNA-186 targeting of Twist1 in prostate cancer. *Gene* **2018**, *654*, 36–42. [CrossRef] [PubMed]

68. Shen, C.J.; Cheng, Y.M.; Wang, C.L. LncRNA PVT1 epigenetically silences miR-195 and modulates EMT and chemoresistance in cervical cancer cells. *J. Drug Target.* **2017**, *25*, 637–644. [CrossRef] [PubMed]

69. Zhang, S.; Zhang, G.; Liu, J. Long noncoding RNA PVT1 promotes cervical cancer progression through epigenetically silencing miR-200b. *APMIS* **2016**, *124*, 649–658. [CrossRef] [PubMed]

70. Wan, L.; Sun, M.; Liu, G.J.; Wei, C.C.; Zhang, E.B.; Kong, R.; Xu, T.P.; Huang, M.D.; Wang, Z.X. Long Noncoding RNA PVT1 Promotes Non-Small Cell Lung Cancer Cell Proliferation through Epigenetically Regulating LATS2 Expression. *Mol. Cancer Ther.* **2016**, *15*, 1082–1094. [CrossRef] [PubMed]

71. Kong, R.; Zhang, E.B.; Yin, D.D.; You, L.H.; Xu, T.P.; Chen, W.M.; Xia, R.; Wan, L.; Sun, M.; Wang, Z.X.; et al. Long noncoding RNA PVT1 indicates a poor prognosis of gastric cancer and promotes cell proliferation through epigenetically regulating p15 and p16. *Mol. Cancer* **2015**, *14*, 82. [CrossRef] [PubMed]

72. Zhou, Q.; Chen, J.; Feng, J.; Wang, J. Long noncoding RNA PVT1 modulates thyroid cancer cell proliferation by recruiting EZH2 and regulating thyroid-stimulating hormone receptor (TSHR). *Tumour Biol.* **2016**, *37*, 3105–3113. [CrossRef] [PubMed]

73. Gou, X.; Zhao, X.; Wang, Z. Long noncoding RNA PVT1 promotes hepatocellular carcinoma progression through regulating miR-214. *Cancer Biomark.* **2017**, *20*, 511–519. [CrossRef] [PubMed]

74. Yang, A.; Wang, H.; Yang, X. Long non-coding RNA PVT1 indicates a poor prognosis of glioma and promotes cell proliferation and invasion via target EZH2. *Biosci. Rep.* **2017**, *37*. [CrossRef] [PubMed]

75. Chen, L.; Ma, D.; Li, Y.; Li, X.; Zhao, L.; Zhang, J.; Song, Y. Effect of long non-coding RNA PVT1 on cell proliferation and migration in melanoma. *Int. J. Mol. Med.* **2018**, *41*, 1275–1282. [CrossRef] [PubMed]

76. Tseng, Y.Y.; Moriarity, B.S.; Gong, W.; Akiyama, R.; Tiwari, A.; Kawakami, H.; Ronning, P.; Reuland, B.; Guenther, K.; Beadnell, T.C.; et al. PVT1 dependence in cancer with MYC copy-number increase. *Nature* **2014**, *512*, 82–86. [CrossRef] [PubMed]

77. Zhang, M.; Wu, W.B.; Wang, Z.W.; Wang, X.H. lncRNA NEAT1 is closely related with progression of breast cancer via promoting proliferation and EMT. *Eur. Rev. Med. Pharmacol. Sci.* **2017**, *21*, 1020–1026. [PubMed]

78. Qian, K.; Liu, G.; Tang, Z.; Hu, Y.; Fang, Y.; Chen, Z.; Xu, X. The long non-coding RNA NEAT1 interacted with miR-101 modulates breast cancer growth by targeting EZH2. *Arch. Biochem. Biophys.* **2016**, *615*, 1–9. [CrossRef] [PubMed]

79. Chen, Q.; Cai, J.; Wang, Q.; Wang, Y.; Liu, M.; Yang, J.; Zhou, J.; Kang, C.; Li, M.; Jiang, C. Long Noncoding RNA NEAT1, Regulated by the EGFR Pathway, Contributes to Glioblastoma Progression Through the WNT/beta-Catenin Pathway by Scaffolding EZH2. *Clin. Cancer Res.* **2018**, *24*, 684–695. [CrossRef] [PubMed]

80. Parasramka, M.; Yan, I.K.; Wang, X.; Nguyen, P.; Matsuda, A.; Maji, S.; Foye, C.; Asmann, Y.; Patel, T. BAP1 dependent expression of long non-coding RNA NEAT-1 contributes to sensitivity to gemcitabine in cholangiocarcinoma. *Mol. Cancer* **2017**, *16*, 22. [CrossRef] [PubMed]

81. Yap, T.A.; Aerts, J.G.; Popat, S.; Fennell, D.A. Novel insights into mesothelioma biology and implications for therapy. *Nat. Rev. Cancer* **2017**, *17*, 475–488. [CrossRef] [PubMed]

82. Schaner, M.E.; Ross, D.T.; Ciaravino, G.; Sorlie, T.; Troyanskaya, O.; Diehn, M.; Wang, Y.C.; Duran, G.E.; Sikic, T.L.; Caldeira, S.; et al. Gene expression patterns in ovarian carcinomas. *Mol. Biol. Cell* **2003**, *14*, 4376–4386. [CrossRef] [PubMed]

83. Di Palma, T.; Lucci, V.; de Cristofaro, T.; Filippone, M.G.; Zannini, M. A role for PAX8 in the tumorigenic phenotype of ovarian cancer cells. *BMC Cancer* **2014**, *14*, 292. [CrossRef] [PubMed]

84. Gardi, N.L.; Deshpande, T.U.; Kamble, S.C.; Budhe, S.R.; Bapat, S.A. Discrete molecular classes of ovarian cancer suggestive of unique mechanisms of transformation and metastases. *Clin. Cancer Res.* **2014**, *20*, 87–99. [CrossRef] [PubMed]

85. Chapel, D.B.; Husain, A.N.; Krausz, T.; McGregor, S.M. PAX8 Expression in a Subset of Malignant Peritoneal Mesotheliomas and Benign Mesothelium has Diagnostic Implications in the Differential Diagnosis of Ovarian Serous Carcinoma. *Am. J. Surg. Pathol.* **2017**, *41*, 1675–1682. [CrossRef] [PubMed]

86. Laury, A.R.; Hornick, J.L.; Perets, R.; Krane, J.F.; Corson, J.; Drapkin, R.; Hirsch, M.S. PAX8 reliably distinguishes ovarian serous tumors from malignant mesothelioma. *Am. J. Surg. Pathol.* **2010**, *34*, 627–635. [CrossRef] [PubMed]

87. Dong, S.; Qu, X.; Li, W.; Zhong, X.; Li, P.; Yang, S.; Chen, X.; Shao, M.; Zhang, L. The long non-coding RNA, GAS5, enhances gefitinib-induced cell death in innate EGFR tyrosine kinase inhibitor-resistant lung adenocarcinoma cells with wide-type EGFR via downregulation of the IGF-1R expression. *J. Hematol. Oncol.* **2015**, *8*, 43. [CrossRef] [PubMed]

88. Xue, D.; Zhou, C.; Lu, H.; Xu, R.; Xu, X.; He, X. LncRNA GAS5 inhibits proliferation and progression of prostate cancer by targeting miR-103 through AKT/mTOR signaling pathway. *Tumour Biol.* **2016**. [CrossRef] [PubMed]

89. Ye, K.; Wang, S.; Zhang, H.; Han, H.; Ma, B.; Nan, W. Long Noncoding RNA GAS5 Suppresses Cell Growth and Epithelial-Mesenchymal Transition in Osteosarcoma by Regulating the miR-221/ARHI Pathway. *J. Cell. Biochem.* **2017**, *118*, 4772–4781. [CrossRef] [PubMed]

90. Rinn, J.L.; Kertesz, M.; Wang, J.K.; Squazzo, S.L.; Xu, X.; Brugmann, S.A.; Goodnough, L.H.; Helms, J.A.; Farnham, P.J.; Segal, E.; et al. Functional demarcation of active and silent chromatin domains in human HOX loci by noncoding RNAs. *Cell* **2007**, *129*, 1311–1323. [CrossRef] [PubMed]

91. Tsai, M.C.; Manor, O.; Wan, Y.; Mosammaparast, N.; Wang, J.K.; Lan, F.; Shi, Y.; Segal, E.; Chang, H.Y. Long noncoding RNA as modular scaffold of histone modification complexes. *Science* **2010**, *329*, 689–693. [CrossRef] [PubMed]

92. Wu, Y.; Zhang, L.; Wang, Y.; Li, H.; Ren, X.; Wei, F.; Yu, W.; Liu, T.; Wang, X.; Zhou, X.; et al. Long non-coding RNA HOTAIR promotes tumor cell invasion and metastasis by recruiting EZH2 and repressing E-cadherin in oral squamous cell carcinoma. *Int. J. Oncol.* **2015**, *46*, 2586–2594. [CrossRef] [PubMed]

93. Battistelli, C.; Cicchini, C.; Santangelo, L.; Tramontano, A.; Grassi, L.; Gonzalez, F.J.; de Nonno, V.; Grassi, G.; Amicone, L.; Tripodi, M. The Snail repressor recruits EZH2 to specific genomic sites through the enrollment of the lncRNA HOTAIR in epithelial-to-mesenchymal transition. *Oncogene* **2017**, *36*, 942–955. [CrossRef] [PubMed]

94. Xia, M.; Yao, L.; Zhang, Q.; Wang, F.; Mei, H.; Guo, X.; Huang, W. Long noncoding RNA HOTAIR promotes metastasis of renal cell carcinoma by up-regulating histone H3K27 demethylase JMJD3. *Oncotarget* **2017**, *8*, 19795–19802. [CrossRef] [PubMed]

95. Liu, Y.W.; Sun, M.; Xia, R.; Zhang, E.B.; Liu, X.H.; Zhang, Z.H.; Xu, T.P.; De, W.; Liu, B.R.; Wang, Z.X. LincHOTAIR epigenetically silences miR34a by binding to PRC2 to promote the epithelial-to-mesenchymal transition in human gastric cancer. *Cell Death Dis.* **2015**, *6*, e1802. [CrossRef] [PubMed]

96. Li, J.T.; Wang, L.F.; Zhao, Y.L.; Yang, T.; Li, W.; Zhao, J.; Yu, F.; Wang, L.; Meng, Y.L.; Liu, N.N.; et al. Nuclear factor of activated T cells 5 maintained by Hotair suppression of miR-568 upregulates S100 calcium binding protein A4 to promote breast cancer metastasis. *Breast Cancer Res.* **2014**, *16*, 454. [CrossRef] [PubMed]

97. Zhang, H.; Cai, K.; Wang, J.; Wang, X.; Cheng, K.; Shi, F.; Jiang, L.; Zhang, Y.; Dou, J. MiR-7, inhibited indirectly by lincRNA HOTAIR, directly inhibits SETDB1 and reverses the EMT of breast cancer stem cells by downregulating the STAT3 pathway. *Stem Cells* **2014**, *32*, 2858–2868. [CrossRef] [PubMed]

98. Ji, P.; Diederichs, S.; Wang, W.; Boing, S.; Metzger, R.; Schneider, P.M.; Tidow, N.; Brandt, B.; Buerger, H.; Bulk, E.; et al. MALAT-1, a novel noncoding RNA, and thymosin beta4 predict metastasis and survival in early-stage non-small cell lung cancer. *Oncogene* **2003**, *22*, 8031–8041. [CrossRef] [PubMed]

99. Chen, Y.; Xiao, Z.; Hu, M.; Luo, X.; Cui, Z. Diagnostic efficacy of long non-coding RNA MALAT-1 in human cancers: A meta-analysis study. *Oncotarget* **2017**, *8*, 102291–102300. [CrossRef] [PubMed]

100. Cheng, Y.; Imanirad, P.; Jutooru, I.; Hedrick, E.; Jin, U.H.; Rodrigues Hoffman, A.; Leal de Araujo, J.; Morpurgo, B.; Golovko, A.; Safe, S. Role of metastasis-associated lung adenocarcinoma transcript-1 (MALAT-1) in pancreatic cancer. *PLoS ONE* **2018**, *13*, e0192264. [CrossRef] [PubMed]

101. Chou, J.; Wang, B.; Zheng, T.; Li, X.; Zheng, L.; Hu, J.; Zhang, Y.; Xing, Y.; Xi, T. MALAT1 induced migration and invasion of human breast cancer cells by competitively binding miR-1 with cdc42. *Biochem. Biophys. Res. Commun.* **2016**, *472*, 262–269. [CrossRef] [PubMed]

102. Xu, S.; Sui, S.; Zhang, J.; Bai, N.; Shi, Q.; Zhang, G.; Gao, S.; You, Z.; Zhan, C.; Liu, F.; et al. Downregulation of long noncoding RNA MALAT1 induces epithelial-to-mesenchymal transition via the PI3K-AKT pathway in breast cancer. *Int. J. Clin. Exp. Pathol.* **2015**, *8*, 4881–4891. [PubMed]

103. Li, J.; Wang, J.; Chen, Y.; Li, S.; Jin, M.; Wang, H.; Chen, Z.; Yu, W. LncRNA MALAT1 exerts oncogenic functions in lung adenocarcinoma by targeting miR-204. *Am. J. Cancer Res.* **2016**, *6*, 1099–1107. [PubMed]

104. Jin, C.; Yan, B.; Lu, Q.; Lin, Y.; Ma, L. Reciprocal regulation of Hsa-miR-1 and long noncoding RNA MALAT1 promotes triple-negative breast cancer development. *Tumour Biol.* **2016**, *37*, 7383–7394. [CrossRef] [PubMed]

105. Hirata, H.; Hinoda, Y.; Shahryari, V.; Deng, G.; Nakajima, K.; Tabatabai, Z.L.; Ishii, N.; Dahiya, R. Long Noncoding RNA MALAT1 Promotes Aggressive Renal Cell Carcinoma through Ezh2 and Interacts with miR-205. *Cancer Res.* **2015**, *75*, 1322–1331. [CrossRef] [PubMed]

106. Li, Q.; Zhang, C.; Chen, R.; Xiong, H.; Qiu, F.; Liu, S.; Zhang, M.; Wang, F.; Wang, Y.; Zhou, X.; et al. Disrupting MALAT1/miR-200c sponge decreases invasion and migration in endometrioid endometrial carcinoma. *Cancer Lett.* **2016**, *383*, 28–40. [CrossRef] [PubMed]

107. Fan, Y.; Shen, B.; Tan, M.; Mu, X.; Qin, Y.; Zhang, F.; Liu, Y. TGF-beta-induced upregulation of malat1 promotes bladder cancer metastasis by associating with suz12. *Clin. Cancer Res.* **2014**, *20*, 1531–1541. [CrossRef] [PubMed]

108. Zhang, Z.C.; Tang, C.; Dong, Y.; Zhang, J.; Yuan, T.; Li, X.L. Targeting LncRNA-MALAT1 suppresses the progression of osteosarcoma by altering the expression and localization of beta-catenin. *J. Cancer* **2018**, *9*, 71–80. [CrossRef] [PubMed]

109. Wang, W.; Zhu, Y.; Li, S.; Chen, X.; Jiang, G.; Shen, Z.; Qiao, Y.; Wang, L.; Zheng, P.; Zhang, Y. Long noncoding RNA MALAT1 promotes malignant development of esophageal squamous cell carcinoma by targeting beta-catenin via Ezh2. *Oncotarget* **2016**, *7*, 25668–25682. [PubMed]

110. Wu, X.S.; Wang, X.A.; Wu, W.G.; Hu, Y.P.; Li, M.L.; Ding, Q.; Weng, H.; Shu, Y.J.; Liu, T.Y.; Jiang, L.; et al. MALAT1 promotes the proliferation and metastasis of gallbladder cancer cells by activating the ERK/MAPK pathway. *Cancer Biol. Ther.* **2014**, *15*, 806–814. [CrossRef] [PubMed]

111. Han, Y.; Wu, Z.; Wu, T.; Huang, Y.; Cheng, Z.; Li, X.; Sun, T.; Xie, X.; Zhou, Y.; Du, Z. Tumor-suppressive function of long noncoding RNA MALAT1 in glioma cells by downregulation of MMP2 and inactivation of ERK/MAPK signaling. *Cell Death Dis.* **2016**, *7*, e2123. [CrossRef] [PubMed]

112. Dong, Y.; Liang, G.; Yuan, B.; Yang, C.; Gao, R.; Zhou, X. MALAT1 promotes the proliferation and metastasis of osteosarcoma cells by activating the PI3K/Akt pathway. *Tumour Biol.* **2015**, *36*, 1477–1486. [CrossRef] [PubMed]

113. Jin, Y.; Feng, S.J.; Qiu, S.; Shao, N.; Zheng, J.H. LncRNA MALAT1 promotes proliferation and metastasis in epithelial ovarian cancer via the PI3K-AKT pathway. *Eur. Rev. Med. Pharmacol. Sci.* **2017**, *21*, 3176–3184. [PubMed]

114. Yan, W.; Wu, Q.; Yao, W.; Li, Y.; Liu, Y.; Yuan, J.; Han, R.; Yang, J.; Ji, X.; Ni, C. MiR-503 modulates epithelial-mesenchymal transition in silica-induced pulmonary fibrosis by targeting PI3K p85 and is sponged by lncRNA MALAT1. *Sci. Rep.* **2017**, *7*, 11313. [CrossRef] [PubMed]

115. Yang, G.; Shen, T.; Yi, X.; Zhang, Z.; Tang, C.; Wang, L.; Zhou, Y.; Zhou, W. Crosstalk between long non-coding RNAs and Wnt/beta-catenin signalling in cancer. *J. Cell. Mol. Med.* **2018**, *22*, 2062–2070. [CrossRef] [PubMed]

116. Liang, J.; Liang, L.; Ouyang, K.; Li, Z.; Yi, X. MALAT1 induces tongue cancer cells' EMT and inhibits apoptosis through Wnt/beta-catenin signaling pathway. *J. Oral Pathol. Med.* **2016**, *48*, 98–105.

117. Zhao, Y.; Yang, Y.; Trovik, J.; Sun, K.; Zhou, L.; Jiang, P.; Lau, T.S.; Hoivik, E.A.; Salvesen, H.B.; Sun, H.; et al. A novel wnt regulatory axis in endometrioid endometrial cancer. *Cancer Res.* **2014**, *74*, 5103–5117. [CrossRef] [PubMed]

118. Liu, S.; Yan, G.; Zhang, J.; Yu, L. Knockdown of Long Noncoding RNA (lncRNA) Metastasis-Associated Lung Adenocarcinoma Transcript 1 (MALAT1) Inhibits Proliferation, Migration, and Invasion and Promoted Apoptosis By Targeting miR-124 in Retinoblastoma. *Oncol. Res.* 2017. [CrossRef] [PubMed]

119. Vassallo, I.; Zinn, P.; Lai, M.; Rajakannu, P.; Hamou, M.F.; Hegi, M.E. WIF1 re-expression in glioblastoma inhibits migration through attenuation of non-canonical WNT signaling by downregulating the lncRNA MALAT1. *Oncogene* **2016**, *35*, 12–21. [CrossRef] [PubMed]

120. Rickman, D.S.; Schulte, J.H.; Eilers, M. The Expanding World of N-MYC-Driven Tumors. *Cancer Discov.* **2018**, *8*, 150–163. [CrossRef] [PubMed]

121. Ma, L.; Young, J.; Prabhala, H.; Pan, E.; Mestdagh, P.; Muth, D.; Teruya-Feldstein, J.; Reinhardt, F.; Onder, T.T.; Valastyan, S.; et al. miR-9, a MYC/MYCN-activated microRNA, regulates E-cadherin and cancer metastasis. *Nat. Cell Biol.* **2010**, *12*, 247–256. [CrossRef] [PubMed]

122. Vadie, N.; Saayman, S.; Lenox, A.; Ackley, A.; Clemson, M.; Burdach, J.; Hart, J.; Vogt, P.K.; Morris, K.V. MYCNOS functions as an antisense RNA regulating MYCN. *RNA Biol.* **2015**, *12*, 893–899. [CrossRef] [PubMed]

123. Jacobs, J.F.; van Bokhoven, H.; van Leeuwen, F.N.; Hulsbergen-van de Kaa, C.A.; de Vries, I.J.; Adema, G.J.; Hoogerbrugge, P.M.; de Brouwer, A.P. Regulation of MYCN expression in human neuroblastoma cells. *BMC Cancer* **2009**, *9*, 239. [CrossRef] [PubMed]

124. Zhao, X.; Li, D.; Pu, J.; Mei, H.; Yang, D.; Xiang, X.; Qu, H.; Huang, K.; Zheng, L.; Tong, Q. CTCF cooperates with noncoding RNA MYCNOS to promote neuroblastoma progression through facilitating MYCN expression. *Oncogene* **2016**, *35*, 3565–3576. [CrossRef] [PubMed]

125. Raveh, E.; Matouk, I.J.; Gilon, M.; Hochberg, A. The H19 Long non-coding RNA in cancer initiation, progression and metastasis—A proposed unifying theory. *Mol. Cancer* **2015**, *14*, 184. [CrossRef] [PubMed]

126. Matouk, I.J.; Halle, D.; Raveh, E.; Gilon, M.; Sorin, V.; Hochberg, A. The role of the oncofetal H19 lncRNA in tumor metastasis: Orchestrating the EMT-MET decision. *Oncotarget* **2016**, *7*, 3748–3765. [CrossRef] [PubMed]

127. Panzitt, K.; Tschernatsch, M.M.; Guelly, C.; Moustafa, T.; Stradner, M.; Strohmaier, H.M.; Buck, C.R.; Denk, H.; Schroeder, R.; Trauner, M.; et al. Characterization of HULC, a novel gene with striking up-regulation in hepatocellular carcinoma, as noncoding RNA. *Gastroenterology* **2007**, *132*, 330–342. [CrossRef] [PubMed]

128. Yu, X.; Zheng, H.; Chan, M.T.; Wu, W.K. HULC: An oncogenic long non-coding RNA in human cancer. *J. Cell. Mol. Med.* **2017**, *21*, 410–417. [CrossRef] [PubMed]

129. Lu, Y.; Li, Y.; Chai, X.; Kang, Q.; Zhao, P.; Xiong, J.; Wang, J. Long noncoding RNA HULC promotes cell proliferation by regulating PI3K/AKT signaling pathway in chronic myeloid leukemia. *Gene* **2017**, *607*, 41–46. [CrossRef] [PubMed]

130. Wang, J.; Ma, W.; Liu, Y. Long non-coding RNA HULC promotes bladder cancer cells proliferation but inhibits apoptosis via regulation of ZIC2 and PI3K/AKT signaling pathway. *Cancer Biomark.* **2017**, *20*, 425–434. [CrossRef] [PubMed]

131. Wu, M.; Lin, Z.; Li, X.; Xin, X.; An, J.; Zheng, Q.; Yang, Y.; Lu, D. HULC cooperates with MALAT1 to aggravate liver cancer stem cells growth through telomere repeat-binding factor 2. *Sci. Rep.* **2016**, *6*, 36045. [CrossRef] [PubMed]

132. Yang, X.J.; Huang, C.Q.; Peng, C.W.; Hou, J.X.; Liu, J.Y. Long noncoding RNA HULC promotes colorectal carcinoma progression through epigenetically repressing NKD2 expression. *Gene* **2016**, *592*, 172–178. [CrossRef] [PubMed]

133. Wang, J.; Liu, X.; Wu, H.; Ni, P.; Gu, Z.; Qiao, Y.; Chen, N.; Sun, F.; Fan, Q. CREB up-regulates long non-coding RNA, HULC expression through interaction with microRNA-372 in liver cancer. *Nucleic Acids Res.* **2010**, *38*, 5366–5383. [CrossRef] [PubMed]

134. Kong, D.; Wang, Y. Knockdown of lncRNA HULC inhibits proliferation, migration, invasion, and promotes apoptosis by sponging miR-122 in osteosarcoma. *J. Cell. Biochem.* **2018**, *119*, 1050–1061. [CrossRef] [PubMed]

135. Zhang, Y.; Li, Z.; Zhong, Q.; Chen, Q.; Zhang, L. Molecular mechanism of HEIH and HULC in the proliferation and invasion of hepatoma cells. *Int. J. Clin. Exp. Med.* **2015**, *8*, 12956–12962. [PubMed]

136. Wang, Y.; Liu, Z.; Yao, B.; Li, Q.; Wang, L.; Wang, C.; Dou, C.; Xu, M.; Liu, Q.; Tu, K. Long non-coding RNA CASC2 suppresses epithelial-mesenchymal transition of hepatocellular carcinoma cells through CASC2/miR-367/FBXW7 axis. *Mol. Cancer* **2017**, *16*, 123. [CrossRef] [PubMed]

137. Zhao, L.; Zhang, Y. Long noncoding RNA CASC2 regulates hepatocellular carcinoma cell oncogenesis through miR-362-5p/Nf-kappaB axis. *J. Cell. Physiol.* 2018. [CrossRef]

138. Askarian-Amiri, M.E.; Crawford, J.; French, J.D.; Smart, C.E.; Smith, M.A.; Clark, M.B.; Ru, K.; Mercer, T.R.; Thompson, E.R.; Lakhani, S.R.; et al. SNORD-host RNA Zfas1 is a regulator of mammary development and a potential marker for breast cancer. *RNA* **2011**, *17*, 878–891. [CrossRef] [PubMed]

139. Gao, K.; Ji, Z.; She, K.; Yang, Q.; Shao, L. Long non-coding RNA ZFAS1 is an unfavourable prognostic factor and promotes glioma cell progression by activation of the Notch signaling pathway. *Biomed. Pharmacother.* **2017**, *87*, 555–560. [CrossRef] [PubMed]

140. Li, T.; Xie, J.; Shen, C.; Cheng, D.; Shi, Y.; Wu, Z.; Deng, X.; Chen, H.; Shen, B.; Peng, C.; et al. Amplification of Long Noncoding RNA ZFAS1 Promotes Metastasis in Hepatocellular Carcinoma. *Cancer Res.* **2015**, *75*, 3181–3191. [CrossRef] [PubMed]

141. Liu, G.; Wang, L.; Han, H.; Li, Y.; Lu, S.; Li, T.; Cheng, C. LncRNA ZFAS1 promotes growth and metastasis by regulating BMI1 and ZEB2 in osteosarcoma. *Am. J. Cancer Res.* **2017**, *7*, 1450–1462. [PubMed]

142. Lv, Q.L.; Chen, S.H.; Zhang, X.; Sun, B.; Hu, L.; Qu, Q.; Huang, Y.T.; Wang, G.H.; Liu, Y.L.; Zhang, Y.Y.; et al. Upregulation of long noncoding RNA zinc finger antisense 1 enhances epithelial-mesenchymal transition in vitro and predicts poor prognosis in glioma. *Tumour Biol.* **2017**, *39*. [CrossRef] [PubMed]

143. Nie, F.; Yu, X.; Huang, M.; Wang, Y.; Xie, M.; Ma, H.; Wang, Z.; De, W.; Sun, M. Long noncoding RNA ZFAS1 promotes gastric cancer cells proliferation by epigenetically repressing KLF2 and NKD2 expression. *Oncotarget* **2017**, *8*, 38227–38238. [CrossRef] [PubMed]

144. Thorenoor, N.; Faltejskova-Vychytilova, P.; Hombach, S.; Mlcochova, J.; Kretz, M.; Svoboda, M.; Slaby, O. Long non-coding RNA ZFAS1 interacts with CDK1 and is involved in p53-dependent cell cycle control and apoptosis in colorectal cancer. *Oncotarget* **2016**, *7*, 622–637. [CrossRef] [PubMed]

145. Tian, F.M.; Meng, F.Q.; Wang, X.B. Overexpression of long-noncoding RNA ZFAS1 decreases survival in human NSCLC patients. *Eur. Rev. Med. Pharmacol. Sci.* **2016**, *20*, 5126–5131. [PubMed]

146. Wang, W.; Xing, C. Upregulation of long noncoding RNA ZFAS1 predicts poor prognosis and prompts invasion and metastasis in colorectal cancer. *Pathol. Res. Pract.* **2016**, *212*, 690–695. [CrossRef] [PubMed]

147. Xia, B.; Hou, Y.; Chen, H.; Yang, S.; Liu, T.; Lin, M.; Lou, G. Long non-coding RNA ZFAS1 interacts with miR-150-5p to regulate Sp1 expression and ovarian cancer cell malignancy. *Oncotarget* **2017**, *8*, 19534–19546. [CrossRef] [PubMed]

148. Xie, S.; Ge, Q.; Wang, X.; Sun, X.; Kang, Y. Long non-coding RNA ZFAS1 sponges miR-484 to promote cell proliferation and invasion in colorectal cancer. *Cell Cycle* **2018**, *17*, 154–161. [CrossRef] [PubMed]

149. Cregan, S.; Breslin, M.; Roche, G.; Wennstedt, S.; MacDonagh, L.; Albadri, C.; Gao, Y.; O'Byrne, K.J.; Cuffe, S.; Finn, S.P. Kdm6a and Kdm6b: Altered expression in malignant pleural mesothelioma. *Int. J. Oncol.* **2017**, *50*, 1044–1052.

150. Rhodes, D.R.; Kalyana-Sundaram, S.; Mahavisno, V.; Varambally, R.; Yu, J.; Briggs, B.B.; Barrette, T.R.; Anstet, M.J.; Kincead-Beal, C.; Kulkarni, P.; et al. Oncomine 3.0: Genes, pathways, and networks in a collection of 18,000 cancer gene expression profiles. *Neoplasia* **2007**, *9*, 166–180. [CrossRef] [PubMed]

151. Oncomine. Available online: www.oncomine.org (accessed on 2 April 2018).

152. cBioportal. Available online: www.cbioportal.org (accessed on 2 April 2018).

153. Gao, J.; Aksoy, B.A.; Dogrusoz, U.; Dresdner, G.; Gross, B.; Sumer, S.O.; Sun, Y.; Jacobsen, A.; Sinha, R.; Larsson, E.; et al. Integrative analysis of complex cancer genomics and clinical profiles using the cBioPortal. *Sci. Signal.* **2013**, *6*, pl1. [CrossRef] [PubMed]

154. Cerami, E.; Gao, J.; Dogrusoz, U.; Gross, B.E.; Sumer, S.O.; Aksoy, B.A.; Jacobsen, A.; Byrne, C.J.; Heuer, M.L.; Larsson, E.; et al. The cBio cancer genomics portal: An open platform for exploring multidimensional cancer genomics data. *Cancer Discov.* **2012**, *2*, 401–404. [CrossRef] [PubMed]

155. Goswami, C.P.; Nakshatri, H. PROGgeneV2: Enhancements on the existing database. *BMC Cancer* **2014**, *14*, 970. [CrossRef] [PubMed]

156. PROGgene. Available online: http://watson.compbio.iupui.edu/chirayu/proggene/database/?url=proggene (accessed on 2 April 2018).

157. Brosseau, S.; Dhalluin, X.; Zalcman, G.; Scherpereel, A. Immunotherapy in relapsed mesothelioma. *Immunotherapy* **2018**, *10*, 77–80. [CrossRef] [PubMed]

International Journal of
Molecular Sciences

MDPI

Case Report

Peritoneal Mesothelioma with Residential Asbestos Exposure. Report of a Case with Long Survival (Seventeen Years) Analyzed by Cgh-Array

Gabriella Serio [1,*], Federica Pezzuto [1], Andrea Marzullo [1], Anna Scattone [2], Domenica Cavone [3], Alessandra Punzi [1], Francesco Fortarezza [1], Mattia Gentile [4], Antonia Lucia Buonadonna [4], Mattia Barbareschi [5] and Luigi Vimercati [3]

[1] Department of Emergency and Organ Transplantation, Division of Pathology, Medical School, University of Bari, 11 G. Cesare Square, 70124 Bari, Italy; pezzuto.federica@libero.it (F.P.); andrea.marzullo@uniba.it (A.M.); alessandra.punzi@hotmail.it (A.P.); francescofortarezza.md@gmail.com (F.F.)
[2] Division of Pathology, IRCCS, National Cancer Institute "Giovanni Paolo II", 70124 Bari, Italy; a.scattoneanatopat@libero.it
[3] Department of Interdisciplinary Medicine, Occupational Health Division, Medical School, University of Bari, 70124 Bari, Italy; domenica.cavone@uniba.it (D.C.); luigi.vimercati@uniba.it (L.V.)
[4] Division of Medical Genetics "Di Venere Hospital", 70124 Bari, Italy; mattiagentile@libero.it (M.G.); albuonadonna@libero.it (A.L.B.)
[5] Department of Pathology, Santa Chiara Hospital, 38121 Trento, Italy; mattiabarbareschi@apss.tn.it
*   Correspondence: gabriella.serio1@uniba.it; Tel.: +39-080-5593-198; Fax: +39-080-5478-280

Received: 27 June 2017; Accepted: 18 August 2017; Published: 22 August 2017

Abstract: Malignant mesothelioma is a rare and aggressive tumor with limited therapeutic options. We report a case of a malignant peritoneal mesothelioma (MPM) epithelioid type, with environmental asbestos exposure, in a 36-year-old man, with a long survival (17 years). The patient received standard treatment which included cytoreductive surgery (CRS) and hyperthermic intraperitoneal chemotherapy (HIPEC). Methods and Results: Molecular analysis with comparative genomic hybridization (CGH)-array was performed on paraffin-embedded tumoral samples. Multiple chromosomal imbalances were detected. The gains were prevalent. Losses at 1q21, 2q11.1→q13, 8p23.1, 9p12→p11, 9q21.33→q33.1, 9q12→q21.33, and 17p12→p11.2 are observed. Chromosome band 3p21 (*BAP1*), 9p21 (*CDKN2A*) and 22q12 (*NF2*) are not affected. *Conclusions:* the defects observed in this case are uncommon in malignant peritoneal mesothelioma. Some chromosomal aberrations that appear to be random here, might actually be relevant events explaining the response to therapy, the long survival and, finally, may be considered useful prognostic factors in peritoneal malignant mesothelioma (PMM).

Keywords: peritoneum; mesothelioma; molecular-array; asbestos

1. Introduction

Peritoneal malignant mesothelioma (PMM) is a rare, lethal malignancy, whose main known cause is occupational or environmental asbestos exposure. Other possible risk factors are radiation exposure and genetic predisposition [1]. The peritoneal mesothelioma incidence is rising continuously without peaking; often the disease is unresectable and the patient's prognosis poor (10–13 months) [2]. In recent years, the prognosis has improved due to therapeutic advancements such as surgery in combination with other treatment modalities such as hyperthermic intraperitoneal chemotherapy or radiation [3]. It has been ascertained that mesothelioma generally occurs in elderly asbestos-exposed men. When the tumor occurs in younger people, generally a genetic predisposition and environmental exposure to asbestos or other mineral fibers are implicated. Genetic susceptibility alone cannot explain mesothelioma in young people, but an inordinate sensitivity to these fibers must be considered. In fact, in familial mesothelioma, germline bands 3p21 (*BAP1*) mutations make the members susceptible to the development of cancer even in the presence of low amounts of inhaled asbestos [4].

Patients younger than 40 years appear to have a significantly longer survival compared to older patients (both in pleural and peritoneal mesothelioma). In a recent report, Thomas et al. [5] suggested that female gender, peritoneal site, histology, surgery, and radiation therapy are the better prognostic factors. However, the authors emphasize the need for genetic study of the tumor to identify recurring changes that can define the disease at a young age. Therefore, the identification of genetic changes will produce the development of new agents targeting these oncogenic abnormalities. Many DNA copy number alterations (CNAs) have been identified in peritoneal malignant mesothelioma [6]. Generally, the genetic alterations observed in malignant mesothelioma (MM) are deletions in 1p21-22, 1p36, 3p21, 4p, 4p12-13, 4q, 4q31-32, 6q14-25, 6q22, 9p21, 10p13-pter, 13q13-14, 14q, 14q12-24, 14q32, 15q15, 17p12-13, 17p12-pter, and 22q12, and amplifications in 1q23, 1q32, 1p36.33, 5p, 5p15.1-pter, 7p, 7p14-15, 7q21, 7q31, 8q, 8q22-23, 8q24, and 15q22-25 shown by comparative genomic hybridization (CGH) analysis [7–15]. Deletion of 9p21 is particularly frequent in MM, especially in the biphasic histotype, although this deletion has been reported to be less frequent in peritoneal than in pleura mesothelioma [15–17]. Also 3p21 and 22q12 losses are less common in peritoneal mesothelioma. Generally, a higher frequency of loss was seen in pleural MM, gains were seen more frequently in peritoneal MM, including regions at 3q, 7q, 8p, 9p, 16p, and 20q [16,18,19].

This study describes the genetic alterations observed in a case of diffuse asbestos-related peritoneal mesothelioma affecting a young man (under 40 years) who has survived and is currently disease-free.

2. Results

The CGH-array analysis revealed multiple chromosomal abnormalities (27 gains and 7 losses). Deletions were found at 1q21, 2q11.1→2q13, 8p23.1, 9p12→9p11, 9q21.33→9q33.1, 9q12→9q21.33, and 17p12→17p11.2. Amplifications were at 1p36.33→1p36.32, 1q31, 3p25.3→3p25.1, 3p22.2→3p22.1, 3q29, 4p16.3→4p16.1, 4q13.1, 5p15.33, 7p22.3→7p22.1, 7q21.11, 8q24.3, 9q34.11→9q34.3, 10q26.3, 11p15.5→11p15.4, 11q13.1→11q13.2, 11q13.3→11q13.4, 12q24.33, 13q33.2, 16q22.1→16q22.2, 17p13.1, 17q24.3→17q25.3, 19p13.3, 20p13→20p12.3, 22q11.21→22q11.23, Xp22.33, Xq22.2, and Xq28. CNAs detected are shown in Figure 1.

3. Discussion

Epidemiological studies report a significant risk of mesothelioma in people exposed to asbestos in non-occupational settings. The prominent characteristic of non-occupational exposure is the considerably younger age at the start of exposure, and hence the risk of a longer exposure and latency. Furthermore, the risk of MM is increased with cumulative dose of asbestos even in analyses limited to non-occupationally exposed subjects [20]. Peritoneal mesothelioma is difficult to diagnose, particularly when the tumor affects women and younger patients. Currently, cytoreductive surgery, combined with hyperthermic intraperitoneal chemoperfusion is the best treatment for young patients. However, clinical trials have shown that this treatment improves median overall survival by 27–46 months [21]. In a multivariate analysis, Thomas et al. [5] report that younger age was a favorable independent prognostic factor, including after multimodal treatments. Because multimodal treatment is associated with a significant perioperative risk it is necessary to consider prognostic markers during patient selection. At present, the morphological growth patterns of the tumor and the mitotic index are the prevalent aspects used to select patient groups to be submitted to treatment options [21,22].

Frequent molecular alterations [deletion/mutation] at *CDKN2A* (9p21), *NF2* (22q12), and *BAP1* (3p21) are considered the predictive prognostic factors for progression-free survival and overall survival in malignant peritoneal mesothelioma (MPM) [23]. The molecular pathway and mechanism still remains unknown due to a lack of large-scale studies. Currently, in oncology, these alterations are not decisive, nor do they affect the choice of chemotherapy. In a younger patient with non-asbestos-related peritoneal mesothelioma, epithelioid type, Sheffield et al. [16] observed that loss of function in *NF2* was correlated with a worse prognosis and no response to chemotherapy. By contrast, a middle-aged woman with non-asbestos-related peritoneal mesothelioma, sarcomatoid type, with mutations in *NF2* (22q12.1), *CDKN2A* (9p21), *p53* (17p12), and *LATS2* (13q12), had an excellent response to treatment. Hence the question, whether it is appropriate to exclude patients with sarcomatoid mesothelioma from aggressive treatment options. *BAP1* was normal in both cases.

Alakus et al. [24], did not observe *CDKN2A* changes when analyzing a series of peritoneal mesothelioma. *BAP1* was lost or inactivated in three of seven cases of peritoneal mesothelioma, suggesting a potential genetic predisposition in these patients. The same results were reported by Borczuk et al. [18] in a series of 32 peritoneal mesothelioma cases. However, when considering *BAP1* in other tumors, its loss is a marker of poor prognosis; in mesotheliomas, the real biologic role of BAP1 (germline or somatic mutations) is unclear and it appears to work with different results [25].

Singhi et al. [26], in a recent report of 86 malignant peritoneal mesothelioma, had observed that the absence of *BAP1* nuclear expression correlated with increased mean age and with the epithelioid subtype; the loss or absence of nuclear *BAP1* expression was not associated to asbestos exposure, incomplete cytoreduction, invasion, or metastasis. *BAP1* immunohistochemical staining, loss vs. preserved, was not associated with clinical outcome, whereas the *CDKN2A* and *NF2* deletions were negative prognostic markers.

Our mesothelioma asbestos-related (environmental exposure), epithelioid subtype, shows a high nuclear immunohistochemical *BAP1* expression, the chromosome bands 3p21 (*BAP1*), 9p21 (*CDKN2A*), and 22q12 (*NF2*) are not altered, and the young patient is disease-free.

The Pin2/TRF1-interacting telomerase inhibitor (*PinX1*) is a novel cloned gene located at human chromosome 8p23, playing a vital role in maintaining telomeres length and chromosome stability [27]. Loss of heterozygosity (LOH) regions, *PinX1* overexpression or inhibition are detected in human malignancies. Thus, *PinX1* might be considered as a putative tumor suppressor gene. The mechanism for *PinX1*-gene inactivation in human cancer is not clear. Recently, it has been demonstrated that *PinX1* expression is directly activated by *p53* in cervical cancer. The authors suggest that *PinX1* inhibition via *p53* transcriptional activity results in the enhancement of telomerase activity [28]. *PinX1* has also been demonstrated to be a new potential cancer therapy target (i.e., fibrosarcoma, hepatocellular carcinoma, breast cancer, gastric carcinoma, lymphoma, etc.). Interestingly, our case presented loss at 8p23.

Gelsolin, a Ca^{2+}-regulated actin filament severing, capping, and nucleating protein, is a ubiquitous, multifunctional regulator of cell structure and metabolism. Gelsolin (*GSN*) can act as a transcriptional cofactor in signal transduction and its own expression and function can be influenced by epigenetic changes. In humans, intracellular (*cGSN*) and extracellular forms (*pGSN*) of *GSN* are encoded by genes on chromosome 9q33. A decreased expression of *cGSN* has been observed in breast, bladder, lung, colorectal, ovarian cancers, etc. and was associated with cell proliferation and survival [29]. *GSN* is involved in the modulation of several signaling pathways, including c-erb-2/EGFR, *p53*, PI3K, Ras-PI3K-Rac, etc. [27,30]. *GSN* represses transactivation of *p53* via inhibition of nuclear translocation of p53, thus inhibiting *p53*-mediated apoptosis in hepatocarcinoma cells [31]. Our case reports 9q33 loss, being the putative site of *GSN*, this tumor suppressor gene.

Finally, Telomerase is a ribonucleoprotein complex mainly composed of a reverse transcriptase catalytic subunit which copies a template region of its RNA subunit to the end of the telomere. Human telomeres function as a protective structure capping the ends of chromosomes. Dysfunction of telomeres plays an important role in cancer initiation and progression. The active human telomerase enzyme is composed of human telomerase reverse transcriptase (*hTERT*), encoded by the TERT gene located at 5p15.33; human telomerase (*hTR*) encoded by the TERC gene at 3q26.3; and dyskerin encoded by the *DKC1* gene located at Xq28. Gains at 5p15.33, 3q26.3, and Xq28 are the regions known to be involved in cancer cells [18,32,33]. Gains at 5p15.33 and Xq38 are shown in our case. So, detection of TERT amplification may be useful to reveal an initial mesothelioma but we should also consider that TERT promoter mutations might be associated with inactivation of many cell cycle regulator genes.

Our study is coherent with the CGH findings reported in the literature for the mesothelioma diagnosis, but failed to identify a compelling alteration to explain the patient's outcomes. So, CGH-array may be a helpful and sometimes definitive approach in the diagnosis of mesothelioma, to evidence the role of TSGs in the tumor pathogenesis, but it is not the method to detect genes-status involved in the tumor progression. Therefore, whole-genome sequencing data represent a crucial milestone not only in the understanding of the mesothelioma pathogenesis, but also to detect the genes that are important for personalized therapy. Next-generation sequencing (NGS) coupled with bioinformatics analysis of the somatic mutations would allow for the screening of tumor-specific mutated proteins, candidate targets for the design of individualized therapy. Our patient, who received a combined chemotherapy with raltitrexed and oxaliplatin, appears to have responded to therapy not merely by chance. The combination of ralitrexed and oxaliplatin would have a synergistic effect on the early inhibition of DNA synthesis promoted by the mutations described [34]. No alteration of the chromosome 18p11.32 (site of the thymidylate synthase gene) was observed in our case.

A smaller total number of losses in the tumoral chromosomes might be related with a longer survival [7,11], but further investigations need to be performed to characterize the recurrent mutations useful to improve the prognosis.

4. Material and Methods

4.1. Case Report: Clinical History an Pathological Features

In May 1999, a 36-year-old man was admitted to the Hospital "Azienda Ospedaliera Universitaria Policlinico" in Bari, Apulia Region—Southern Italy, for ascites, abdominal pain, and weight loss. His past medical history was unremarkable; he had been a smoker (20 cigarettes per day) since he was 13 years old. A CT-scan revealed that there was irregular thickening (thickness > 2 mm) of the peritoneum and suspect nodules but no abdominal viscera invasion. Routine laboratory tests were unremarkable. Tumor markers were negative. An exploratory laparotomy with biopsy was performed. Histology showed MM with a solid epithelial (60%, predominant subtype) (Figure 1a) and pseudotubular (40%, subdominant) pattern. The neoplastic cells were generally polygonal, cuboidal or low columnar, with a pale-to-eosinophilic, abundant cytoplasm; cellular atypia ranged from mild to moderate. Nucleoli were prominent and eosinophilic. Nuclear pseudoinclusions and psammoma bodies were also seen (Figure 1a). The mitotic count was $2/mm^2$. Necrotic foci and vascular invasion were not found. Immunohistochemical (IHC) analysis showed intense expression of 5/6 cytokeratins, calretinin, WT-1, HBME-1, and vimentin. Ki67-index was 5%. In addition, IHC detection of BAP-1 was performed using a primary anti-human *BAP1* antibody (C-4, Santa Cruz Biotechnology, Santa Cruz, CA, USA). Tumor tissue showed a strong/moderate *BAP1* nuclear expression in pseudo-tubular and solid epithelioid pattern (Figure 1b).

Patient opted for surgery at a specialist center (Institute Gustave Roussy, France); where he underwent cytoreductive surgery (CRS) and hyperthermic intraperitoneal chemotherapy with oxaliplatin (HIPEC). Complete cytoreduction of all macroscopic tumors (>2 mm) was combined with peritonectomy, omentectomy and organ resection (cholecystectomy, half the distal small intestine and right colon). In accordance with the French hospital protocol, preoperatively the patient underwent six cycles of chemotherapy (raltitrexed (Tomudex®) and oxaliplatin combined). The patient was enrolled in the Apulia Regional Mesothelioma Register by the pathology department, in compliance with Italian law for the compulsory notification of new cases of MM. According to the standardized mesothelioma register questionnaire to detect asbestos exposure, lifestyle habits and work history, including a possible asbestos exposure during military service, were investigated. The work history of cohabiting family members and a family history of cancer were also evaluated. Exposure during leisure, travel or hobby activities was excluded, as well as any exposure to ionizing radiation. The ascertained asbestos exposure was residential: he had lived near a source of asbestos pollution, in-situ in buildings, at a distance of less than 15 m for 36 years. In fact, the subject had always lived, since birth, in an apartment overlooking military barracks, built in the period 1920–1930, and also played football there. The presence of asbestos (eternit) in the barracks had been ascertained in roofs and chimneys. In 2001, reclaiming of the asbestos roofing began, and ended in 2006. At follow-up in November 2016 (Positron Emission Tomography (PET)/CT-scan performed in July 2016 was negative), more than 17 years after diagnosis, the patient is alive and disease-free without recurrence. Currently, he has chronic diarrhea and chronic abdominal pain.

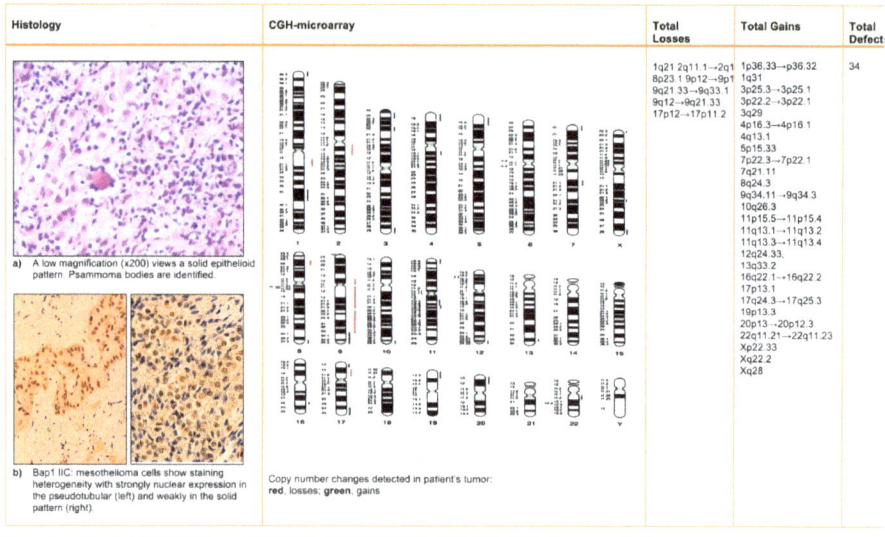

Figure 1. Malignant peritoneal mesothelioma: histology, immunohistochemical (IHC) and comparative genomic hybridization (CGH)-array results.

4.2. CGH-Array Analysis

Informed written consent to the use of histological samples for additional studies was obtained. Molecular analysis with CGH-array was performed on paraffin-embedded tissue. Genomic DNA was extracted from 5-μm sections of paraffin-embedded tissue with the Dneasy Tissue Kit (Qiagen, Hilden, Germany) according to the manufacturer's instructions. Normal sex-matched DNA was extracted from peripheral blood lymphocytes according to standard hybridization procedures (Nucleon BACC3, Amersham Pharmacia Biotech, Bucks, UK). Array-CGH with a genomic resolution of about 0.5 Mb, increasing to 0.25 Mb in the subtelomeric regions, was carried out using the Cytochip V3 genomeARRAY slide (Techno Genetics Srl-Bouty Spa, Milan, Italy), containing 5.380 BAC clones, according to the manufacturer's instructions. Slides were scanned at 633 nm (Cy3) and 543 nm (Cy5) using the Scan ArrayGx (PerkinElmer, Waltham, MA, USA). Image analysis was done using BlueFuse software (Bluegnome Limited, Cambridge, UK). Once the positions of the biological sample were known, a powerful quantification algorithm was used to calculate the amount of signal at each spot location. For each clone, a log2 of the ratio Cy3/Cy5 fluorescent intensity was calculated. The raw results delivered by quantification were subjected to a series of post-processing stages including normalization, data exclusion, and identification of copy number change regions considering the replicate standard deviation values, the internal controls, the degree of confidence, and the median of the log_2 ratio of clones in the regions. Data points lying beyond three standard deviations were considered to be part of a change analysis region. Regions exceeding the ratio thresholds of log 0.3 and log −0.3 and containing at least one clone were considered to be amplifications or deletions, respectively. The results of experiments were visualized on the copy number panel. Full reports, including an ISCN summary of regions of change, were provided as Excel spreadsheets in the results directory.

5. Conclusions

On the basis of the evidence of genomic abnormalities found in our case, related to asbestos exposure in a patient with a supposed genetic cancer predisposition, these may be very important in the choice of therapy, but systematic molecular analyses are needed to understand the complexity and heterogeneity of the chromosomal aberrations that characterize such malignancies. Chromosomal regions of common allelic loss or gain may contain more than one gene associated with clinical-pathological features such as age, histological type, therapy response, or survival. Some chromosomal aberrations that appear to be random here might actually be relevant events explaining the response to therapy, the long survival and, finally, may be considered useful prognostic factors in PMM.

Acknowledgments: The authors are grateful to Mary V. Pragnell, for language assistance. Also, a special thanks to patient, R.F., for allowing the writing of this article. Finally, the authors are most grateful to colleagues of the Department of Surgery, "Gustave Roussy" hospital, for surgical and oncological assistance.

Author Contributions: Gabriella Serio and Andrea Marzullo: principal investigators, planned and designed the study, performed histological analysis, drafted the manuscript; Domenica Cavone: administered questionnaires; Federica Pezzuto, Anna Scattone, Alessandra Punzi, and Francesco Fortarezza: prepared samples and carried out histological analisys; Mattia Barbareschi carried out immunohistochemical analysis; Antonia Lucia Buonadonna and Mattia Gentile: performed molecular analysis; Luigi Vimercati: revised the manuscript.

Conflicts of Interest: The authors declare no conflict of interest.

Ethical Approval: Written informed consent was obtained from the patient.

References

1. Dogan, A.U.; Baris, Y.I.; Dogan, M.; Emri, S.; Steele, I.; Elmishad, A.G.; Carbone, M. Genetic predisposition to fiber carcinogenesis causes a mesothelioma epidemic in Turkey. *Cancer Res.* **2006**, *66*, 5063–5082. [CrossRef] [PubMed]
2. Reid, A.; de Klerk, N.H.; Magnani, C.; Ferrante, D.; Berry, G.; Musk, A.W.; Merler, E. Mesothelioma risk after 40 years since first exposure to asbestos: A pooled analysis. *Thorax* **2014**, *69*, 843–850. [CrossRef] [PubMed]
3. Yan, T.D.; Welch, L.; Black, D.; Sugarbaker, P.H. A systematic review on the efficacy of cytoreductive surgery combined with perioperative intraperitoneal chemotherapy for diffuse malignancy peritoneal mesothelioma. *Ann. Oncol.* **2007**, *18*, 827–834. [CrossRef] [PubMed]
4. Baumann, F.; Flores, E.; Napolitano, A.; Kanodia, S.; Taioli, E.; Pass, H.; Yang, H.; Carbone, M. Mesothelioma patients with germline BAP1 mutations have 7-fold improved long-term survival. *Carcinogenesis* **2015**, *36*, 76–81. [CrossRef] [PubMed]
5. Thomas, A.; Chen, Y.; Yu, T.; Gill, A.; Prasad, V. Distinctive clinical characteristics of malignant mesothelioma in young patients. *Oncotarget* **2015**, *30*, 16766–16773. [CrossRef] [PubMed]
6. Chirac, P.; Maillet, D.; Leprêtre, F.; Isaac, S.; Glehen, O.; Figeac, M.; Villeneuve, L.; Péron, J.; Gibson, F.; Galateau-Sallé, F.; et al. Genomic copy number alterations in 33 malignant peritoneal mesothelioma analyzed by comparative genomic hybridization array. *Hum. Pathol.* **2016**, *55*, 72–82. [CrossRef] [PubMed]
7. Serio, G.; Scattone, A.; Gentile, M.; Nazzaro, P.; Pennella, A.; Buonadonna, A.L.; Pollice, L.; Musti, M. Familial pleural mesothelioma with environmental asbestos exposure: Losses of DNA sequences by comparative genomic hybridization (CGH). *Histopathology* **2004**, *45*, 643–645. [CrossRef] [PubMed]
8. Musti, M.; Kettunen, E.; Dragonieri, S.; Lindholm, P.; Cavone, D.; Serio, G.; Knuutila, S. Cytogenetic and molecular genetic changes in malignant mesothelioma. *Cancer Genet. Cytogenet.* **2006**, *170*, 9–15. [CrossRef] [PubMed]

9. Lindholm, P.M.; Salmenkivi, K.; Vauhkonen, H.; Nicholson, A.G.; Anttila, S.; Kinnula, V.L.; Knuutila, S. Gene copy number analysis in malignant pleural mesothelioma using oligonucleotide array CGH. *Cytogenet. Genome Res.* **2007**, *119*, 46–52. [CrossRef] [PubMed]

10. Taniguchi, T.; Karnan, S.; Fukui, T.; Yokoyama, T.; Tagawa, H.; Yokoi, K.; Ueda, Y.; Mitsudomi, T.; Horio, Y.; Hida, T.; et al. Genomic profiling of malignant pleural mesothelioma with array-based comparative genomic hybridization shows frequent non-random chromosomal alteration regions including JUN amplification on 1p32. *Cancer Sci.* **2007**, *98*, 438–446. [CrossRef] [PubMed]

11. Serio, G.; Gentile, M.; Pennella, A.; Marzullo, A.; Buonadonna, A.L.; Nazzaro, P.; Testini, M.; Musti, M.; Scattone, A. Characterization of a complex chromosome aberration in two cases of peritoneal mesothelioma arising primarily in the hernial sac. *Pathol. Int.* **2009**, *59*, 415–421. [CrossRef] [PubMed]

12. Jean, D.; Daubriac, J.; Le Pimpec-Barthes, F.; Galateau Salle, F.; Jaurand, M.C. Molecular changes in mesothelioma with an impact on prognosis and treatment. *Arch. Pathol. Lab. Med.* **2012**, *136*, 277–293. [CrossRef] [PubMed]

13. Takeda, M.; Kasai, T.; Enomoto, Y.; Takano, M.; Morita, K.; Nakai, T.; Iizuka, N.; Maruyama, H.; Ohbayashi, C. Comparison of genomic abnormality in malignant mesothelioma by the site of origin. *J. Clin. Pathol.* **2014**, *67*, 1038–1043. [CrossRef] [PubMed]

14. Kato, S.; Tomson, B.N.; Buys, T.P.; Elkin, S.K.; Carter, J.L.; Kurzrock, R.; Lindholm, P.M.; Salmenkivi, K.; Vauhkonen, H. Genomic landscape of malignant mesotheliomas. *Mol. Cancer Ther.* **2016**, *15*, 2498–2507. [CrossRef] [PubMed]

15. Bueno, R.; Stawiski, E.W.; Goldstein, L.D.; Durinck, S.; De Rienzo, A.; Modrusan, Z.; Gnad, F.; Nguyen, T.T.; Jaiswal, B.S.; Chirieac, L.R.; et al. Comprehensive genomic analysis of malignant pleural mesotelioma identifies recurrent mutations, gene fusions and splicing alterations. *Nat. Genet.* **2016**, *48*, 407–416. [CrossRef] [PubMed]

16. Sheffield, B.S.; Tinker, A.V.; Shen, Y.; Hwang, H.; Li-Chang, H.H.; Pleasance, E.; Ch'ng, C.; Lum, A.; Lorette, J.; McConnell, Y.J.; et al. Personalized oncogenomics: Clinical experience with malignant peritoneal mesothelioma using whole genome sequencing. *PLoS ONE* **2015**, *10*, e0119689. [CrossRef] [PubMed]

17. Ugurluer, G.; Chang, K.; Gamez, M.E.; Arnett, A.L.; Jayakrishnan, R.; Miller, R.C.; Sio, T.T. Genome-based mutational analysis by Next Generation Sequencing in patients with malignant pleural and peritoneal mesothelioma. *Anticancer Res.* **2016**, *36*, 2331–2338. [CrossRef] [PubMed]

18. Borczuk, A.C.; Pei, J.; Taub, R.N.; Levy, B.; Nahum, O.; Chen, J.; Chen, K.; Testa, J.R. Genome-wide analysis of abdominal and pleural malignant mesothelioma with DNA arrays reveals both common and distinct regions of copy number alteration. *Cancer Biol. Ther.* **2016**, *17*, 328–335. [CrossRef] [PubMed]

19. Joseph, N.M.; Chen, Y.Y.; Nasr, A.; Yeh, I.; Talevich, E.; Onodera, C.; Bastian, B.C.; Rabban, J.T.; Garg, K.; Zaloudek, C.; et al. Genomic profiling of malignant peritoneal mesothelioma reveals recurrent alterations in epigenetic regulatory genes BAP1, SETD2, and DDX3X. *Mod. Pathol.* **2017**, *30*, 246–254. [CrossRef] [PubMed]

20. Kanarek, M.S.; Mandich, M.K. Peritoneal mesothelioma and asbestos: Clarifying the relationship by epidemiology. *Epidemiology* **2016**, *6*, 1000233. [CrossRef]

21. Valente, K.; Blackham, A.U.; Levine, E.; Russell, G.; Votanopoulos, K.I.; Stewart, J.H.; Shen, P.; Geisinger, K.R.; Sirintrapun, S.J. A histomorphologic grading system that predicts overall survival in diffuse malignant peritoneal mesothelioma with epithelioid subtype. *Am. J. Surg. Pathol.* **2016**, *40*, 1243–1248. [CrossRef] [PubMed]

22. Krasinskas, A.M.; Borczuk, A.C.; Hartman, D.J.; Chabot, J.A.; Taub, R.N.; Mogal, A.; Pingpank, J.; Bartlett, D.; Dacic, S. Prognostic significance of morphological growth patterns and mitotic index of epithelioid malignant peritoneal mesothelioma. *Histopathology* **2016**, *68*, 729–737. [CrossRef] [PubMed]

23. Travis, D.V.; Brambilla, E.; Burke, A.P.; Marx, A.; Nicholson, A.G. *WHO Classification of Tumours of the Lung, Pleura, Thymus and Heart*; International Agency for Research on Cancer: Lyon, France, 2015.

24. Alakus, H.; Yost, S.E.; Woo, B.; French, R.; Lin, G.Y.; Jepsen, K.; Frazer, K.A.; Lowy, A.M.; Harismendy, O. BAP1 mutation is a frequent somatic event in peritoneal malignant mesothelioma. *J. Transl. Med.* **2015**, *16*, 122. [CrossRef] [PubMed]

25. Luchini, C.; Veronese, N.; Yachida, S.; Cheng, L.; Nottegar, A.; Stubbs, B.; Solmi, M.; Capelli, P.; Pea, A.; Barbareschi, M.; et al. Different prognostic roles of tumor suppressor gene BAP1 in cancer: A systematic review with meta-analysis. *Genes Chromosomes Cancer* **2016**, *55*, 741–749. [CrossRef] [PubMed]

26. Singhi, A.D.; Krasinskas, A.M.; Choudry, H.A.; Barlett, D.L.; Pingpank, J.F.; Zeh, H.J.; Luvison, A.; Fuhrer, K.; Bahary, N.; Seethala, R.R.; et al. The prognostic significance of BAP1, NF2, and CDKN2A in malignant peritoneal mesothelioma. *Mod. Pathol.* **2016**, *29*, 14–24. [CrossRef] [PubMed]

27. Li, H.L.; Song, J.; Yong, H.M.; Hou, P.F.; Chen, Y.S.; Song, W.B.; Bai, J.; Zheng, J.N. PinX1: Structure, regulation and its functions in cancer. *Oncotarget* **2016**, *7*, 66267–66275. [CrossRef] [PubMed]

28. Wu, G.; Liu, D.; Jiang, K.; Zhang, L.; Zeng, Y.; Zhou, P.; Zhong, D.; Gao, M.; He, F.; Zheng, Y. PinX1, a novel target gene of p53, is suppressed by HPV16 E6 in cervical cancer cells. *Biochim. Biophys. Acta* **2014**, *1839*, 88–96. [CrossRef] [PubMed]

29. Stock, A.M.; Klee, F.; Edlund, K.; Grinberg, M.; Hammad, S.; Marchan, R.; Cadenas, C.; Niggemann, B.; Zänker, K.S.; Rahnenführer, J.; et al. Gelsolin is associated with longer metastasis-free survival and reduced cell migration in estrogen receptor-positive breast cancer. *Anticancer Res.* **2015**, *35*, 5277–5285. [PubMed]

30. An, J.H.; Kim, J.W.; Jang, S.M.; Kim, C.H.; Kang, E.J.; Choi, K.H. Gelsolin negatively regulates the activity of tumor suppressor p53 through their physical interaction in hepatocarcinoma HepG2 cells. *Biochem. Biophys. Res. Commun.* **2011**, *412*, 44–49. [CrossRef] [PubMed]

31. Liu, J.; Zhang, C.; Feng, Z. Tumor suppressor p53 and its gain-of-function mutants in cancer. *Acta Biochim. Biophys. Sin.* **2014**, *46*, 170–179. [CrossRef] [PubMed]

32. Knuutila, S.; Armengol, G.; Björkqvist, A.M.; el-Rifai, W.; Larramendy, M.L.; Monni, O.; Szymanska, J. Comparative genomic hybridization study on pooled DNAs from tumors of one clinical-pathological entity. *Cancer Genet. Cytogenet.* **1998**, *100*, 25–30. [CrossRef]

33. Cao, Y.; Bryan, T.M.; Reddel, R.R. Increased copy number of the TERT and TERC telomerase subunit genes in cancer cells. *Cancer Sci.* **2008**, *99*, 1092–1099. [CrossRef] [PubMed]

34. Fizazi, K.; Caliandro, R.; Soulié, P.; Fandi, A.; Daniel, C.; Bedin, A.; Doubre, H.; Viala, J.; Rodier, J.; Trandafir, L.; et al. Combination raltitrexed (Tomudex®)-oxaliplatin: A step forward in the struggle against mesothelioma? The Institute Gustave Roussy experience with chemotherapy and chemo-immunotherapy in mesothelioma. *Eur. J. Cancer* **2000**, *36*, 1514–1521. [CrossRef]

International Journal of
Molecular Sciences

MDPI

Review

Mesotheliomas in Genetically Engineered Mice Unravel Mechanism of Mesothelial Carcinogenesis

Didier Jean [1,2,3,4] and Marie-Claude Jaurand [1,2,3,4,*]

[1] Inserm, UMR-1162, Génomique Fonctionnelle des Tumeurs Solides, F-75010 Paris, France; didier.jean@inserm.fr
[2] Université Paris Descartes, Labex Immuno-Oncologie, Sorbonne Paris Cité, F-75000 Paris, France
[3] Institut Universitaire d'Hématologie, Université Paris Diderot, Sorbonne Paris Cité, F-75010 Paris, France
[4] Université Paris 13, Sorbonne Paris Cité, F-93206 Saint-Denis, France
* Correspondence: marie-claude.jaurand@inserm.fr; Tel.: +33-(0)1-7263-9350

Received: 5 July 2018; Accepted: 23 July 2018; Published: 27 July 2018

Abstract: Malignant mesothelioma (MM), a rare and severe cancer, mainly caused as a result of past-asbestos exposure, is presently a public health concern. Current molecular studies aim to improve the outcome of the disease, providing efficient therapies based on the principles of precision medicine. To model the molecular profile of human malignant mesothelioma, animal models have been developed in rodents, wild type animals and genetically engineered mice harbouring mutations in tumour suppressor genes, especially selecting genes known to be inactivated in human malignant mesothelioma. Animals were either exposed or not exposed to asbestos or to other carcinogenic fibres, to understand the mechanism of action of fibres at the molecular level, and the role of the selected genes in mesothelial carcinogenesis. The aim of the manuscript was to compare mesothelioma models to human malignant mesothelioma and to specify the clue genes playing a role in mesothelial carcinogenesis. Collectively, MM models recapitulate the clinical features of human MM. At least two altered genes are needed to induce malignant mesothelioma in mice. Two pathways regulated by *Cdkn2a* and *Trp53* seem independent key players in mesothelial carcinogenesis. Other genes and pathways appear as bona fide modulators of the neoplastic transformation.

Keywords: malignant mesothelioma; mesothelium; mineral fibres; gene mutations; tumor suppressor genes; signalling pathways; carcinogenesis

1. Introduction

Human malignant mesothelioma (HMM) is a cancer with current poor outcome, which is diagnosed with advanced non-curable disease. HMM has a strong association with asbestos exposure, a natural mineral fibre. The present researches mainly aim to find efficient therapeutics. Many of the current studies focus on target therapy to counteract the physio-pathological mechanisms allowing mesothelioma cells to grow in and invade their microenvironment, and to escape from the immune survey. For that purpose, mesotheliomas are developed in so called "mesothelioma models", which include orthotopic or heterotopic xenografts of human mesothelioma cell lines and patient-derived xenografts in immunodeficient mice [1,2]. Moreover, experimental mesotheliomas models have been developed for different purposes. Malignant mesotheliomas (MM) models have been generated to understand the carcinogenic mechanism induced by asbestos fibres or to identify the most relevant genes and important signalling pathways associated to mesothelial cell transformation. This aim was developed with both in vitro studies on mammalian cells, including mesothelial cells, and in vivo studies in animals [3]. Efforts have been also made to generate MM in animals treated or not treated with asbestos fibres. More recently, recombinant inbred mouse lines were designed to determine the genetic bases of the disease. In this context, genetically engineered mice (GEM) carrying genes modified to mimic the human disease were chosen and exposed or not to carcinogenic fibres. These experiments allow comparison between mesotheliomas developed in different genetic context and possibly emphasise specific clinical and molecular features.

The application of target therapy needs a deep knowledge of the tumour microenvironment characteristics to permit an appropriate way to suppress tumour cell proliferation, survival, migration, invasiveness and impair the interactions with the microenvironment as final outcome to eradicate the tumour. The different animal models may bring some relevant knowledge of the specific molecular pattern of the tumours and of the disease. In this review, we will discuss the features of mesothelioma induced in animals and to what extent they are close to the HMM.

2. Human Malignant Mesothelioma

2.1. Human Malignant Pleural Mesothelioma

The clinical and pathological features of pleural MM will be briefly summarised here. Several reviews can be suggested to the reader [4–6].

2.1.1. Natural History

The major risk factor for malignant pleural mesothelioma (MPM) is a past exposure to asbestos fibres, and more than 80% of MM are located in the pleura as a result of inhalation exposure. MPM occurs after a long delay, up to 40 years, after the beginning of exposure. However, malignant peritoneal mesothelioma (MPeM) is also found in asbestos-exposed patients, exceptionally in the testis [7]. MPM can be found in populations not exposed for occupational reasons, but showing domestic or environmental exposures [8–11]

2.1.2. Histological Classification

On the basis of histological morphology, MM is divided into three major histologic types, epithelioid, sarcomatoid, or mixed (biphasic) categories. Epithelioid and sarcomatoid categories have several secondary growth patterns as reported by Hussain et al. [12].

2.1.3. Physiopathology

Mesothelial cells form a monolayer at the surface of the mesothelium. Their cellular morphology is not uniform, depending on the regional location with flattened, intermediate, cuboidal and microvilli-rich mesothelial cells. Mesothelial cells play an important role in maintaining pleural homeostasis [13]. Pleuro-lymphatic communication is made through stomas [14]. In human, stomas open at the mesothelial surface and extend into a lymphatic capillary connected to the submesothelial lymphatics [15]. Inhaled asbestos fibres are deposited in the respiratory airways, reach the lung and are translocated into the pleura. The presence of fibres has been demonstrated both in the human pleura and in animals [16,17].

2.1.4. Molecular Alterations in MPM

Many publications have reported molecular alterations in MPM (see for review [6,18]). They concern copy number alterations (CNAs) of chromosome regions, gene mutations and epigenetic modifications. One recurrent finding is the numerous chromosome rearrangements, with several specific chromosomal gains on 1, 5, 7 and 17 or losses on 1, 3, 4, 6, 9, 13 and 22 [19]. Losses in 3p21, 9p21, 14q and part or whole chromosome 22 were recurrently observed. These loci contain many tumour suppressor genes (TSGs) such as *BAP1*, *CDKN2B*, *CDKN2A*, and *NF2* which are frequently inactivated. Other genes of interest, *LATS2*, *SETD2* and *TP53* are inactivated at a lower extent [20–22]. A loss on the chromosome region 14q11.2–q21 was the only difference detected between patients exposed (loss) and not exposed (no loss) to asbestos [23,24]. Gene alterations consisted in base substitution, apuric or apurinic base losses, deletion of one or several exons, or the whole gene. Gene fusions and splice alterations were also described mostly in *NF2*, *BAP1* and *SETD2* genes [22]. So far, no recurrent oncogene was found altered in MM, but an oncogenic hotspot mutation was reported in the promoter of *TERT* in 15% MPM [25]. However amplification of oncogenes such as *PDGFRB*, *MYC* or *VEGFR* could play a role in mesothelial neoplastic transformation [26–28].

Investigation of epigenetic changes demonstrated changes in gene methylation, and differential expression in non-coding RNAs such as microRNAs and long non-coding RNAs in comparison with normal cell [29]. It is known that miRNAs interact with the regulation of oncogenes and TSGs and can work either as oncogenes or TSGs [30]. Methylome analyses have shown a variety of methylation profiles in MPM, and an association with asbestos exposure [31]. Analysis of promoter methylation of cell cycle control genes showed that the number of methylated genes was a predictor of asbestos exposure [32]. MiRNome analyses also revealed differential expression between MPM and normal counterparts, between MPM and reactive pleural cells and between histological categories [33–35].

2.1.5. Alterations in Regulatory Pathways

Whole genomic and transcriptomic analyses have emphasised the regulatory pathways activated or inactivated in MPM. Hippo and PI3K/AKT/mTOR are deregulated either because of the mutation in critical genes of the pathway and/or inappropriate activation of members of the pathway [36–38]. Other regulatory pathways that play a role in development and cell differentiation are reported to be differentially activated in comparison with normal cells, Hedgehog that is associated with the maintenance of cancer stem cells, and Wnt, a pathway, which plays a role in intracellular signal traffic [39–41]. Important deregulation of the mitotic spindle assembly checkpoint pathway (MSAC) and microtubule network has been reported in MPM, although no mutation was detected in these genes [42]. Highest levels of expression of genes of the MSAC pathway, notably in sarcomatoid MPM [42].

2.1.6. Molecular Classification of MPM

In addition to histological classification, molecular classification of MPM was performed from trancriptomic analyses. Studying primary MPM cultures and tumour samples by transcriptomic microarray resulted in the definition of two molecular classes (C1 and C2) [43]. Gene mutations were investigated in selected genes *BAP1*, *CDKN2A*, *CDKN2B*, *NF2* and *TP53*. Briefly, *BAP1* alterations were more frequent in C1 and epithelioid MPM were found in both groups, with a worse survival prognosis in the C2 subgroup. Pathway analysis revealed that EMT was differentially regulated between MPM subgroups [43].

In an extensive study, transcriptomes, whole exomes (*n* = 99) and targeted exomes were analysed in MPM tumours [22]. Using RNA-seq data, four molecular subtypes were defined, sarcomatoid, epithelioid, biphasic-epithelioid (biphasic-E) and biphasic-sarcomatoid (biphasic-S). In this study, genes significantly mutated were identified: *BAP1*, *NF2*, *TP53*, *SETD2*, *DDX3X*, *ULK2*, *RYR2*, *CFAP45*, *SETDB1* and *DDX51*, and a multitude of mutations in several genes. These mutations result in the alteration of several signal pathways such as Hippo, mTOR, histone methylation, RNA helicase and p53 pathways. Hippo pathway was altered in all molecular subtypes, mTOR more in biphasic-S. Histone methylation and *BAP1* alteration were more frequent in epithelioid MPM. Six mutation signatures were identified, but none was associated to asbestos exposure [22].

Gene expression was also investigated to differentiate MPM cells and benign mesothelial hyperplasia (MH) using NanoString technologies in tumour tissues [33]. One hundred and seventeen genes were selected. An unsupervised cluster analysis defined two clusters, one composed only of MPM and one only of MH samples. Interestingly, this approach identified already known mesothelioma genes, *BAP1* and *NF2* being downregulated, and *MSLN*, which encodes mesothelin, upregulated in MPM in comparison with MH. In contrast, *CDKN2A* was not statistically deregulated in MPM in comparison with MH [33]. This suggests different roles of these genes in the neoplastic progression of mesothelial cells.

2.1.7. MPM Response to Treatments

There is agreement that globally, MPM survival is dependent on the histological subtype; epithelioid mesothelioma having better prognosis that sarcomatoid mesothelioma. The recent molecular analyses have shown that the outcome of MPM is also related to the molecular group with differential outcome within epithelioid mesothelioma [43].

2.2. Malignant Peritoneal Mesothelioma

MPeM also found as a result of asbestos inhalation, is reported as slightly different from MPM. As in MPM, the major histologic types of MPeM as in MPM are found, with the epithelioid type being the most frequent. Histologic variants comprise heterologous (osteosarcomatous, chondrosarcomatous, and rhabdomyosarcomatous) elements and desmoplastic mesothelioma [44]. However, MPeM shows differences with MPM in terms of survival, which is longer than MPM. The main risk factor remains asbestos exposure in about 50% of the cases, lower than in MPM [45].

Genome wide analysis of epithelioid MPeM and MPM showed similarities in CNAs [24]. Overall, regions of copy number gain were more common in MPeM, whereas losses were more common in pleural MPM. Losses occurring in 3p, 9p and 22q genomic regions carrying the TSGs *BAP1*, *CDKN2A* and *NF2*, respectively were seen at a statistically significant higher rate MPM than in MPeM [24]. The authors studied CNAs in groups of different exposures and found different results. Patients with history of medical radiation exposure showed multiple regions of gain, including 1q, 3p, 3q and 5p. Region of losses in 6q, 14q, 17p and 22q and gains 7q, 10p, 10q, 17q were found in tumours from asbestos-exposed patients [24]. Reccurent mutations are also found in similar genes than MPM [46], even if specific alterations were described in subgroup of MPeM such as *ALK* rearrangement [47].

2.3. Conclusions on Human Malignant Mesothelioma Biology

HMM appears to have a spectrum of different features. First, MM can grow in the serosa of the pleura, peritoneum, pericardium or tunica vaginalis. The MM tumour morphology is heterogeneous. MM cells in different tumours differ by their physiological and genomic status, and relationship with their microenvironment. Although some physiological and molecular alterations are recurrently found in mesothelioma cells, sometimes at a high rate, given tumours have specific features that need to be known to more precisely define groups of tumours and perform precise therapeutics. In the following, it is discussed to what extent models of MM are close to HMM.

3. Models of Malignant Mesothelioma

Mesotheliomas have been developed in rodents by injection of asbestos fibres in wild type (WT) rats or in mice and GEM mice, exposed or non-exposed to asbestos, refractory ceramic fibres (RCF) or carbon nanotubes (CNT).

3.1. Spontaneous Mesotheliomas in Wild Type Rodents

Spontaneous mesotheliomas that occur in control or sham cohorts in toxicological studies using rats are rare events. An incidence of 4.3% (7/395) of genital and serosal mesotheliomas, and only one pleural mesothelioma has been reported in male rats, with a variety of morphological patterns [48]. More recently, 0.2–5% mesotheliomas of the tunica vaginalis (MTV) were classified as epithelioid, sarcomatoid of mixed, consistent with the histologic classification in HMM [49]. Spontaneous mesotheliomas were reported in male F344/N rats controls in a summary over 5 decades from 2-year National Toxicology Program carcinogenicity bioassays. The frequency was 0.2–5% MTV [49]. Spontaneous mesothelioma is also rare in mice [50–52].

3.2. Mesothelioma in Animal Experiments

3.2.1. Asbestos-Induced Mesotheliomas in WT Animals

These studies were carried out mainly in rats, less in mice. The aim was to investigate the carcinogenicity of different types of fibres [53]. Rats were exposed by inhalation, intra-tracheal instillation or intra-serosal administration. Lung tumours and mesotheliomas were observed at different rates, depending on the route of exposure and fibre type [54]. The natural history of mesotheliomas showed similarities with HMM, they occurred after a long delay and ascites developed after exposure via the intra-peritoneal route. Histological analyses reported similar features as found in HMM, but epithelioid is not the most frequent histologic category. For instance, after administration in the pleural cavity of rats, reported histologic types were tubulo-papillary (a category of pleural epithelioid, 8.2%), mixed (74.8%) and spindle (16.9%) MM [55].

Recently, a whole exome sequencing of asbestos-induced murine mesotheliomas (MuMM) was performed in 3 different strains of WT mice stains, BALB/c, CBA and C57BL/6, and 15 MM cell lines were analysed, obtained from 4, 4 and 6 ascites, respectively [56]. In all but one cell line, recurrent genomic changes included homozygous (Hom) loss of *Cdkn2a* (this gene encodes two proteins, p16$^{\text{Ink4a}}$ and p19^{Arf}) and deletion in *Lats2* and *Setd2*, but no mutation in *Bap1* or *Nf2*. Hom loss of *Trp53* was found in one cell line. Mutation signature was principally C to T, as found in HMM, and G to A transitions, but transversions were also found. BALB/c cell lines carried more mutations than the others. Several pathways were affected by mutations such as Wnt, Hedgehog, Notch, mTOR, MAPK and p53 pathways, but not Hippo [56]. These results suggest a unique key role of *Cdkn2a* in murine mesothelial carcinogenesis. Moreover, mesotheliomas arose in the absence of alteration of *Bap1* and *Nf2*, as in HMM, consistent with a role of other pathways affected by the genes mutated at low frequency, or epigenetic mechanism.

An epigenetic mechanism of inactivation of *Cdkn2a* locus was suggested to be an initial step of MuMM induction, leading later to allelic deletion of Arf, in WT mice exposed to CNT by intrapleural instillation [57].

3.2.2. Mesothelioma in GEM

Spontaneous MuMM

GEM heterozygous (Htz) or homozygous (Hom) in *Nf2*, *Bap1*, *Cdkn2a* (*Ink4a* and/or *Arf*), *Trp53* or *Bap1* genes, either alone or in combination, were generated, based on the knowledge of the TSGs genes playing a role in mesothelial carcinogenesis. GEM in *Rb*, *Tsc1* and *Pten* were also generated despite the absence of mutation in HMM [58–61]. Tables 1 and 2 summarize the different studies carried out with GEM.

One MuMM was reported (6%) in *Nf2*$^{\text{KO3/+}}$ carrying the loss of *Nf2* exon 3 [8]. Jongsma et al. [59] injected AdCre in the pleural cavity of mice carrying conditional mutant alleles in *Nf2*, *Cdkn2a*, *Trp53* or *Rb*, and Htz *Ink4a* mutant [59]. The highest rate of thoracic MuMM was observed in double mutants *Nf2* and *Cdkn2a*, *Trp53* or *Rb* and triple mutants *Nf2*, *Trp53* and *Ink4a*. Mutations in *Cdkn2a*, *Ink4a* or *Trp53* were the most pejorative in term of MuMM incidence. *Rb* inactivation induces the lowest incidence of MuMM. Hom *Nf2* enhanced tumours rate in *Rb* mutants [59]. A majority of epithelioid mutants was only found in Hom *Nf2*/Htz *Trp53* mice. Guo et al. [58] injected AdCre in the peritoneal cavity or in the bladder, in conditional mutants *Trp53* and *Tsc1*. High rate of MuMM was found in double Hom *Trp53*/*Tsc1* mutants, but none in Htz/Hom mutants. MM were mostly of epithelioid type [58]. Hom *Pten* leads to MuMM with a frequency of 7% in mice, but when coupled with Hom *Trp53*, 56% of mice developed pleural MuMM. The histologies of Hom *Pten* and *Trp53* MuMM were sarcomatoid and biphasic [61].

Three types of Htz *Bap1* mutants were generated in mice, one was knockout in exons 6 and 7 of *Bap1*, and the two others with point mutations identical to germline mutations found in two human families (W and L, respectively) with a *BAP1* cancer syndrome presenting mesothelioma in several family members [60]. Htz germline mutations in *BAP1* predispose to a range of benign and malignant tumours, including mesothelioma. In Htz mice, although numerous types of cancers were developed, mesothelioma was absent or rare (2/93 Htz mice) and developed after a long delay (19 and 29 months). The tumour type with the highest incidence was ovarian sex cord stromal tumours, 38 of 60 (63%) in *Bap1* mutant mice.

Collectively, the results show a differential role of the altered genes. Data from Jongsma et al. [59] suggest a prominent role of *Cdkn2a* and *Trp53*, compared to *Nf2*, as mice harbouring Hom *Nf2* and either Htz *Cdkn2a* or Htz *Trp53* have longer survival than mice with Hom *Cdkn2a* or *Trp53* and Htz *Nf2*. However, Htz *Trp53* in association with Htz or Hom *Tcs1* did not induce MM, contrary to its association with *Nf2*, but consistent with a bona fine role of *Nf2 in MM* [58,59]. Results also showed that *Bap1* Htz mutations are not sufficient to induce MuMM [60]. All histologic types of mesotheliomas were observed, with a majority of mixed and sarcomatoid types, with the exception of epithelioid type in *Tsc1/Trp53* mice. Despite the different genetic background of mice, these studies underline several key genes for MM, consistent with findings in HMM, and that MM can develop with a variety TGS mutations, and likely with more than one TSG.

MuMM in Mice Exposed to Carcinogenic Fibres

Mice harbouring Htz genes *Nf2*, *CdKn2a* (*Ink4a* and/or *Arf*), *Trp53* or *Bap1* and their WT counterparts were exposed to carcinogenic fibres administered intra-peritoneally [60,62–68]. In one study, both *Nf2* and *Cdkn2a* were HTz [69]. MuMM arose in both WT and Htz mice, more frequently and with a shorter survival in Htz mice than in their WT counterparts, showing the role of these genes in enhancing sensitivity to fibres. MuMM generally arise after a long delay, often preceded by the occurrence of ascetic fluid. MM were detected several months after exposure, 18 and 27 weeks In Htz *Nf2* mice [66,69] and 21 to 37 weeks in Htz *Cdkn2a*, *Ink4a* or *Arf* [63]. Median survivals were around one year or more. From the number of MuMM or lag time after fibre exposure in different genetic situations, it is difficult to establish a hierarchy between genes, because of the variety of protocols between studies (mice strains, dose and schedule of exposure, fibre type). Htz *Trp53* mice were also developed high rate of MuMM when exposed to asbestos or to CNT [68,70].

Additionally, genes other than TSG such as *Asc*, relevant of inflammatory process, was also Hom- or Htz-inactivated in GEM [71]. Inactivation of *Asc* in GEM non-significantly reduced the percentage of mice with MuMM, but the disease-free survival was significantly lowered. These results suggest a role of inflammation in disease progression and the authors showed a relation with IL1b/IL1R signalling [71].

Asbestos induces MuMM in MexTAg transgenic mice that carry a fragment of the Simian Virus 40 (SV40) TAg open reading frame [72]. These MuMM replicate many aspects of MM at the molecular level, but MuMM development was not dependent on *Cdkn2a*, likely attributable to the Tag expression [73].

3.2.3. Mutation Profiles in MuMM of Mice Exposed to Fibres

Genetic alterations have been studied in MM cells cultured from ascitic fluids in fibre-exposed GEM. In MM cells from Htz *Nf2* mice, a loss of heterozygosity (LOH) of *Nf2* was found in all or a majority of MM cell lines from Htz *Nf2* mice, 85% (6/7), 83.5% (10/12) and 100%, respectively [64–66]. Inactivation of *Cdkn2a* and *Cdkn2b* was predominant, and resulted from biallelic deletions. Otherwise, co-deletion of *Cdkn2a* (*Ink4a* and *Arf*) and *Cdkn2b* was predominant [64,65,74]. Rates of *Trp53* mutations were less frequent, about 20% as in HMM [64,65,74]. Two cell lines with alteration of *Trp53* were *Cdkn2a* (*Ink4a* and *Arf*) and *Cdkn2b* WT, suggesting two different pathways of carcinogenesis [74]. A role of the hippo pathway is suggested by the activation of Yap/Taz in tissue from asbestos-exposed Htz *Nf2* mice, as shown by its nuclear localisation [75].

Altomare et al. [62] reported biallelic inactivation of *Arf* in all cell lines from Htz *Arf* mice, in 3/7 from WT mice, and no deletion of *Ink4a* or *Ink4b* (*Cdkn2b*) in all but one cell line from these mice, and no loss of p53 protein. However, one WT cell line showed loss of *Trp53* and p53 and retention of both *Cdkn2a* and *Cdkn2b*. Most of MM cells from Htz *Arf* mice showed hemizygous loss of *Faf1* and down-regulation of its protein, which regulated TNF-α-mediated NF-κB signalling pathway in these cells. Accordingly, in Htz *Cdkn2a* (*Ink4a* and *Arf*) mice, a biallelic loss of both *Ink4a* and *Arf* was observed, with protein loss of p16^{Ink4a} and p19^{ARF}, and in Htz *Ink4a* mice, there was a biallelic inactivation of *Ink4a*, loss of p16^{Ink4a} or p53, and frequent loss p15^{Inkba} and p19^{Arf}, but one cell line from Htz *Ink4a* mice expressed p19^{Arf} but did not express p53 [63]. In the three configurations of Htz *Ink4a*, *Arf* or *Cdkn2a* (*Ink4a* and *Arf*), nearly all cell lines expressed N*f2* and p53 [63]. The reciprocity between retention and loss of *Cdkn2a* (*Ink4a* and *Arf*) and *Trp53* expression of p53 consistent with an alternative role of the p53 pathway independently of hippo pathway and Ink4a regulation. These results suggest a major role of Arf in a context of fibre exposure and the role of alternative pathways in mesothelial carcinogenesis, as suggested above from the results obtained in Htz *Nf2* mice.

Molecular analyses of cell lines from Htz *Bap1* mice showed *Bap1* LOH, but no alteration of *Ink4a*, *Ink4b* and *Arf*, in contrast to WT mice that retain WT Bap1, but were deleted in *Ink4a*, *Ink4b* and *Arf*, suggesting two alternative mechanisms of MM development despite the fact that *CDKN2A* and *BAP1* mutations are not exclusive in HMM [76]. Rb protein was down regulated in cells from Htz *Bap1* mice due to aberrant epigenetic of the *Rb* promoter, suggesting a role of *Bap1* on *Rb* expression [76]. Fifty per cent of MM cell lines from ascites in asbestos-exposed Htz *Trp53* mice had loss of the WT allele. In addition while cell lines with no loss of WT allele were diploid, those with LOH were tetraploid, consistent with a genetic instability related to checkpoint.

In tissues from asbestos-exposed Htz *Nf2* mice, Rehrauer et al. [75] reported a higher number of mutations determined by RNA-seq, with an increase in A to G mutations, but not T to C, as compared to sham. This may be due to hydrolytic deamination of adenosine (*Ada*), as *Ada* expression is significantly increased, and linked to an *Adar* downstream activity [75].

Table 1. Induction of murine mesotheliomas (MuMM) in genetically engineered mice (GEM) (Injection of AdCre in GEM [1]).

Gene(s) Affected	Gene(s) Status	MuMM %	Epi [2] %	Sarco [2] %	Mixed %	Survival [3] Weeks	Reference
Nf2 / Ink4a/Arf	Htz / Hom	34[4]	28.6	21.4	50	58[5]	[ ]
Nf2 / Ink4a/Arf	Htz / Hom	34.6[4]	22.2	27.8	50	71[5]	[ ]
Nf2 / Ink4a/Arf	Htz / Hom	79[4]	2.2	68.9	28.9	31[5]	[ ]
Nf2 / Rb	Htz / Hom	5.9[4]	0	Primarily sarco	Some mixed	ND[7]	[ ]
Nf2 / Rb	Hom / Hom	13.3[4]	0	Primarily sarco	Some mixed	ND	[ ]
Nf2 / Rb	Htz / Hom	26.3[4]	0	Primarily sarco	Some mixed	ND	[ ]
Nf2 / Rb	Htz / Hom	6.75[6]	0	Primarily sarco	Some mixed	ND	[ ]
Nf2 / Rb	Htz / Hom	13.3[6]	0	Primarily sarco	Some mixed	ND	[ ]
Nf2 / Rb	Htz / Hom	20[6]	0	Primarily sarco	Some mixed	ND	[ ]
Nf2 / Trp53	Htz / Hom	59[4]	25	25	50	29[5]	[ ]
Nf2 / Trp53	Htz / Hom	25[4]	60	40	0	86[5]	[ ]
Nf2 / Trp53	Hom / Hom	82[4]	15.5	46.7	37.8	19[5]	[ ]
Nf2 / Trp53	Hom / Hom	93.7[4]	0	40	60	ND	[ ]
Nf2 / Trp53 / Ink4a	Htz / Hom / Hom	91.1[4]	0	76.6	23.4	11	[ ]
Tsc1 / Trp53	Hom / Hom	85[6]	Mostly			37	[ ]
Tsc1 / Trp53	WT / WT	0[6]	NA[7]			>57	[ ]
Tsc1 / Trp53	WT / WT	0[6]	NA			>57	[ ]
Tsc1 / Trp53	Hom / WT	0[6]	NA			>57	[ ]
Tsc1 / Trp53	Hom / Hom	73[8]	Mostly			44	[ ]
Tsc1 / Trp53	WT / WT	0[8]	NA			>57	[ ]
Tsc1 / Trp53	Hom / WT	0[8]	NA			>57	[ ]
Tsc1 / Trp53	WT / WT	0[8]	NA			>57	[ ]

[1] Strain of mice: FVB/N [59]; Hybrids [58]; [2] Epi.: Epithelioid; Sarco.: Sarcomatoid; [3] Median survival of the series; [4] After intrathoracic injection of AdCre; [5] Mice with thoracic tumours; [6] After intraperitoneal injection of AdCre; [7] ND: Not done; NA: Not applicable; [8] After injection of AdCre in the bladder.

Table 2. Induction of MuMM in GEM (Induction of MuMM by injection of fibres).

Mice Strain	Gene(s) Affected	Gene(s) Status	Treatment	MuMM %	Epi [2] %	Sarco [2] %	Mixed %	Survival [3] Weeks	Reference
FVB/N	Nf2	Htz	Asbestos	47	30.4 Htz + WT	65.2 Htz + WT	4.3 Htz + WT		[66]
FVB/N	None	WT	Asbestos	15	30.4 Htz + WT	65.2 Htz + WT	4.3 Htz + WT		[66]
FVB/N	Nf2	Htz	Saline	0	NA	NA	NA		[66]
FVB/N	None	WT	Saline	0	NA	NA	NA		[66]
129sv/Jae	Nf2	Htz	Asbestos	85	6.25	18.75	75	43	[64]
129sv/Jae	None	WT	Asbestos	59	31	27.6	41.4	52	[64]
FVB/N	Nf2	Htz	RCF	55	27	38.4	34.6	68	[65]
FVB/N	None	WT	RCF	7.1	0	0	100	80	[65]
C57/Bl6	Nf2	Htz	Asbestos	10	ND	ND	ND	ND	[75]
C57/Bl6	Arf	Htz	Asbestos	96.2	68	12	20	42	[62]
C57/Bl6	None	WT	Asbestos	81.5	68.2	18.2	13.6	56	[62]
Hybrids	Ink4a/Arf	Htz	Asbestos	88	Occasional	Prevalent	Occasional	29.6	[63]
Hybrids	Ink4a/Arf	Htz	TiO2	0	NA	NA	NA	NA	[63]
Hybrids	Ink4a	Htz	Asbestos	66	Occasional	Prevalent	Occasional	34.6	[63]
Hybrids	Arf	WT	Asbestos	65	Occasional	Prevalent	Occasional	38	[63]
Hybrids	None	Htz	Asbestos	50	Occasional	Prevalent	Occasional	49.4	[63]
Hybrids	Nf2	Htz	Asbestos	ND	ND	Most sarcomatous	ND	38	[69]
Hybrids	Nf2Ink4a/Arf	HtzHtz	Asbestos	ND	ND	ND	ND	24	[69]
Hybrids	None	WT	Asbestos	ND	ND	ND	ND	45	[69]
129/Sv on a 75% background	Trp53	Htz	Asbestos	76 (after 44 weeks)	ND	ND	ND		[68]
C57/Bl6 background	Trp53	Hom	Asbestos	ND	ND	ND	ND		[68]
129/Sv on a 75% background	None	WT	Asbestos	32 (after 67 weeks)	ND	ND	ND		[68]
FVB	Bap1	Htz	Asbestos	73	ND	ND	ND	43	[76]
FVB	None	WT	Asbestos	32	ND	ND	ND	55	[76]
FVB	Bap1	Htz (L)	Asbestos	71	ND	ND	ND	46	[60]
FVB	Bap1	Htz (W)	Asbestos	74	ND	ND	ND	48	[60]
FVB	None	WT	Asbestos	35	ND	ND	ND	60	[60]
C57BL/6	Bap1	Htz	Asbestos low dose	36	all or part	all or part	ND	57	[67]
C57BL/6	None	WT	Asbestos low dose	10				57	[67]
C57BL/6	None	WT	Saline	0	NA	NA		NA	[67]
C57BL/6	Bap1	Htz	Asbestos std dose	60	all or part	all or part	ND	39	[67]
C57BL/6	Bap1	WT	Asbestos std dose	28	all or part	all or part		57	[67]
C57BL/6	Asc	Hom	Asbestos	55	0	75	25	66.2	[71]
C57BL/6	Asc	Htz	Asbestos	65	0	68	32	69.4	[71]
C57BL/6	None	WT	Asbestos	80	0	67	33	OK	[71]

[2] Epi.: Epithelioid; Sarco.: Sarcomatoid; [3] Median survival of the series.

4. Discussion

Literature data on MM in rodents led us to consider several issues concerning the molecular mechanism of mesothelial cell transformation, and its relationship with exposure to mineral and synthetic fibres. Most studies showed remarkable similarities between human and rodent MM. In both species, MM is a rare spontaneous cancer that is found in the similar sites, pleura, peritoneum and tunica vaginalis. When exposed to carcinogenic fibres, MM occurs after a long delay post-exposure, and all histological categories are observed. From studies carried out in GEM, no single gene predisposes to MM since MuMM are only in fibre-exposed mice, but asbestos is a powerful agent to facilitate the development of MM. MuMM were developed in mice harbouring Htz and Hom inactivation of TSG, or Hom and Hom inactivation.

In WT animals, exposure to fibres induces a significant incidence of MPM or MPeM, depending on the route of administration, respectively, in both rats and mice. The animals were symptomatic, showing ascites after intra-peritoneal administration of fibres. When reported, early MM appeared after several months, and were further detected during the whole life time of the species. In mice, the median survival in animals was about more than one year, except in Hom *Nf2/Trp53* and Hom *Nf2/Trp53/Ink4a*. The survival was lower in asbestos-exposed GEM mice.

Although no precise quantitative data in the distribution of histological categories are given, the epithelioid type is not the most frequent in WT rodents and in GEM. In GEM the most frequent categories are sarcomatoid or mixed MPeM. In contrast, the epithelioid type is the most frequent human MPeM. However, a prevalence of epithelioid MPeM was reported in GEM Hom *Trp53/Tsc1* not exposed to asbestos [58], and in both WT and Htz *Arf* asbestos-exposed mice [62].

Investigations of spontaneous MM in GEM harbouring co-mutations in TSG showed that two genes, *Cdkn2a* and *Trp53* are predominant for MM development, as biallelic inactivation generates the highest rate of MM [58,59]. This is found despite the biallelic inactivation of *Nf2*, suggesting a cooperative but not predominant role of this gene [59]. Accordingly, in asbestos-exposed Htz *Nf2* mice, *Nf2* LOH is associated to loss of *Cdkn2a* and/or *Cdkn2b*. A key role of *Cdkn2a* and *Trp53* is also seen when using cell cultured from ascites fluids from Htz *Cdkn2a*, *Ink4a* and *Arf* GEM. Among genes encoded at the *Ink4* locus (*Ink4a*, *Ink4b* and *Arf*), *Trp53* biallelic inactivation is an alternative mechanism to carcinogenesis via genes inactivated at the *Ink4* locus. Of note, TP53 mutations are found in 11% of HMM (Cosmic database v85, https://cancer.sanger.ac.uk). Interestingly, in Htz *Bap1* mice, *Cdkn2a* or *Cdkn2a*b are not inactivated in MM, in contrast to MM with WT Bap1 where both genes are lost, but *Rb* down regulation was evidenced [76]. Independently of the inactivation of TSG already known to be involved in MM, mutations in genes involved in other regulatory pathways act as complementary mechanism accounting for mesothelial carcinogenesis.

As a whole, these studies brought information on the molecular changes in MM. A few genes are key players in the carcinogenic process. Others are bona fide modulators, which may be requested to favour the progression of the tumour, due to their involvement in signal or metabolic pathways. The diversity of mutated genes, the complex combination of altered genes, and the variety of associated deregulated pathways, lead to the heterogeneity of the tumour molecular profiles and is in agreement with the inter-tumour heterogeneity observed in HMM.

5. Conclusions

The data on asbestos-exposed mice do not bring significant information on the mechanism of genotoxicity of asbestos fibres. A better identification of the mutation signatures, characterisation of deleted regions and break points localisation and epigenetic changes, in both MM tumours and MM cell lines could help understanding the mechanism genome damage [77]. Inflammation is thought to act as a contributor, but it is not known whether it is the driving force for DNA damaging at lower doses than required in experiments [78]. Events entailing gene deletion and rearrangements should be considered. The contribution of gene methylation is not enough documented, but *Rb* is regulated by DNA methylation in Htz *Bap1* mice [76]. Jongsma et al. [59] reported that epigenetic inactivation of *Ink4a*, although enhancing the malignancy, does not contribute to the development of pleural MuMM in Htz *Nf2/Trp53*, in agreement with the evidence of deletions of this gene demonstrated in several studies [59]. Nevertheless, hypermethylation of *Cdkn2a* locus preceding allelic *Arf* deletion was suggested to be a mesothelial carcinogenesis step in pleura of mice exposed to CNT [57]. These studies have emphasised the diversity of the molecular events entailing the development of MM in experimental animals, and their consistency with the molecular status of HMM, in term of key genes and pathways, and potent modulators of tumour progression.

Author Contributions: Both D.J. and M.-C.J. equally contributed to the redaction of the manuscript.

Conflicts of Interest: The authors declare no conflict of interest.

Abbreviations

CNT	Carbon Nanotubes
GEM	Genetically Engineered Mice
HMM	Human Malignant Mesothelioma
MM	Malignant Mesothelioma
MPeM	Malignant Peritoneal Mesothelioma
MuMM	Murine Malignant Mesothelioma

References

1. Boyer, A.; Pasquier, E.; Tomasini, P.; Ciccolini, J.; Greillier, L.; Andre, N.; Barlesi, F.; Mascaux, C. Drug repurposing in malignant pleural mesothelioma: A breath of fresh air? *Eur. Respir. Rev.* **2018**, *27*, 170098. [CrossRef] [PubMed]

2. Wu, L.; Allo, G.; John, T.; Li, M.; Tagawa, T.; Opitz, I.; Anraku, M.; Yun, Z.; Pintilie, M.; Pitcher, B.; et al. Patient-Derived Xenograft Establishment from Human Malignant Pleural Mesothelioma. *Clin. Cancer Res.* **2017**, *23*, 1060–1067. [CrossRef] [PubMed]

3. Singh, A.; Pruett, N.; Hoang, C.D. In vitro experimental models of mesothelioma revisited. *Transl. Lung Cancer Res.* **2017**, *6*, 248–258. [CrossRef] [PubMed]

4. Katzman, D.; Sterman, D.H. Updates in the diagnosis and treatment of malignant pleural mesothelioma. *Curr. Opin. Pulm. Med.* **2018**, *24*, 319–326. [CrossRef] [PubMed]

5. Lacourt, A.; Leveque, E.; Goldberg, M.; Leffondre, K. Dose-time response association between occupational asbestos exposure and pleural mesothelioma: Authors' response. *Occup. Environ. Med.* **2018**, *75*, 161–162. [CrossRef] [PubMed]

6. McCambridge, A.J.; Napolitano, A.; Mansfield, A.S.; Fennell, D.A.; Sekido, Y.; Nowak, A.K.; Reungwetwattana, T.; Mao, W.; Pass, H.I.; Carbone, M.; et al. Progress in the Management of Malignant Pleural Mesothelioma in 2017. *J. Thorac. Oncol.* **2018**, *13*, 606–623. [CrossRef] [PubMed]

7. Serio, G.; Pagliarule, V.; Marzullo, A.; Pezzuto, F.; Gentile, M.; Pennella, A.; Nazzaro, P.; Buonadonna, A.L.; Covelli, C.; Lettini, T.; et al. Molecular changes of malignant mesothelioma in the testis and their impact on prognosis: Analyses of two cases. *Int. J. Clin. Exp. Pathol.* **2016**, *9*, 7658–7667.

8. Goswami, E.; Craven, V.; Dahlstrom, D.L.; Alexander, D.; Mowat, F. Domestic asbestos exposure: A review of epidemiologic and exposure data. *Int. J. Environ. Res. Public Health* **2013**, *10*, 5629–5670. [CrossRef] [PubMed]

9. Noonan, C.W. Environmental asbestos exposure and risk of mesothelioma. *Ann. Transl. Med.* **2017**, *5*, 234. [CrossRef] [PubMed]

10. Serio, G.; Pezzuto, F.; Marzullo, A.; Scattone, A.; Cavone, D.; Punzi, A.; Fortarezza, F.; Gentile, M.; Buonadonna, A.L.; Barbareschi, M.; et al. Peritoneal Mesothelioma with Residential Asbestos Exposure. Report of a Case with Long Survival (Seventeen Years) Analyzed by Cgh-Array. *Int. J. Mol. Sci.* **2017**, *18*, 1818. [CrossRef] [PubMed]

11. Vimercati, L.; Cavone, D.; Lovreglio, P.; De Maria, L.; Caputi, A.; Ferri, G.M.; Serio, G. Environmental asbestos exposure and mesothelioma cases in Bari, Apulia region, southern Italy: A national interest site for land reclamation. *Environ. Sci. Pollut.* **2018**, *25*, 15692–15701. [CrossRef] [PubMed]

12. Husain, A.N.; Colby, T.V.; Ordonez, N.G.; Allen, T.C.; Attanoos, R.L.; Beasley, M.B.; Butnor, K.J.; Chirieac, L.R.; Churg, A.M.; Dacic, S.; et al. Guidelines for Pathologic Diagnosis of Malignant Mesothelioma 2017 Update of the Consensus Statement From the International Mesothelioma Interest Group. *Arch. Pathol. Lab. Med.* **2018**, *142*, 89–108. [CrossRef] [PubMed]
13. Mutsaers, S.E.; Jaurand, M.C.; GaryLee, Y.C.; Prele, C.M. Mesothelial cells and pleural immunology. In *Textbook of Pleural Diseases*, 3nd ed.; Light, R.W., GaryLee, Y.C., Eds.; CRC Press: Boca Raton, FL, USA, 2016; pp. 27–44.
14. Wang, N.S. Anatomy of the pleura. *Clin. Chest Med.* **1998**, *19*, 229–240. [CrossRef]
15. Oshiro, H.; Miura, M.; Iobe, H.; Kudo, T.; Shimazu, Y.; Aoba, T.; Okudela, K.; Nagahama, K.; Sakamaki, K.; Yoshida, M.; et al. Lymphatic Stomata in the Adult Human Pulmonary Ligament. *Lymphat. Res. Biol.* **2015**, *13*, 137–145. [CrossRef] [PubMed]
16. Broaddus, V.C.; Everitt, J.I.; Black, B.; Kane, A.B. Non-neoplastic and neoplastic pleural endpoints following fiber exposure. *J. Toxicol. Environ. Health B Crit. Rev.* **2011**, *14*, 153–178. [CrossRef] [PubMed]
17. Fleury Feith, J.; Jaurand, M.C. Pleural lymphatics and pleural diseases related to fibres. *Rev. Pneumol. Clin.* **2013**, *69*, 358–362. [CrossRef] [PubMed]
18. Oehl, K.; Vrugt, B.; Opitz, I.; Meerang, M. Heterogeneity in Malignant Pleural Mesothelioma. *Int. J. Mol. Sci.* **2018**, *19*, 1603. [CrossRef] [PubMed]
19. Hylebos, M.; Van Camp, G.; Vandeweyer, G.; Fransen, E.; Beyens, M.; Cornelissen, R.; Suls, A.; Pauwels, P.; van Meerbeeck, J.P.; Op de Beeck, K. Large-scale copy number analysis reveals variations in genes not previously associated with malignant pleural mesothelioma. *Oncotarget* **2017**, *8*, 113673–113686. [CrossRef] [PubMed]
20. Jean, D.; Daubriac, J.; Le Pimpec-Barthes, F.; Galateau-Salle, F.; Jaurand, M.C. Molecular changes in mesothelioma with an impact on prognosis and treatment. *Arch. Pathol. Lab. Med.* **2012**, *136*, 277–293. [CrossRef] [PubMed]
21. Tranchant, R.; Quetel, L.; Tallet, A.; Meiller, C.; Renier, A.; de Koning, L.; de Reynies, A.; Le Pimpec-Barthes, F.; Zucman-Rossi, J.; Jaurand, M.C.; et al. Co-occurring Mutations of Tumor Suppressor Genes, LATS2 and NF2, in Malignant Pleural Mesothelioma. *Clin. Cancer Res.* **2017**, *23*, 3191–3202. [CrossRef] [PubMed]
22. Bueno, R.; Stawiski, E.W.; Goldstein, L.D.; Durinck, S.; De Rienzo, A.; Modrusan, Z.; Gnad, F.; Nguyen, T.T.; Jaiswal, B.S.; Chirieac, L.R.; et al. Comprehensive genomic analysis of malignant pleural mesothelioma identifies recurrent mutations, gene fusions and splicing alterations. *Nat. Genet.* **2016**, *48*, 407–416. [CrossRef] [PubMed]
23. Jean, D.; Thomas, E.; Renier, A.; de Reynies, A.; Lecomte, C.; Andujar, P.; Fleury-Feith, J.; Giovannini, M.; Zucman-Rossi, J.; Stern, M.H.; et al. Syntenic relationships between genomic profiles of fiber-induced murine and human malignant mesothelioma. *Am. J. Pathol.* **2011**, *176*, 881–894. [CrossRef] [PubMed]
24. Borczuk, A.C.; Pei, J.; Taub, R.N.; Levy, B.; Nahum, O.; Chen, J.; Chen, K.; Testa, J.R. Genome-wide analysis of abdominal and pleural malignant mesothelioma with DNA arrays reveals both common and distinct regions of copy number alteration. *Cancer Biol. Ther.* **2016**, *17*, 328–335. [CrossRef] [PubMed]
25. Tallet, A.; Nault, J.C.; Renier, A.; Hysi, I.; Galateau-Salle, F.; Cazes, A.; Copin, M.C.; Hofman, P.; Andujar, P.; Le Pimpec-Barthes, F.; et al. Overexpression and promoter mutation of the TERT gene in malignant pleural mesothelioma. *Oncogene* **2014**, *33*, 3748. [CrossRef] [PubMed]
26. Barone, E.; Gemignani, F.; Landi, S. Overexpressed genes in malignant pleural mesothelioma: Implications in clinical management. *J. Thorac. Dis.* **2018**, *10*, S369–S382. [CrossRef] [PubMed]
27. Riquelme, E.; Suraokar, M.B.; Rodriguez, J.; Mino, B.; Lin, H.Y.; Rice, D.C.; Tsao, A.; Wistuba, I.I. Frequent coamplification and cooperation between C-MYC and PVT1 oncogenes promote malignant pleural mesothelioma. *J. Thorac. Oncol.* **2014**, *9*, 998–1007. [CrossRef] [PubMed]
28. Tsao, A.S.; Harun, N.; Fujimoto, J.; Devito, V.; Lee, J.J.; Kuhn, E.; Mehran, R.; Rice, D.; Moran, C.; Hong, W.K.; et al. Elevated PDGFRB gene copy number gain is prognostic for improved survival outcomes in resected malignant pleural mesothelioma. *Ann. Diagn. Pathol.* **2014**, *18*, 140–145. [CrossRef] [PubMed]
29. Quinn, L.; Finn, S.P.; Cuffe, S.; Gray, S.G. Non-coding RNA repertoires in malignant pleural mesothelioma. *Lung Cancer* **2015**, *90*, 417–426. [CrossRef] [PubMed]
30. Vannini, I.; Fanini, F.; Fabbri, M. Emerging roles of microRNAs in cancer. *Curr. Opin. Genet. Dev.* **2018**, *48*, 128–133. [CrossRef] [PubMed]

31. Christensen, B.C.; Houseman, E.A.; Godleski, J.J.; Marsit, C.J.; Longacker, J.L.; Roelofs, C.R.; Karagas, M.R.; Wrensch, M.R.; Yeh, R.F.; Nelson, H.H.; et al. Epigenetic profiles distinguish pleural mesothelioma from normal pleura and predict lung asbestos burden and clinical outcome. *Cancer Res.* **2009**, *69*, 227–234. [CrossRef] [PubMed]

32. Christensen, B.C.; Godleski, J.J.; Marsit, C.J.; Houseman, E.A.; Lopez-Fagundo, C.Y.; Longacker, J.L.; Bueno, R.; Sugarbaker, D.J.; Nelson, H.H.; Kelsey, K.T. Asbestos exposure predicts cell cycle control gene promoter methylation in pleural mesothelioma. *Carcinogenesis* **2008**, *29*, 1555–1559. [CrossRef] [PubMed]

33. Bruno, R.; Ali, G.; Giannini, R.; Proietti, A.; Lucchi, M.; Chella, A.; Melfi, F.; Mussi, A.; Fontanini, G. Malignant pleural mesothelioma and mesothelial hyperplasia: A new molecular tool for the differential diagnosis. *Oncotarget* **2017**, *8*, 2758–2770. [CrossRef] [PubMed]

34. Busacca, S.; Germano, S.; De Cecco, L.; Rinaldi, M.; Comoglio, F.; Favero, F.; Murer, B.; Mutti, L.; Pierotti, M.; Gaudino, G. MicroRNA signature of malignant mesothelioma with potential diagnostic and prognostic implications. *Am. J. Respir. Cell Mol. Biol.* **2010**, *42*, 312–319. [CrossRef] [PubMed]

35. Reid, G. MicroRNAs in mesothelioma: From tumour suppressors and biomarkers to therapeutic targets. *J. Thorac. Dis.* **2015**, *7*, 1031–1040. [PubMed]

36. De Santi, C.; Melaiu, O.; Bonotti, A.; Cascione, L.; Di Leva, G.; Foddis, R.; Cristaudo, A.; Lucchi, M.; Mora, M.; Truini, A.; et al. Deregulation of miRNAs in malignant pleural mesothelioma is associated with prognosis and suggests an alteration of cell metabolism. *Sci. Rep.* **2017**, *7*, 3140. [CrossRef] [PubMed]

37. Jaurand, M.C.; Jean, D. Biomolecular Pathways and Malignant Pleural Mesothelioma. In *Malignant Pleural Mesothelioma: Present Status and Future Directions*; Mineo, T.C., Ed.; Bentham Science Publishers: Sharjah, UAE, 2015; pp. 173–196.

38. Sekido, Y. Molecular pathogenesis of malignant mesothelioma. *Carcinogenesis* **2013**, *34*, 1413–1419. [CrossRef] [PubMed]

39. Felley-Bosco, E.; Opitz, I.; Meerang, M. Hedgehog Signaling in Malignant Pleural Mesothelioma. *Genes* **2015**, *6*, 500–511. [CrossRef] [PubMed]

40. Gradilla, A.C.; Sanchez-Hernandez, D.; Brunt, L.; Scholpp, S. From top to bottom: Cell polarity in Hedgehog and Wnt trafficking. *BMC Biol.* **2018**, *16*, 37. [CrossRef] [PubMed]

41. Leprieur, E.G.; Jablons, D.M.; He, B. Old Sonic Hedgehog, new tricks: A new paradigm in thoracic malignancies. *Oncotarget* **2018**, *9*, 14680–14691. [CrossRef] [PubMed]

42. Suraokar, M.B.; Nunez, M.I.; Diao, L.; Chow, C.W.; Kim, D.; Behrens, C.; Lin, H.; Lee, S.; Raso, G.; Moran, C.; et al. Expression profiling stratifies mesothelioma tumors and signifies deregulation of spindle checkpoint pathway and microtubule network with therapeutic implications. *Ann. Oncol.* **2014**, *25*, 1184–1192. [CrossRef] [PubMed]

43. De Reynies, A.; Jaurand, M.C.; Renier, A.; Couchy, G.; Hysi, I.; Elarouci, N.; Galateau-Salle, F.; Copin, M.C.; Hofman, P.; Cazes, A.; et al. Molecular classification of malignant pleural mesothelioma: Identification of a poor prognosis subgroup linked to the epithelial-to-mesenchymal transition. *Clin. Cancer Res.* **2014**, *20*, 1323–1334. [CrossRef] [PubMed]

44. Churg, A.; Attanoos, R.; Borczuk, A.C.; Chirieac, L.R.; Galateau-Salle, F.; Gibbs, A.; Henderson, D.; Roggli, V.; Rusch, V.; Judge, M.J.; et al. Dataset for Reporting of Malignant Mesothelioma of the Pleura or Peritoneum: Recommendations From the International Collaboration on Cancer Reporting (ICCR). *Arch. Pathol. Lab. Med.* **2016**, *140*, 1104–1110. [CrossRef] [PubMed]

45. Bridda, A.; Padoan, I.; Mencarelli, R.; Frego, M. Peritoneal mesothelioma: A review. *MedGenMed* **2007**, *9*, 32. [PubMed]

46. Joseph, N.M.; Chen, Y.Y.; Nasr, A.; Yeh, I.; Talevich, E.; Onodera, C.; Bastian, B.C.; Rabban, J.T.; Garg, K.; Zaloudek, C.; et al. Genomic profiling of malignant peritoneal mesothelioma reveals recurrent alterations in epigenetic regulatory genes BAP1, SETD2, and DDX3X. *Mod. Pathol.* **2017**, *30*, 246–254. [CrossRef] [PubMed]

47. Hung, Y.P.; Dong, F.; Watkins, J.C.; Nardi, V.; Bueno, R.; Dal Cin, P.; Godleski, J.J.; Crum, C.P.; Chirieac, L.R. Identification of ALK Rearrangements in Malignant Peritoneal Mesothelioma. *JAMA Oncol.* **2018**, *4*, 235–238. [CrossRef] [PubMed]

48. Tanigawa, H.; Onodera, H.; Maekawa, A. Spontaneous Mesotheliomas in Fischer Rats—A Histological and Electron Microscopic Study. *Toxicol. Pathol.* **1987**, *15*, 157–163. [CrossRef] [PubMed]

49. Maronpot, R.R.; Nyska, A.; Foreman, J.E.; Ramot, Y. The legacy of the F344 rat as a cancer bioassay model (a retrospective summary of three common F344 rat neoplasms). *Crit. Rev. Toxicol.* **2016**, *46*, 641–675. [CrossRef] [PubMed]

50. Bomhard, E. Frequency of spontaneous tumours in NMRI mice in 21-month studies. *Exp. Toxicol. Pathol.* **1993**, *45*, 269–289. [CrossRef]

51. Peters, R.L.; Rabstein, L.S.; Spahn, G.J.; Madison, R.M.; Huebner, R.J. Incidence of spontaneous neoplasms in breeding and retired breeder BALB-cCr mice throughout the natural life span. *Int. J. Cancer* **1972**, *10*, 273–282. [CrossRef] [PubMed]

52. Shirai, M.; Maejima, T.; Tanimoto, T.; Kumagai, K.; Makino, T.; Kai, K.; Teranishi, M.; Sanbuissho, A. Mixed Type of Malignant Mesothelioma in an Aged Male ICR Mouse. *J. Toxicol. Pathol.* **2011**, *24*, 169–172. [CrossRef] [PubMed]

53. IARC. Man-made vitreous fibres and Radon. *IARC Monogr. Eval. Carcinog. Risks Hum.* **1988**, *43*, 1–300.

54. Boulanger, G.; Andujar, P.; Pairon, J.C.; Billon-Galland, M.A.; Dion, C.; Dumortier, P.; Brochard, P.; Sobaszek, A.; Bartsch, P.; Paris, C.; et al. Quantification of short and long asbestos fibers to assess asbestos exposure: A review of fiber size toxicity. *Environ. Health* **2014**, *13*, 59. [CrossRef] [PubMed]

55. Wagner, J.C.; Berry, G. Mesotheliomas in rats following inoculation with asbestos. *Br. J. Cancer* **1969**, *23*, 567–581. [CrossRef] [PubMed]

56. Sneddon, S.; Patch, A.M.; Dick, I.M.; Kazakoff, S.; Pearson, J.V.; Waddell, N.; Allcock, R.J.N.; Holt, R.A.; Robinson, B.W.S.; Creaney, J. Whole exome sequencing of an asbestos-induced wild-type murine model of malignant mesothelioma. *BMC Cancer* **2017**, *17*, 396. [CrossRef] [PubMed]

57. Chernova, T.; Murphy, F.A.; Galavotti, S.; Sun, X.M.; Powley, I.R.; Grosso, S.; Schinwald, A.; Zacarias-Cabeza, J.; Dudek, K.M.; Dinsdale, D.; et al. Long-Fiber Carbon Nanotubes Replicate Asbestos-Induced Mesothelioma with Disruption of the Tumor Suppressor Gene *Cdkn2a* (*Ink4a/Arf*). *Curr. Biol.* **2017**, *27*, 3302–3314.e6. [CrossRef] [PubMed]

58. Guo, Y.; Chirieac, L.R.; Bueno, R.; Pass, H.; Wu, W.; Malinowska, I.A.; Kwiatkowski, D.J. Tsc1-Tp53 loss induces mesothelioma in mice, and evidence for this mechanism in human mesothelioma. *Oncogene* **2014**, *33*, 3151–3160. [CrossRef] [PubMed]

59. Jongsma, J.; van Montfort, E.; Vooijs, M.; Zevenhoven, J.; Krimpenfort, P.; van der Valk, M.; van de Vijver, M.; Berns, A. A conditional mouse model for malignant mesothelioma. *Cancer Cell* **2008**, *13*, 261–271. [CrossRef] [PubMed]

60. Kadariya, Y.; Cheung, M.; Xu, J.; Pei, J.; Sementino, E.; Menges, C.W.; Cai, K.Q.; Rauscher, F.J.; Klein-Szanto, A.J.; Testa, J.R. Bap1 Is a Bona Fide Tumor Suppressor: Genetic Evidence from Mouse Models Carrying Heterozygous Germline Bap1 Mutations. *Cancer Res.* **2016**, *76*, 2836–2844. [CrossRef] [PubMed]

61. Sementino, E.; Menges, C.W.; Kadariya, Y.; Peri, S.; Xu, J.; Liu, Z.; Wilkes, R.G.; Cai, K.Q.; Rauscher, F.J., 3rd; Klein-Szanto, A.J.; et al. Inactivation of *Tp53* and *Pten* drives rapid development of pleural and peritoneal malignant mesotheliomas. *J. Cell. Physiol.* **2018**. [CrossRef] [PubMed]

62. Altomare, D.A.; Menges, C.W.; Pei, J.; Zhang, L.; Skele-Stump, K.L.; Carbone, M.; Kane, A.B.; Testa, J.R. Activated TNF-α/NF-κB signaling via down-regulation of Fas-associated factor 1 in asbestos-induced mesotheliomas from Arf knockout mice. *Proc. Natl. Acad. Sci. USA* **2009**, *106*, 3420–3425. [CrossRef] [PubMed]

63. Altomare, D.A.; Menges, C.W.; Xu, J.; Pei, J.; Zhang, L.; Tadevosyan, A.; Neumann-Domer, E.; Liu, Z.; Carbone, M.; Chudoba, I.; et al. Losses of both products of the *Cdkn2a/Arf* locus contribute to asbestos-induced mesothelioma development and cooperate to accelerate tumorigenesis. *PLoS ONE* **2011**, *6*, e18828. [CrossRef] [PubMed]

64. Altomare, D.A.; Vaslet, C.A.; Skele, K.L.; De Rienzo, A.; Devarajan, K.; Jhanwar, S.C.; McClatchey, A.I.; Kane, A.B.; Testa, J.R. A mouse model recapitulating molecular features of human mesothelioma. *Cancer Res.* **2005**, *65*, 8090–8095. [CrossRef] [PubMed]

65. Andujar, P.; Lecomte, C.; Renier, A.; Fleury-Feith, J.; Kheuang, L.; Daubriac, J.; Janin, A.; Jaurand, M.C. Clinico-pathological features and somatic gene alterations in refractory ceramic fibre-induced murine mesothelioma reveal mineral fibre-induced mesothelioma identities. *Carcinogenesis* **2007**, *28*, 1599–1605. [CrossRef] [PubMed]

66. Fleury-Feith, J.; Lecomte, C.; Renier, A.; Matrat, M.; Kheuang, L.; Abramowski, V.; Levy, F.; Janin, A.; Giovannini, M.; Jaurand, M.C. Hemizygosity of *Nf2* is associated with increased susceptibility to asbestos-induced peritoneal tumours. *Oncogene* **2003**, *22*, 3799–3805. [CrossRef] [PubMed]

67. Napolitano, A.; Pellegrini, L.; Dey, A.; Larson, D.; Tanji, M.; Flores, E.G.; Kendrick, B.; Lapid, D.; Powers, A.; Kanodia, S.; et al. Minimal asbestos exposure in germline BAP1 heterozygous mice is associated with deregulated inflammatory response and increased risk of mesothelioma. *Oncogene* **2016**, *35*, 1996–2002. [CrossRef] [PubMed]

68. Vaslet, C.A.; Messier, N.J.; Kane, A.B. Accelerated progression of asbestos-induced mesotheliomas in heterozygous *p53*$^{+/-}$ mice. *Toxicol. Sci.* **2002**, *68*, 331–338. [CrossRef] [PubMed]

69. Menges, C.W.; Kadariya, Y.; Altomare, D.; Talarchek, J.; Neumann-Domer, E.; Wu, Y.; Xiao, G.H.; Shapiro, I.M.; Kolev, V.N.; Pachter, J.A.; et al. Tumor suppressor alterations cooperate to drive aggressive mesotheliomas with enriched cancer stem cells via a p53-miR-34a-c-Met axis. *Cancer Res.* **2014**, *74*, 1261–1271. [CrossRef] [PubMed]

70. Takagi, A.; Hirose, A.; Nishimura, T.; Fukumori, N.; Ogata, A.; Ohashi, N.; Kitajima, S.; Kanno, J. Induction of mesothelioma in *p53*$^{+/-}$ mouse by intraperitoneal application of multi-wall carbon nanotube. *J. Toxicol. Sci.* **2008**, *33*, 105–116. [CrossRef] [PubMed]

71. Kadariya, Y.; Menges, C.W.; Talarchek, J.; Cai, K.Q.; Klein-Szanto, A.J.; Pietrofesa, R.A.; Christofidou -Solomidou, M.; Cheung, M.; Mossman, B.T.; Shukla, A.; et al. Inflammation-Related IL1beta/IL1R Signaling Promotes the Development of Asbestos-Induced Malignant Mesothelioma. *Cancer Prev. Res.* **2016**, *9*, 406–414. [CrossRef] [PubMed]

72. Robinson, C.; van Bruggen, I.; Segal, A.; Dunham, M.; Sherwood, A.; Koentgen, F.; Robinson, B.W.; Lake, R.A. A novel SV40 TAg transgenic model of asbestos-induced mesothelioma: Malignant transformation is dose dependent. *Cancer Res.* **2006**, *66*, 10786–10794. [CrossRef] [PubMed]

73. Robinson, C.; Dick, I.M.; Wise, M.J.; Holloway, A.; Diyagama, D.; Robinson, B.W.; Creaney, J.; Lake, R.A. Consistent gene expression profiles in MexTAg transgenic mouse and wild type mouse asbestos-induced mesothelioma. *BMC Cancer* **2015**, *15*, 983. [CrossRef] [PubMed]

74. Lecomte, C.; Andujar, P.; Renier, A.; Kheuang, L.; Abramowski, V.; Mellottee, L.; Fleury-Feith, J.; Zucman-Rossi, J.; Giovannini, M.; Jaurand, M.C. Similar tumor suppressor gene alteration profiles in asbestos-induced murine and human mesothelioma. *Cell Cycle* **2005**, *4*, 1862–1869. [CrossRef] [PubMed]

75. Rehrauer, H.; Wu, L.; Blum, W.; Pecze, L.; Henzi, T.; Serre-Beinier, V.; Aquino, C.; Vrugt, B.; de Perrot, M.; Schwaller, B.; et al. How asbestos drives the tissue towards tumors: YAP activation, macrophage and mesothelial precursor recruitment, RNA editing, and somatic mutations. *Oncogene* **2018**, *37*, 2645–2659. [CrossRef] [PubMed]

76. Xu, J.; Kadariya, Y.; Cheung, M.; Pei, J.; Talarchek, J.; Sementino, E.; Tan, Y.; Menges, C.W.; Cai, K.Q.; Litwin, S.; et al. Germline mutation of Bap1 accelerates development of asbestos-induced malignant mesothelioma. *Cancer Res.* **2014**, *74*, 4388–4397. [CrossRef] [PubMed]

77. Bignell, G.R.; Greenman, C.D.; Davies, H.; Butler, A.P.; Edkins, S.; Andrews, J.M.; Buck, G.; Chen, L.; Beare, D.; Latimer, C.; et al. Signatures of mutation and selection in the cancer genome. *Nature* **2010**, *463*, 893–898. [CrossRef] [PubMed]

78. Moller, P.; Danielsen, P.H.; Jantzen, K.; Roursgaard, M.; Loft, S. Oxidatively damaged DNA in animals exposed to particles. *Crit. Rev. Toxicol.* **2013**, *43*, 96–118. [CrossRef] [PubMed]

International Journal of
Molecular Sciences

MDPI

Article

Experimental Model of Human Malignant Mesothelioma in Athymic Mice

Didier J. Colin [1,†], David Cottet-Dumoulin [2,†], Anna Faivre [2], Stéphane Germain [1], Frédéric Triponez [2] and Véronique Serre-Beinier [2,*]

1   MicroPET/SPECT/CT Imaging Laboratory, Centre for BioMedical Imaging (CIBM), University Hospitals and University of Geneva, 1211 Geneva 4, Switzerland; didier.colin@unige.ch (D.J.C.); stephane.germain@hcuge.ch (S.G.)
2   Department of Thoracic and Endocrine Surgery, University Hospitals and University of Geneva, 1211 Geneva 4, Switzerland; David.Cottet-Dumoulin@unige.ch (D.C.-D.); anna.faivre@unige.ch (A.F.); frederic.triponez@hcuge.ch (F.T.)
*   Correspondence: Veronique.Serre-Beinier@hcuge.ch; Tel.: +41-22-379-5107
†   These authors contributed equally to this work.

Received: 12 June 2018; Accepted: 25 June 2018; Published: 26 June 2018

Abstract: Malignant pleural mesothelioma (MPM) is a thoracic aggressive cancer caused by asbestos exposure, which is difficult to diagnose and treat. Here, we characterized an in vivo orthotopic xenograft model consisting of human mesothelioma cells (designed as H2052/484) derived from a pleural NCI-H2052 tumor injected in partially immunodeficient athymic mice. We assessed tumor formation and tumor-dependent patterns of inflammation. H2052/484 cells conserved their mesothelioma phenotype and most characteristics from the parental NCI-H2052 cells. After intra-thoracic injection of H2052/484 cells, thoracic tumors developed in nearly all mice (86%) within 14 days, faster than from parental NCI-H2052 cells. When the mice were euthanized, the pleural lavage fluid was examined for immune cell profiles. The pleural immune cell population increased with tumor development. Interestingly, the proportion of myeloid-derived suppressor cell and macrophage (especially CD206+ M2 macrophages) populations increased in the pleural fluid of mice with large mesothelioma development, as previously observed in immunocompetent mice. This reliable orthotopic model recapitulates human mesothelioma and may be used for the study of new treatment strategies.

Keywords: cancer; pleura; mesothelioma; orthotopic xenotransplantation; athymic mouse; immune cells

1. Introduction

Malignant pleural mesothelioma (MPM) is an aggressive tumor that develops in the lining of the lungs. This cancer is causally associated with asbestos exposure. Although asbestos use is banned in many of the world's industrialized countries, the incidence of mesothelioma has overall not decreased for the last twenty years in most occidental countries [1]. Surgery is an option for early-stage MPM patients but not for most patients with advanced invasive disease [2] for whom treatment consists of palliative chemotherapy combining cisplatin with pemetrexed. While this treatment may relieve symptoms, it provides only modest survival, since the median survival average is 9–18 months from the time of diagnosis. Therefore, there is an urgent need for more effective treatments. Previous results from our laboratory, mostly obtained from in vitro experiments, suggest that inhibition of the Macrophage migration inhibitory factor (MIF)/CD74 pathway decreases the development of MPM [3]. These data need validation in a reliable in vivo preclinical model. Several murine mesothelioma models have been developed, and the selection of an appropriate model depends upon the experimental aims. Asbestos-induced and genetically engineered mesothelioma mouse models recapitulate the phenotypic and genetic heterogeneity as well as the carcinogenesis steps of human mesothelioma. They have also a strong predictive power for drug response and resistance, but their use to validate new therapies is limited by a low take rate, a long latency in tumor development, and a high cost. Syngeneic transplantation of murine mesothelioma cell lines could be an alternative with a high take rate and a rapid tumor development. Nevertheless, murine and human cells present fundamental phenotypic and functional differences. For example, there are two CD74 isoforms in mice (p31 and p41) and four in humans (p33, p35, p41, and p43) [4]. Up to now, the role of these different isoforms has not been clearly identified. Therefore, the effect of a treatment on murine mesothelial tumors could be not reproducible in human mesothelioma. Finally, preclinical studies on MPM mostly rely on subcutaneous or peritoneal xenotransplants of human mesothelioma cell lines in immunodeficient mice. These models provide reliable data and allow for rapid clinical translation. The major limitation is that the tumor environment is different from the in situ thoracic pleural mesothelioma environment. Transplantation in the orthotopic site offers a tumor microenvironment close to that of the original human tumor. To date, the use of the orthotopic thoracic site for xenografting has not been widespread, which is largely due to the technical difficulties in reaching and monitoring tumor development in this location. Here, we present a reliable orthotopic model of human MPM obtained after injection of a human mesothelioma cell line into the pleural space of athymic mice. Athymic mice have the advantage to be only partially immunodeficient, since they lack the thymus but produce most other immune cell types. Phenotypical and molecular characterizations of the tumor masses are described. For the first time in this orthotopic xenograft mesothelioma model, immune cell populations in the pleural environment of human mesothelioma-bearing mice are assessed.

2. Results

2.1. Selection and Characterization of the Human H2052/484 Cell Line

We were interested in the human MPM H2052 cells. Previous data from our lab showed that H2052 cells expressed MIF and its receptor CD74 [3] and that reduction of MIF and CD74 expression decreased the growth of H2052 cells. We previously observed that H2052 cells injected into the pleural cavity of athymic nude mice formed extensive pleural tumors in nearly all injected mice (5/6). Nevertheless, H2052 tumors developed slowly in the thoracic cavity. Using 2-deoxy-2-[^{18}F]fluoro-D-glucose ([^{18}F]FDG)-PET/CT analyses, H2052 pleural tumors were identified in the thoracic cavity starting on day 69 after cell injection, and their development continued until 102 days after cell injection, time at which the mice were sacrificed because of the extent of tumor development [3]. In order to evaluate whether preliminary in vivo engraftment of H2052 cells increased their tumorigenicity, thoracic tumors were mechanically dissociated into cell suspensions. Among many cell populations which could be maintained in monolayer culture, one cell line named H2052/484 was further characterized.

First, we compared the vitality and the multiplication of H2052/484 cells to those of the parental H2052 cells and of three other MPM cell lines using a mitochondrial activity assay (3-(4,5-dimethylthiazol-2-yl)-2,5-diphenyltetrazolium bromide, MTT) and crystal violet assay, respectively.

We observed similar vitality and multiplication rates for H2052/484 and the parental H2052 cells (Figure 1). Nutritional supplementation with fetal bovine serum (FBS) did not change the vitality of H28, H2052/484, and H2052 cells after 48 h of culture (Figure 1, upper panels); however, the multiplication of these three cell lines increased dose-dependently with FBS concentrations after 48 h of culture. After 48 h, the density of cells cultured with 10% FBS compared to cells cultured with 0% FBS, estimated by the absorbance level, was 2.81 ± 0.94 times higher for H28 ($n = 3$), 2.99 ± 0.80 times higher for H2052 ($n = 7$), and 6.53 ± 3.10 times higher for H2052/484 ($n = 7$). FBS supplementation dose-dependently increased the vitality of JL-1 and MSTO-211H as well as their multiplication (Figure 1, left lower panels). After 48 h, the vitality of cells cultured with 10% FBS compared to cells cultured with 0% FBS was 1.65 ± 0.23 times higher for JL1 ($n = 4$) and 1.79 ± 0.25 times higher for MSTO-211H ($n = 3$). The density of cells cultured for 48 h with 10% FBS compared to cells cultured with 0% FBS, estimated by the absorbance level, was 7.60 ± 0.07 times higher for JL1 ($n = 3$) and 12.23 ± 0.60 times higher for MSTO-211H ($n = 3$).

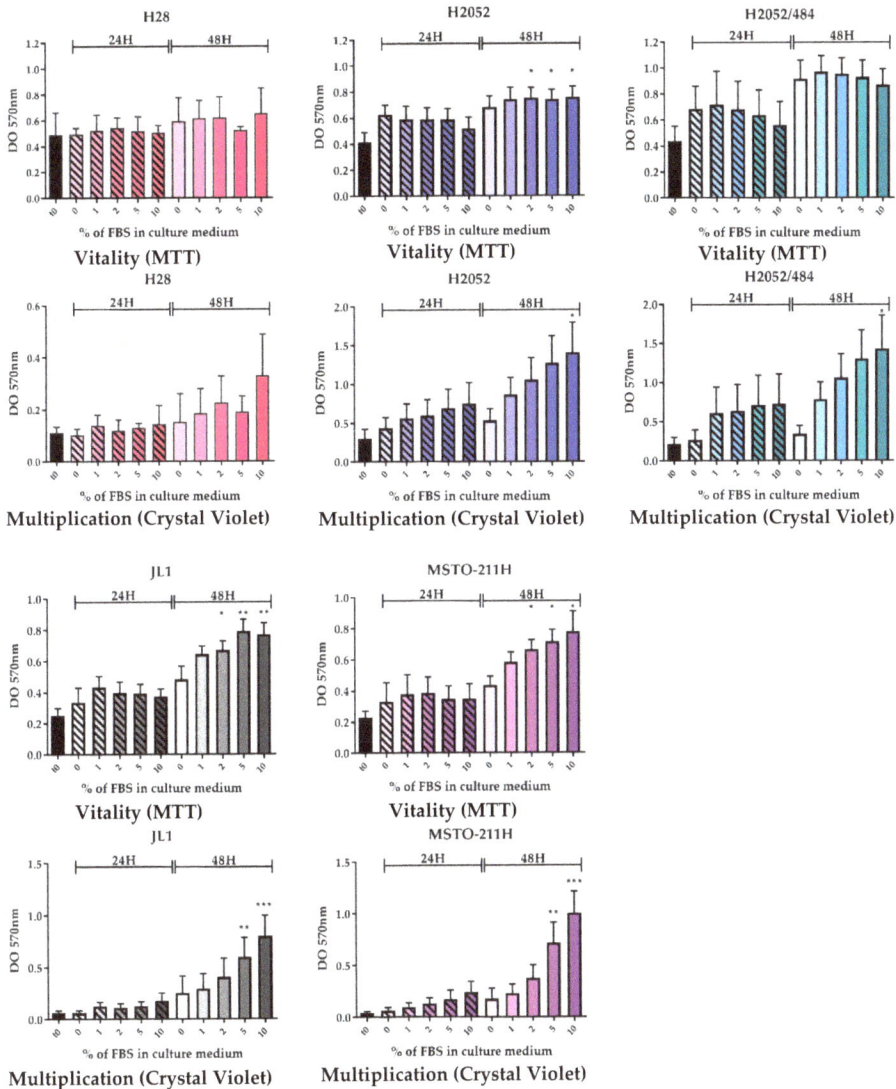

Figure 1. Cell vitality (3-(4,5-dimethylthiazol-2-yl)-2,5-diphenyltetrazolium bromide, MTT) and multiplication (Crystal Violet) of H2052/484 cells (blue-green) are similar to those of the parental H2052 cells (blue). The vitality and multiplication of the five malignant pleural mesothelioma (MPM) cell lines (H28 in pink; H2052 in blue; H2052/484 in blue-green; JL-1 in grey, and MSTO-211H in purple) were evaluated after the cells were cultured for 24 h (hashed bar) and 48 h (full bar) in medium supplemented with different percentages of fetal bovine serum (FBS). DO, optical density. The bars are mean values ($\pm$SEM) for $n = 3$–7 experiments. Kruskal–Wallis test between FBS concentrations and 0%: * $p < 0.05$, ** $p < 0.01$, *** $p < 0.001$.

Then, we compared the phenotype of H2052/484 cells to that of the parental H2052 cells and of three other MPM cell lines by studying the expression of different epithelial-to-mesenchymal (EMT) markers. Compared to parental H2052 cells, H2052/484 cells expressed 1.9 higher mRNA levels of the epithelial marker E-cadherin (CDH1) (Figure 2) and higher mRNA levels of the transcription factors SNAIL2 (3.3-fold change), ZEB1 (1.9-fold change), and ZEB2 (1.4-fold change), which are considered mesenchymal markers.

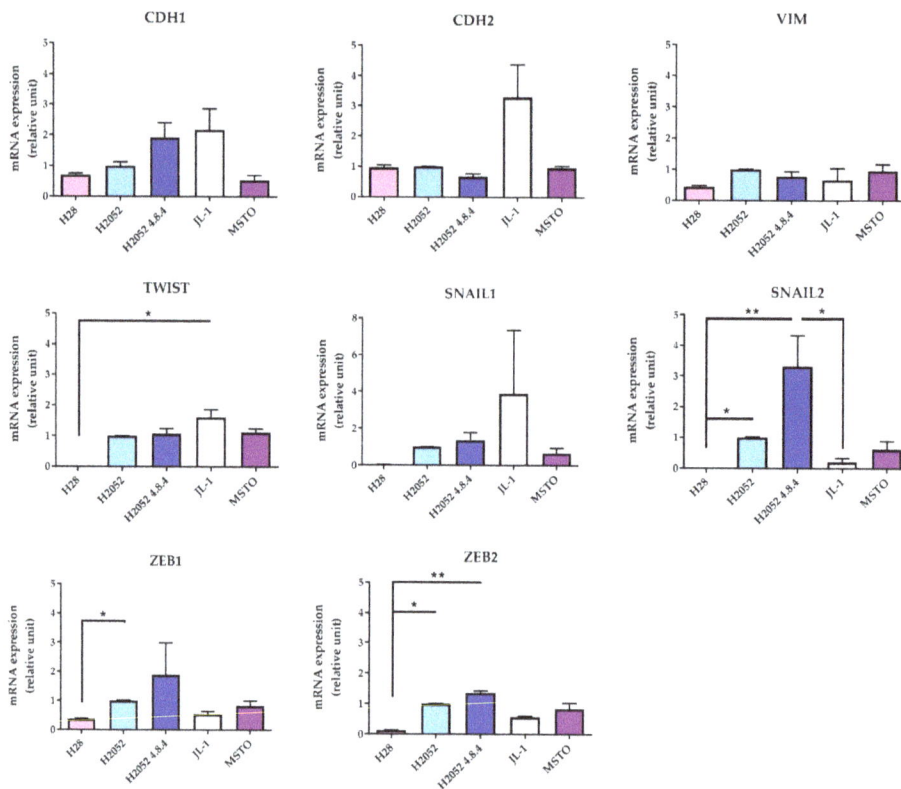

Figure 2. H2052/484 MPM cells express high levels of epithelial–to-mesenchymal (EMT) transcription factors. The mRNA levels of the EMT markers were measured in parental H2052 cells, in H2052/484 cells, and in three other MPM cell lines (H28, JL-1, and MSTO). The relative mRNA expression levels were measured by RT-qPCR and are presented as a ratio to the mRNA levels in parental H2052 cells. The data represent the mean values ($\pm$SD) of three independent experiments. Kruskal–Wallis test between MPM cell lines: * $p < 0.05$, ** $p < 0.01$.

The mRNA expression levels of these three transcription factors were higher in H2052/484 cells compared to the three other MPM cell lines (H28, JL-1, and MSTO). These differences were not statistically significant. Interestingly, H28 cells failed to form tumors in vivo [3] and expressed the lowest mRNA levels of ZEB1, ZEB2, SNAIL1, SNAIL2, and TWIST. H2052/484 cells expressed the lowest level of N-cadherin mRNA (CDH2). Western blot analyses of EMT markers in H2052/484, JL-1, and MSTO cell lines confirmed the highest expression levels of Snail (SNAIL1) and Slug (SNAIL2) and the lowest expression of N-cadherin in H2052/484 cells (Figure 3). We did not detect E-cadherin protein expression in any of the studied MPM cell lines. MIF and CD74 mRNA levels in H2052/484 cells were similar to the levels in parental H2052 (for MIF: 1.39 ± 0.07, $n = 3$, for H2052/484; 1.31 ± 0.05, $n = 3$, for H2052; for CD74: 1.14 ± 0.07, $n = 3$, for H2052/484; 1.22 ± 0.22, $n = 3$, for H2052).

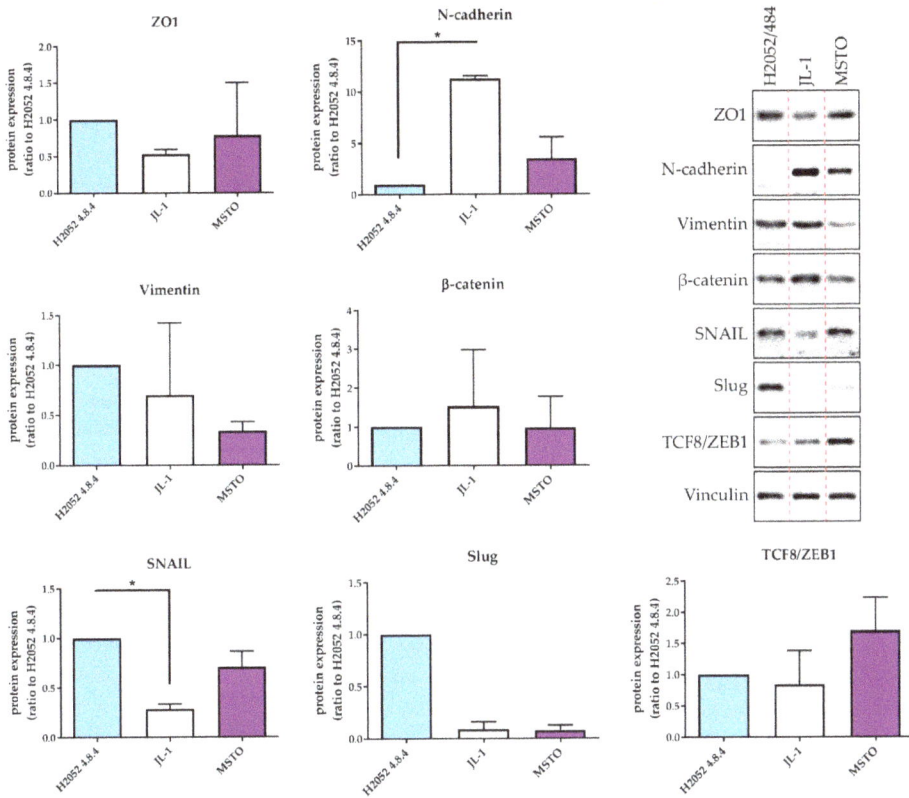

Figure 3. H2052/484 cells express epithelial and mesenchymal markers. Protein expression of EMT markers was measured in H2052/484 cells and two other MPM cell lines (JL-1 and MSTO) by western blotting. Representative western blot results are shown; the dashed red lines indicate the manual cropping of the bands detected for samples run on the same gels and identically exposed. Protein expression levels are presented as the ratio to the respective protein level in H2052/484 cells. The data represent the mean values ($\pm$SD) of three independent experiments. Kruskal–Wallis test between MPM cell lines: * $p < 0.05$.

2.2. Characterization of Orthotopic Tumor Masses Generated by Human MPM H2052/484 Cells

Intrapleural (i.pl.) injection of H2052/484 cells into athymic nude mice yielded sizable tumor masses identifiable by ([^{18}F]FDG)-PET/CT imaging within 2 weeks. H2052/484 tumors developed in nearly all injected mice (24/28). The tumors were distributed freely in the thoracic cavity or attached to the aortic arch (close to the thymus rudiment), to the inferior vena cava, to thoracic muscles, or to the lungs (Figure 4a).

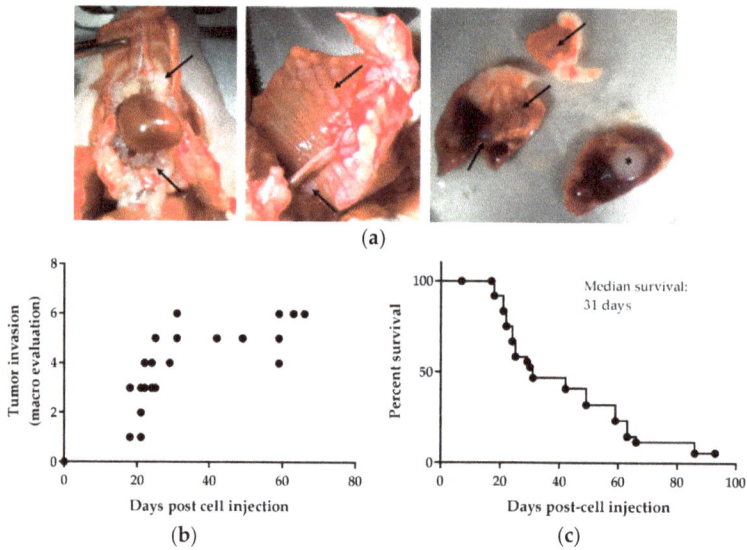

Figure 4. H2052/484 cells formed pleural mesothelioma in athymic mice. H2052/484 cells were injected intrapleurally (i.pl.) (1×10^6 cells) into athymic nude mice (single experiment; n = 28). (**a**) H2052/484 tumors were free in the thoracic cavity or attached to the aortic arch (close to the thymus rudiment, left panel, black arrows), the inferior vena cava (left panel), thoracic muscles (middle panel), or to lungs (right panel). (**b**) The mice were sacrificed at different time points (end-point criteria), and the tumor scores were evaluated post-mortem following criteria described in Table 1. (**c**) Survival of athymic mice after intrapleural injection of H2052/484 MPM cells.

The tumor attached to the left lung (Figure 4, right panel, asterisk) was localized close to the injection site. These tumors were poorly invasive and did not often penetrate deep into intercostal tissues or into the lung to which they were attached. There was no evidence of metastases, as we found no tumors in distant organs. The mice were followed using positron-emission tomography/computerised tomography (PET/CT) imaging until they were sacrificed, once tumor development reached euthanasia endpoints (described in Material and Methods) such as large size or signs of unacceptable pain and clinical distress. Macroscopic evaluation of tumor size and extent was performed when the mice were sacrificed (Figure 4b). The scoring of tumor development is detailed in Table 1. Between days 18 to 29, tumors of different size and extent were observed, ranging from score 1 to 5 and indicating an active phase of tumor growth (Figure 4b). Between days 29 to 66, all tumors reached the maximum scores of 5 or 6 (Table 1). Half of the mice were sacrificed during the active phase of tumor growth, and a median survival of 31 days was observed (Figure 4c).

Figure 5a shows haematoxylin–eosin (HE) staining of explanted and formalin-fixed H2052/484 pleural tumors. Necrotic (N) areas were observed in big tumors (mouse 2, right panel, Figure 5a). A meshwork of capillaries and vessels was detected inside the tumors by identifying the red blood cells on HE-stained slices (Figure 5b) or after immunostaining with anti-CD31 antibody (Figure 5b). Ki67 labelling showed a high proliferation rate of tumoral cells (Figure 5c). Apoptotic cells were also clearly identified after immunolabelling of γ-H2AX histone that showed strong homogeneous nuclear labelling (Figures 5c). The excised H2052/484 tumors were tested for retention of classical MPM markers. H2052/484 tumoral cells expressed calretinin and mesothelin (Figure 6a), confirming their MPM behaviour after in vivo orthotopic engraftment. Finally, cytoplasmic MIF was clearly observed in all H2052/484 tumor cells (Figure 6b). MIF receptor CD74 and co-receptor CD44 were also detected in the cytoplasm and the membrane of MPM cells (Figure 6b), suggesting an active MIF/CD74 pathway.

Table 1. Score of H2052/484 tumor development.

Score	Macroscopic Observations
1	Tumor limited to the thoracic surface of the left lung (injection site)
2	Tumor(s) on the left lung Tumor(s) 2 mm or less along - the pulmonary veins and arteries - the inferior vena cava (mediastinal pleura) No tumor on the aortic arch, the thoracic muscle, the diaphragm, and the pericardium
3	Tumors (more than 2 mm or more than 5 tumors) - on the left and right lungs or/and - along the pulmonary veins and arteries or/and - along the inferior vena cava (mediastinal pleura) Scattered foci of tumor (2 mm or less) on the thoracic muscle or/and the aortic arch No tumor on the diaphragm and the pericardium
4	Tumors (more than 2 mm or more than 5 tumors) on - the lungs or/and - the pulmonary vascular trunk or/and - the mediastinal pleura or/and - the thoracic muscles and/or the aortic arch Scattered foci of tumor (1 mm or less) on the diaphragm or the pericardium
5	Tumors (more than 2 mm or more than 5 tumors) on - the lungs or/and - the pulmonary vascular trunk or/and - the mediastinal pleura or/and - the thoracic muscles and/or the aortic arch or/and - on the diaphragm or the pericardium OR - extension of tumor from the visceral pleura into the underlying pulmonary parenchyma
6	Advanced tumors involving each of the pleural surfaces (parietal, mediastinal, diaphragmatic, and visceral pleura) with confluent pleural tumors.

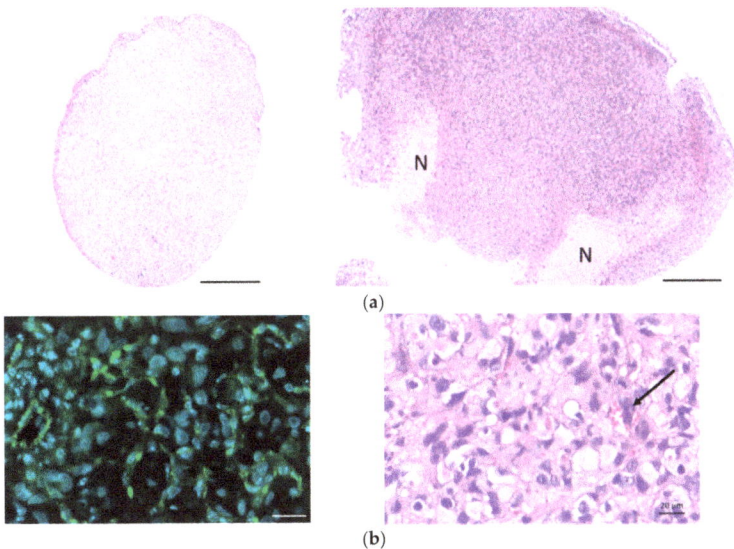

(a)

(b)

Figure 5. *Cont.*

(c)

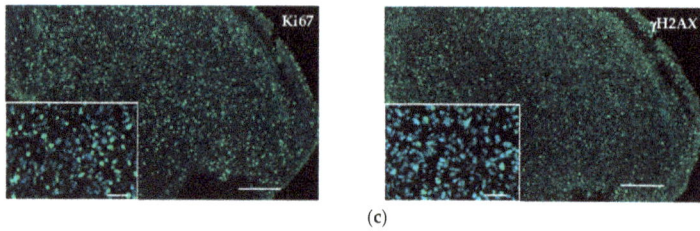

Figure 5. Characterization of H2052/484-derived orthotopic tumors. (**a**) Haematoxylin–eosin (HE) sections of two representative pleural MPM generated after intrathoracic injection of H2052/484 cells in athymic nude mice at day 46 after cell injection. Necrotic areas (N) were identified in the largest tumor (right panel). Scale bars: 500 μm. (**b**) Vascularization of H2052/484 tumors: left, CD31 (green, endothelial marker) expressed in tumors; right, HE staining showing red blood cells in vessels (black arrow) in the tumor. Scale bars: 20 μm. (**c**) Cell proliferation and DNA damage representative of cell apoptosis were identified in H2052/484 tumors. The tumor slices were stained (green) with anti-Ki67 (cell proliferation), anti-γ-H2AX (DNA damage and cell apoptosis), and DAPI (nuclear counterstaining, blue). Scale bars: 500 μm and 50 μm.

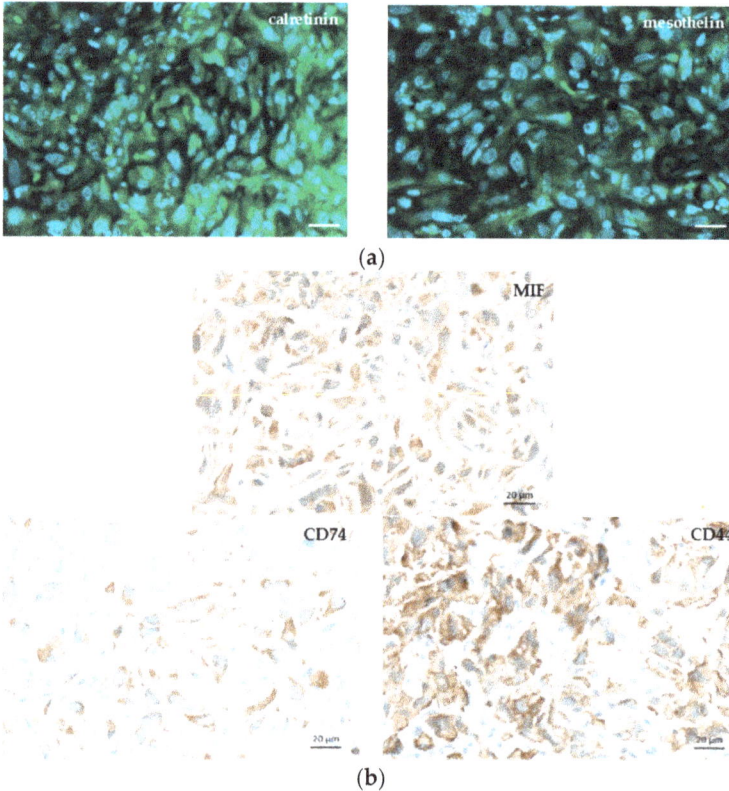

(a)

(b)

Figure 6. MPM markers and macrophage migration inhibitory factor (MIF), CD74, and CD44 expressions in orthotopic H2052/484 tumors. Representative photomicrographs of a H2052/484 intrathoracic tumor stained with (**a**) anti-calretinin, and anti-mesothelin, (**b**) anti-MIF, anti-CD74, and anti-CD44 antibodies. Scale bars: 20 μm.

2.3. Local Immune Cell Response of Athymic Nude Mice Developing Orthotopic H2052/484 Tumors

Due to the lack of thymus maturation, a T-cell deficiency was observed in athymic nude mice. B cells, dendritic cells, and granulocytes were all relatively intact, and there was a compensatory increase in both natural killer (NK)-cell activity and tumoricidal macrophages in these mice (as reviewed in reference [5]). Therefore, an immune response could be expected after malignant cell injection. Using flow cytometry, the immune cell populations (lymphocytes, neutrophils, monocytes, and macrophages) were characterized in the thoracic cavity of mice injected i.pl with H2052/484 cells (Figure 7).

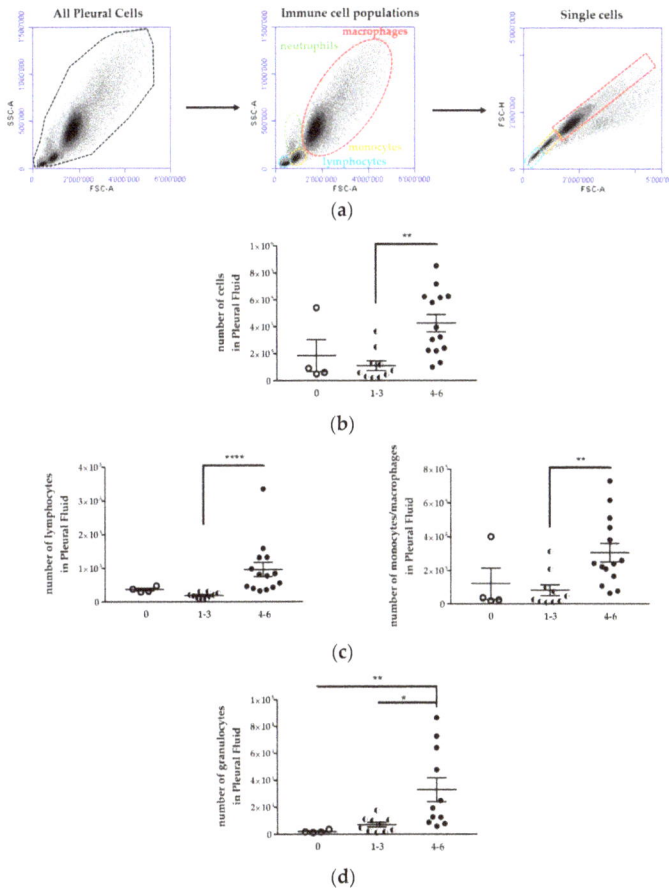

Figure 7. Immune cell populations increased in the pleural fluid of H2052/484 tumor-bearing mice. (**a**) Viable immune cell populations (lymphocytes (blue), monocytes (yellow), macrophages (red) and granulocytes (green)) were identified first according to side scatter (SSC-A) vs. forward scatter (FSC-A). Doublets were excluded using a pulse geometry gate FSC-H × FSC-A plot. Representative flow cytometry dot plots and scatter plots of pleural cells are shown. Comparisons of the number of (**b**) total cells, (**c**) lymphocytes, monocytes/macrophages, and (**d**) granulocytes in the pleural fluid of mice with no tumor (empty dots) and in tumor-bearing mice scored 1 to 3 (1–3; half dots) and 4 to 6 (4–6, full dots). The data represent the mean values ± SEM. Comparisons were made using Kruskal–Wallis test; * $p < 0.05$, ** $p < 0.01$, **** $p < 0.0001$.

First, the cell populations were identified according to their sizes (forward scatter, FSC) and internal structures (side scatter, SSC). Second, labelling of immune cells with specific antibodies was performed as follows: $CD19^+B220^+$ for B lymphocytes, $CD19^-CD11b^+$ for monocytes/macrophages, $CD19^-CD11b^+ F4/80^+$ for macrophages, $CD11b^+F4/80^+CD206^+$ for M2 macrophages, $CD49b^+$ for NK cells, $CD11b^+Gr1^+$ for myeloid-derived suppressor cells (MDSC).

Mice with a high tumor development score (4 to 6) showed an increase of cell number in the pleura compared to mice with a low tumor development score (1 to 3) or mice without tumor (Figure 7b, Table 2). The number of cells for each cell population, i.e., lymphocytes, granulocytes (neutrophils), monocytes, and macrophages, increased in the pleural cavity of mice with a high tumor development score (Figure 7; Table 2).

Table 2. Immune cell number in the pleural fluids from mice with differential H2052/484 MPM development.

Tumor Development Score	0	1–3	4–6
Total pleural cells	$185{,}942 \pm 118{,}481$ $n = 4$	$111{,}068 \pm 35{,}687$ $n = 10$ >0.9999	$425{,}358 \pm 63{,}830$ $n = 14$ 0.1472
p vs. 0			
p vs. 1–3			***0.0021***
Lymphocyte number	$36{,}376 \pm 4135$ $n = 4$	$18{,}779 \pm 2638$ $n = 10$ 0.2943	$96{,}651 \pm 21{,}303$ $n = 14$ 0.3602
p vs. 0			
p vs. 1–3			***<0.0001***
Granulocyte number	2085 ± 515 $n = 4$	7025 ± 1686 $n = 10$ 0.6575	$33{,}012 \pm 8879$ $n = 11$ ***0.0037***
p vs. 0			
p vs. 1–3			***0.0238***
Monocyte/macrophage number	$121{,}038 \pm 93{,}659$ $n = 4$	$81{,}626 \pm 32{,}076$ $n = 10$ >0.9999	$305{,}638 \pm 54{,}706$ $n = 14$ 0.1696
p vs. 0			
p vs. 1–3			***0.0044***

Despite the increased number of lymphocytes in the pleural cavity, the percentage of total lymphocytes in the pleural cell population tended to decrease with the increase of the tumor development score (Figure 8a; Table 3). The percentage of lymphocytes in mice with tumor score 4–6 was 1.7 lower than in mice without tumor. This lymphocyte population identified by the FSC and SSC cytometric parameters included T and B lymphocytes and NK cells. The percentage of $CD19^+B220^+$ B lymphocytes was relatively stable in mice with different tumor development scores (Table 3), representing 6.8 to 17.5% of total cells in the thoracic cavity. The monocytes/macrophages population represented 53.8 to 68.3% of pleural cells (Table 3). This percentage was not different in mice with or without tumor (Figure 8b). This population contains several cell types, including MDSC, dendritic cells, monocytes, macrophages. The percentage of $CD11b^+F4/80^+$ macrophages increased in the thoracic cavity of mice with high tumor development score (Figure 8c, left panel) from $22.9 \pm 8.0\%$ (mice without tumor) to $45.3 \pm 7.0\%$ (mice with tumor development score 4–6) (Table 3).

We observed an increase of the percentage of $CD11b^+F4/80^+CD206^+$ M2 immunosuppressive macrophages in the pleural cell population of mice with higher tumor scores from 1.6% (mice without tumor) to 5.6% (mice with tumor development score 4–6) (Figure 8c, right panel; Table 3). This result is in concordance with previous reports of Jackaman C et al. [6].

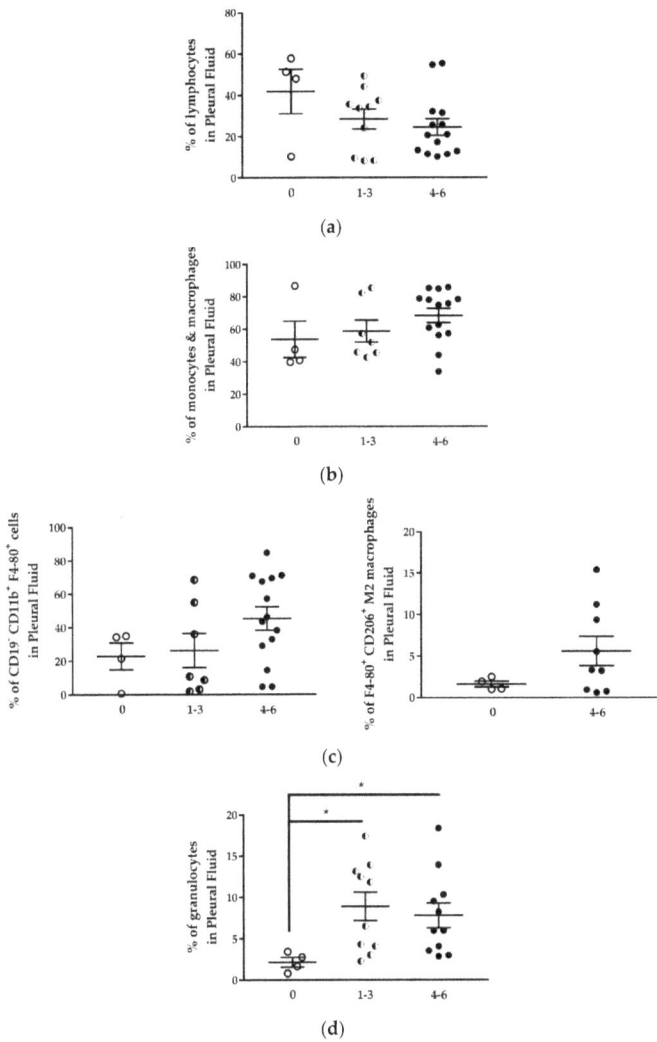

Figure 8. The proportion of immune cell populations in the pleural fluid of H2052/484 tumor-bearing mice changed with the tumor score. Comparisons of percentages of (**a**) total lymphocytes, (**b**) total monocytes/macrophages, (**c**) $CD19^-CD11b^+F4-80^+$ macrophages and $CD206^+$ M2 macrophages, and (**d**) granulocytes in the pleural fluid of mice with no tumor (empty dots) and in tumor-bearing mice with development scores 1–3 (half dots) and 4–6 (full dots). The percentages represent the number of each cell population in the total pleural cell number as determined by flow cytometry. The data represent the mean values ± SEM. Comparisons were made using Kruskal–Wallis test; $^* p < 0.05$.

Using an immunocompetent orthotopic mouse model of MPM, it was shown that the proportion of $CD11b^+F4/80^+$ tumor-associated macrophages increased significantly with MPM progression. Large tumors contained more macrophages of the M3 subset (a macrophage subset expressing a mixed M1 and M2 phenotype) and MDSCs. In this model, the percentage of $CD11b^+Gr1^+$ MDSC was two-fold higher in mice with tumors compared to mice without tumors (Table 3). Finally, the proportion of neutrophils in the pleural cell population was approximately fourfold higher in mice with tumors than in mice without tumors, and this independently of the tumor development score (Figure 8d; Table 3). While the proportion of total lymphocytes decreased with tumor development, we observed a sixfold increase in the neutrophil-to-lymphocyte ratio in mice with tumor development compared to mice without tumors (Table 3).

In summary, a high score (4–6) of H2052/484 tumor development in athymic mice was associated with an increase of immune cells in the thoracic cavity. Two-thirds of this population were CD11b$^+$ cells, with a high proportion of F4/80$^+$ macrophages and Gr1$^+$ MDSC. Tumor development was also associated with a decrease of the percentage of total lymphocytes and an increase of the percentage of neutrophils.

Table 3. Immune cell distribution in the pleural fluids of mice with differential H2052/484 MPM development. MDSC, myeloid-derived-suppressor cells

Tumor Development Score	0	1–3	4–6
Lymphocytes (% of total cells)	41.9 ± 10.7 $n = 4$	28.5 ± 4.8 $n = 10$	24.6 ± 4.0 $n = 14$
p vs. 0		*0.6644*	*0.3602*
p vs. 1–3			*>0.9999*
CD19$^+$ B220$^+$ lymphocytes (% of total cells)	8.7 ± 2.7 $n = 4$	17.5 ± 5.3 $n = 5$	6.8 ± 1.9 $n = 11$
p vs. 0		*>0.9999*	*>0.9999*
p vs. 1–3			*0.2621*
monocytes/macrophages (% of total cells)	53.8 ± 11.2 $n = 4$	58.7 ± 6.8 $n = 7$	68.3 ± 4.3 $n = 14$
p vs. 0		*>0.9999*	*0.6247*
p vs. 1–3			*>0.9999*
CD19$^-$ CD11b$^+$ monocytes/macrophages (% of total cells)	52.1 ± 10.6 $n = 4$	53.3 ± 7.1 $n = 7$	65.5 ± 4.7 $n = 14$
p vs. 0		*>0.9999*	*0.9253*
p vs. 1–3			*0.4993*
CD19$^-$ CD11b$^+$ F4/80$^+$ Macrophages (% of total cells)	22.9 ± 8.0 $n = 4$	26.3 ± 10.2 $n = 7$	45.3 ± 7.0 $n = 14$
p vs. 0		*>0.9999*	*0.2956*
p vs. 1–3			*0.3193*
CD11b$^+$ F4/80$^+$ CD206$^+$ M2 macrophages (% of total cells)	1.6 $n = 4$		5.6 $n = 9$
p vs. 0			*0.4140*
Neutrophils (% of total cells)	2.2 ± 0.6 $n = 4$	8.9 ± 1.7 $n = 10$	7.8 ± 1.5 $n = 14$
p vs. 0		***0.0224***	***0.0456***
p vs. 1–3			*>0.9999*
Neutrophil-to-lymphocyte ratio	0.06 ± 0.01 $n = 4$	0.38 ± 0.06 $n = 10$	0.36 ± 0.06 $n = 11$
p vs. 0		***0.0095***	***0.0206***
p vs. 1–3			*>0.9999*
CD49b$^+$ NK cells (% of total cells)	5.5 ± 1.9 $n = 4$	7.6 ± 1.3 $n = 7$	7.0 ± 1.7 $n = 12$
p vs. 0		*>0.9999*	*>0.9999*
p vs. 1–3			*>0.9999*
CD11b$^+$ Gr1$^+$ MDSC (% of total cells)	27.7 ± 6.7 $n = 4$		51.1 ± 8.1 $n = 9$
p vs. 0			*0.1063*

3. Discussion

In order to assess the effect of the MIF/CD74 pathway in the development of MPM, we derived a new human MPM cell line expressing MIF, CD74, and CD44 and able to generate orthotopic intra-thoracic tumors. H2052/484 cells were obtained from the dissociation of a pleural tumor obtained after NCI-H2052 cell injection. Furthermore, H2052/484 cells conserved their mesothelioma phenotype and most characteristics of the parental H2052 cells. They demonstrated faster tumor growth than parental H2052 cells after intrathoracic injection in athymic mice. This higher tumor development may be related to higher levels of the EMT transcription factors Snail 2, Zeb1, and Zeb2. The activation of the EMT program is commonly observed in human cancers and is closely related to tumor invasiveness and progression [7]. H2052/484 cells were modestly virulent in vivo, and the mice were found to tolerate a certain level of tumor burden (1×10^6 cells) over a two-week time course, without euthanasia requirements due to distress. Thus, this model represents a reproducible mean to test new therapies targeting the MIF/CD74 pathway as well as other pathways that promote the growth of MPM. This model provides a large time window to evaluate the anticancer effects of new treatments and possible tumor relapse and resistance due to subpopulations of cells that might escape therapy. We used athymic mice as hosts, given that our study objective was to evaluate the effects of the MIF/CD74 pathway inhibition on human MPM development in vivo. These mice are partially immunodeficient because of the lack of thymus [8], which leads to a very poor response to thymic-dependent antigens [5]. Except for the lack of T lymphocytes, most other immune cell types are present in these mice, and we observed an increase in the immune cell population in the pleural fluid with the increase of tumor development. $CD19^+B220^+$ B lymphocytes and NK cells ($CD49b^+$) were identified in the pleural fluid. No expansion of these populations was observed in tumor-bearing mice compared to mice without tumors. Myeloid-derived suppressor cells ($CD11^+Gr1^+$), monocytes ($CD19^-CD11b^+$), and macrophages ($CD19^-CD11b^+F4/80^+$) were also detected in the pleural fluid. Interestingly, MDSC and macrophages (especially $CD206^+$ M2 macrophages) expanded during MPM development as previously shown in an immunocompetent mouse model of mesothelioma [6]. The lack of T cells in nude mice is not an obstacle to study the relationship between inflammatory cells and mesothelioma development. Indeed, Jackaman et al. have shown [6], using an immunocompetent mouse model, that the suppressive role of regulatory T cells is important during the early stages of mesothelioma tumor evolution, but, in advanced-stage mesothelioma, myeloid cells and macrophages are major regulatory cells, as confirmed in our study. Both cell types have been shown to promote tumor growth, recurrence, and tumor burden in multiple ways, including promotion of angiogenesis and immunosuppressive activity [9,10]. Several studies reported a role for macrophage migration inhibitory factor (MIF) in promoting MDSC and macrophage accumulation and immunosuppressive activity in several cancers [11,12]. We previously showed that human MPM expresses MIF and its receptor CD74 and that this pathway is important for MPM cells proliferation [3]. In this model, MIF secretion by H2052/484 tumor cells may attract immunosuppressive cells such as MDSC and polarized macrophages toward an immunosuppressive M2 phenotype, thus promoting tumor growth. We plan to evaluate the effect of MIF inhibitors on H2052/484 development. The data obtained on the extent of tumor development and the immune cell types present in the local (pleural) tumor environment should help us to design new therapies. Then, these new agents should be validated on other mouse models such as humanized mouse models. In these models, immunocompromised mice (generally non-obese diabetic (NOD) scid gamma and NOD Rag gamma mice characterized by a great immunodeficiency) are immunologically reconstituted with human immune cells. The effects of MIF inhibitors on the total human immune cell populations could be characterized.

In summary, this study shows that the orthotopic xenotransplantation model of H2052/484 MPM cells in nude mice is a reproducible model to study the functional and mechanistic effects of new treatments for MPM. This model can be used to test the therapeutic effects of MIF inhibition on human MPM development and possibly to develop new therapies for this fatal disease.

4. Materials and Methods

4.1. Isolation of H2052/484 Cells and Cell Culture

The MPM cell lines H28 (NCI-H-28), H2052 (NCI-H2052), and MSTO (MSTO211H) were purchased from American Type Culture Collection (Manassas, Virginie, VA, USA). The MPM cell lines JL-1 was established and characterized in our laboratory from human biopsies [13]. H2052/484 cells were subcultured after mechanical dissociation of an orthotopic tumor explanted 102 days after an intrapleural implantation of 1×10^6 NCI-H2052 cells into an athymic Nude-*Foxn1nu* nu/nu. All cells were routinely cultured in RPMI 1640 medium containing 10% (v/v) fetal bovine serum (complete RPMI, Life Technologies, Carlsbad, CA, USA). The cultures were grown at 37 °C in 5% CO_2.

4.2. Total RNA Isolation and Real-Time RT-PCR

The expressions of *CDH1*, *CDH2*, *VIM*, *TWIST*, *SNAIL1*, *SNAIL2*, *ZEB1*, *ZEB2*, *MIF*, *CD74*, *GAPDH*, *GUSB*, *EEFLA1*, and *TBP* mRNAs were evaluated by quantitative RT-PCR analysis. Total mRNA from each cell lines was extracted by InViTrap® Spin Universal RNA Mini kit (Stratec, Birkenfeld, Germany) according to the manufacturer's instructions.

Reverse transcription and Real-time RT-PCR was performed at the same time using ONE-step kit Converter (Takyon-Eurogentec UF-RTAD-D0701, Eurogentec, Liège, Belgium) and No ROX SYBR MasterMix blue dTTP (Takyon Eurogentec UF-NSMT-B0101, Eurogentec, Liège, Belgium). Real-time RT-PCR was performed on each sample in duplicate with 50 ng cDNA per condition, using a Biorad CFX Connect Real Time system. SYBR green primer sequences for the targeted human genes are available upon request. The results were normalized to the expression levels of housekeeping genes, including *GAPDH*, *GUSB*, *EEFLA1*, and *TBP* genes.

4.3. Cell Growth and Vitality

Five thousand cells per well were cultured into 96-well microplates in complete RPMI. After cell adhesion, the medium was replaced, and the cells were cultured in RPMI without or with increasing concentrations of FBS for 24 h and 48 h at 37 °C. Cell growth was determined by crystal violet staining. Briefly, after fixation with formalin 10%, the cells were stained with 0.1% crystal violet (Sigma-Aldrich Corp., St. Louis, MO, USA) for 30 min at room temperature. The cells were lysed in a 10% acetic acid solution for 30 min at room temperature. The absorbance was read in a spectrophotometer at 570 nm.

Cell vitality was determined by the reduction of 3-(4,5-dimethylthiazol-2-yl)-2,5-diphenyltetrazolium bromide (MTT, Sigma-Aldrich Corp., St. Louis, MO, USA). The MTT solution (500 μg/mL in RPMI) was added for 2 h at 37 °C. The cells were lysed with dimethyl sulfoxide. The absorbance was read in a spectrophotometer at 570 nm.

4.4. Cell Lysis and Western Blotting Analysis

The samples were lysed in RIPA buffer (50 mM Tris, 150 mM NaCl, 0.1% SDS, 0.5% Sodium deoxycholate, 1% Igepal CA630, 2 mM EDTA, 50 mM NaF, pH 8) supplemented with a protease inhibitor cocktail (Roche Molecular Diagnostics, Pleasanton, California, CA, USA) and titrated using the DC Protein Assay (Bio-Rad Laboratories, Hercules, California, CA, USA). Amounts of 5 to 20 μg of proteins were separated by SDS-PAGE and transferred onto nitrocellulose membranes (Amersham, Little Chalfont, UK). The membranes were blocked 1 h at room temperature, incubated overnight at 4 °C with primary antibodies (EMT sampler kit, 9782 Cell signaling (Danvers, Massachusetts, MA, USA) and B-Actin, A2066 Sigma-Aldrich (Sigma-Aldrich Corp., St. Louis, MO, USA), incubated for 1 h with secondary antibodies (Bio-Rad Laboratories, Hercules, California, CA, USA), and developed using a standard ECL protocol. The quantifications were performed by using a ChemiDoc MP and the Image Lab software (Bio-Rad Laboratories, Hercules, California, CA, USA).

4.5. In Vivo Development of H2052/484 MPM Cells in Nude Mice

The mice were anaesthetized with isoflurane. Buprenorphin (analgesic) was injected subcutaneously. (0.05 mg/kg). MPM H2052/484 cells were injected into the left pleural cavity (1×10^6 tumor cells suspended in 50 µL of RPMI) of 19-week-old athymic female nude mice nu/nu (Envigo, Huntingdon, UK) ($n = 28$). Briefly, the mouse was placed on the left side (left lateral decubitus). A 0.5 to 1 cm incision of the skin was made to expose the ribs. An amount of 50 µL of cell suspension was slowly injected into the intercostal space on the right dorsal mid-axyllary line just below the inferior border of the scapula. The wound was closed with three to four absorbable sutures. Anesthesia was stopped. The mouse was placed under a heat lamp until awake.

2-deoxy-2-[^{18}F]fluoro-D-glucose ([^{18}F]FDG)-PET/computed tomography (CT) scans were used to follow the intrapleural tumor growth. PET/CT was performed using a Triumph PET/SPECT/CT system (Trifoil, Chatsworth, CA, USA) on mice fasted for 12 h. An amount of 5–6 MBq of [^{18}F]FDG was i.v. injected retro-orbitally on anesthetized mice. The mice were then left awake at RT during an uptake time of 60 min. Subsequently, 700 µL of 132 mg/mL meglumine ioxitalamate (Telebrix, 6% m/v iodide, Guerbet AG, Zürich, Switzerland) was injected intraperitoneally in mice to delineate the abdominal region, and the mice were subjected to CT scans. Images were obtained at 80 kVp, 160 µA, and 1024 projections were acquired during the 360° rotation, with a field of view of 71.3 mm (1.7× magnification). Sixty min after the [^{18}F]FDG injection, PET scans were started for a duration of 20 min. PET scans were reconstructed with the built-in LabPET software (Triumph-Adler, Nuremberg, Germany), using an OSEM3D (20 iterations) algorithm, and the images were calibrated in Bq/mL by scanning a phantom cylinder. Reconstruction of the CT scans was performed with the Triumph XO software (Triumph-Adler, Nuremberg, Germany) that uses a matrix of 512 and a voxel size of 0.135 mm. CT scans and PET scans were co-registered using the plugin Vivid (Trifoil) for Amira (FEI, Hillsboro, OR, USA) and exported as dicom files. The software Osirix (Pixmeo, Bernex, Switzerland) was used to quantitatively analyse the datasets and generate pictures.

Two weeks after cell injection, the mice were checked every other day by observers for signs of morbidity. Euthanasia endpoints were chosen to minimize the distress of the transplanted mice. They were defined in a previous pilot study in which body weight, body condition scoring, appearance, and behavioral assessments were used to evaluate morbidity in this orthotopic mouse model of mesothelioma. The following criteria were determined: significant tumor growth and/or malignant pleural effusion in the thoracic cavity (detected by PET/CT imaging), labored breathing, abnormal posture, dehydration, and weight loss of 15% within a few days. This study was conducted under protocols revised and approved by the institutional animal care and use committee and by Geneva's veterinary state office.

When committee-approved endpoints (authorization GE/106/16 approved by the "Direction Générale de la Santé", Republic of Geneva, 19 July 2016, 25291) were achieved, the mice were euthanized and closely examined for the presence of thoracic tumors. For each euthanized mouse, blood was drawn from the heart. For histology, spleen, lung, and tumors were explanted and fixed in 10% neutral buffered formalin.

4.6. Immunohistochemistry Analysis

The MPM samples fixed in formalin were embedded in paraffin. Four-µm-thick MPM tumor sections were cut and stained with haematoxilin-eosin (HE) or analyzed by immunohistochemistry. Labeling with anti-MIF (gift of Thierry Roger, Lausanne, Switzerland), anti-CD74 (HPA010592, Sigma-Aldrich Corp., St. Louis, MO, USA), and anti-CD44 antibodies (HPA005785, Sigma-Aldrich Corp., St. Louis, MO, USA) was performed using the Ventana Discovery automated staining system (Ventana Medical Systems, Tucson, AZ, USA). Ventana reagents for the entire procedure were used. Antigen retrieval was performed by heating the slides in CC1 cell conditioning solution for 20 min (EDTA antigen retrieval solution pH 8.4; 20 min for CD74 and CD44, 36 min for MIF). The slides were incubated 30 min at 37 °C with primary antibodies diluted at 1/300 (MIF), 1/1000 (CD74),

Int. J. Mol. Sci. **2018**, *19*, 1881

and 1/500 (CD44) in an antibody diluent from Dako (S2022, Agilent technology, Santa Clara, CA, USA). Anti-MIF, anti-CD74, and anti-CD44 labeling was detected using the rabbit OmniMap kit (760-149). Immunostaining with anti-Ki67 (9027, Cell signaling technology, Danvers, MA, USA), γ-H2AX (sc-101696, Santa Cruz Biotechnology, Dallas, TX, USA), CD31 (ab28364, Abcam, Cambridge, UK), calretinin (18-0211, Invitrogen, Carlsbad, CA, USA), and mesothelin (HPA017172, Sigma-Aldrich Corp., St. Louis, MO, USA) was performed after EDTA antigen retrieval for 15 min. After a 20 min blocking step in PBS 0.2%/Triton X100 (PBST), the sections were incubated with primary antibodies in blocking buffer overnight at 4 °C (Ki67, 1/500; γ-H2AX, 1/500; CD31, 1/50; calretinin, 1/80; mesothelin, 1/80). The sections were washed with PBST and incubated for 1 h with DAPI 0.4 μg/mL and secondary A488-conjugated anti-rabbit antibodies in blocking buffer (A21206, Molecular Probes, ThermoFisher Scientific, Waltham, MA, USA). The slides were washed and mounted in fluorescence mounting medium (Dako, Agilent technology, Santa Clara, CA, USA). The slides were scanned with an Axioscan.Z1 and analyzed with Zen (Zen 2.3, Carl Zeiss, Oberkochen, Germany). Specific binding of all antibodies was previously checked by running controls without primary antibodies (see Supplementary Materials figure).

4.7. Collection of Pleural Fluid and Flow Cytometry Staining

Following euthanization, the thoracic cavity of each mice was opened and washed with 1 mL of cold sterile PBS supplemented with 3% FBS. The pleural fluid (PF) was aspirated and placed on ice before centrifugation at $300 \times g$ for 5 min. The supernatant was removed and stored at -80 °C. The cell pellet was washed with 10 mL of PBS/3% FBS, resuspended in 2 mL PBS/3% FBS, and carefully layered upon 2 mL of Ficoll-Paque Plus separation medium (GE Healthcare, Munich, Germany). The Ficoll gradient was centrifuged for 20 min at $400 \times g$ without brake. The mononuclear cells were collected, washed in PBS/3% FBS, and resuspended in 400 μL of PBS-3% FBS-1mM EDTA (FACS buffer). The cells were incubated with Fc-blocking reagent (TrueStain, Biolegend, San Diego, CA, USA) for 5 min and subsequently stained for 30 min at 4 °C with the relevant antibody. The lymphocytes were characterized with APC anti-CD19 (6D5) and FITC anti-CD45R/B220 (RA3-6B2). The antibodies used to analyze monocytes, macrophages, and MDSC were PE anti-CD11b (M1/70), PE/Cy7 anti-F4-80 (BM8), APC anti-CD206 (C068C2), and FITC anti-Gr1 (RB6-8C5). NK cells were characterized with PE anti-CD49b (DX5) (BioLegend, San Diego, CA, USA). The controls received equivalent concentrations of isotype-matched IgG. All samples were acquired with a BD Accuri C6 flow cytometer and analyzed with BD Accuri C6 and FlowJo software (FlowJo V10-CL, Tree Star Inc., Ashland, OR, USA) Monocytes/macrophages, lymphocytes, and granulocytes were first gated according to a SSC-A vs. FSC-A scatter plot, and doublets were excluded using a pulse geometry gate FSC-H × FSC-A plot.

4.8. Statistics

The results are presented as means ± SEM or SD as indicated. Kruskal–Wallis test was used to examine statistical differences among three or more groups. Differences between pairs of groups were examined for statistical significance using the unpaired Mann–Whitney U test. A p value < 0.05 was considered statistically significant.

Supplementary Materials: Supplementary materials can be found at http://www.mdpi.com/1422-0067/19/7/1881/s1.

Author Contributions: D.J.C. and V.S.-B. conceived and designed the experiments; D.J.C., D.C.-D., A.F., S.G. and V.S.-B. performed the experiments; D.J.C., D.C.-D., A.F. and V.S.-B. analyzed the data; V.S.-B. wrote the paper. F.T. provided financial support for technician, research equipment, and maintenance costs and performed a critical review of the manuscript.

Funding: This research was funded by Ligue Genevoise contre le Cancer (to V.S.-B.) and by the Centre d'Imagerie BioMédicale (CIBM) of the Universities and Hospitals of Geneva and Lausanne (to D.J.C. and S.G.).

Acknowledgments: We thank Mélanie Matthey-Doret and Romain Baechler for their excellent technical assistance. We thank S. Sadowski for her careful and critical reading of the manuscript.

Int. J. Mol. Sci. **2018**, *19*, 1881

Conflicts of Interest: The authors declare no conflict of interest. The founding sponsors had no role in the design of the study; in the collection, analyses, or interpretation of data; in the writing of the manuscript, and in the decision to publish the results.

Abbreviations

MPM	Malignant pleural mesothelioma
MIF	Macrophage migration inhibitory factor
MTT	3-(4,5-dimethylthiazol-2-yl)-2,5-diphenyltetrazolium bromide
FBS	Fetal bovine serum
FDG	Fluoro D glucose
PET/CT	Positron emission tomography/computed tomography
HE	Haematoxylin-eosin
NK	Natural killer
MDSC	Myeloid-derived-suppressor cell
FSC	Forward scatter
SC	Side scatter

References

1. Ogunseitan, O.A. The asbestos paradox: Global gaps in the translational science of disease prevention. *Bull. World Health Organ.* **2015**, *93*, 359–360. [CrossRef] [PubMed]
2. Opitz, I. Management of malignant pleural mesothelioma-The European experience. *J. Thorac. Dis.* **2014**, *6*, S238–S252. [PubMed]
3. D'Amato-Brito, C.; Cipriano, D.; Colin, D.J.; Germain, S.; Seimbille, Y.; Robert, J.H.; Triponez, F.; Serre-Beinier, V. Role of MIF/CD74 signaling pathway in the development of pleural mesothelioma. *Oncotarget* **2016**, *7*, 11512–11525. [CrossRef] [PubMed]
4. Borghese, F.; Clanchy, F.I. CD74: An emerging opportunity as a therapeutic target in cancer and autoimmune disease. *Expert Opin. Ther. Targets* **2011**, *15*, 237–251. [CrossRef] [PubMed]
5. Belizario, J.E. Immunodeficient mouse models: An overview. *Open Immunol. J.* **2009**, *2*, 79–85. [CrossRef]
6. Jackaman, C.; Yeoh, T.L.; Acuil, M.L.; Gardner, J.K.; Nelson, D.J. Murine mesothelioma induces locally-proliferating IL-10$^+$TNF-α^+CD206$^-$CX3CR1$^+$ M3 macrophages that can be selectively depleted by chemotherapy or immunotherapy. *Oncoimmunology* **2016**, *5*, e1173299. [CrossRef] [PubMed]
7. Marcucci, F.; Stassi, G.; De Maria, R. Epithelial-mesenchymal transition: A new target in anticancer drug discovery. *Nat. Rev. Drug Discov.* **2016**, *15*, 311–325. [CrossRef] [PubMed]
8. Pantelouris, E.M. Absence of thymus in a mouse mutant. *Nature* **1968**, *217*, 370–371. [CrossRef] [PubMed]
9. Gabrilovich, D.I.; Ostrand-Rosenberg, S.; Bronte, V. Coordinated regulation of myeloid cells by tumours. *Nat. Rev. Immunol.* **2012**, *12*, 253–268. [CrossRef] [PubMed]
10. Mantovani, A.; Sica, A. Macrophages, innate immunity and cancer: Balance, tolerance, and diversity. *Curr. Opin. Immunol.* **2010**, *22*, 231–237. [CrossRef] [PubMed]
11. Simpson, K.D.; Templeton, D.J.; Cross, J.V. Macrophage migration inhibitory factor promotes tumor growth and metastasis by inducing myeloid-derived suppressor cells in the tumor microenvironment. *J. Immunol.* **2012**, *189*, 5533–5540. [CrossRef] [PubMed]
12. Yaddanapudi, K.; Putty, K.; Rendon, B.E.; Lamont, G.J.; Faughn, J.D.; Satoskar, A.; Lasnik, A.; Eaton, J.W.; Mitchell, R.A. Control of tumor-associated macrophage alternative activation by macrophage migration inhibitory factor. *J. Immunol.* **2013**, *190*, 2984–2993. [CrossRef] [PubMed]
13. Philippeaux, M.M.; Pache, J.C.; Dahoun, S.; Barnet, M.; Robert, J.H.; Mauel, J.; Spiliopoulos, A. Establishment of permanent cell lines purified from human mesothelioma: Morphological aspects, new marker expression and karyotypic analysis. *Histochem. Cell Biol.* **2004**, *122*, 249–260. [CrossRef] [PubMed]

International Journal of
Molecular Sciences

MDPI

Communication

Non-Coding Transcript Heterogeneity in Mesothelioma: Insights from Asbestos-Exposed Mice

Emanuela Felley-Bosco [1],* and Hubert Rehrauer [2]

[1]   Laboratory of Molecular Oncology, Lungen- und Thoraxonkologie Zentrum, University Hospital Zurich, Sternwartstrasse 14, 8091 Zürich, Switzerland
[2]   Functional Genomics Center Zurich, ETH Zurich and University of Zurich, 8057 Zurich, Switzerland; hubert.rehrauer@fgcz.ethz.ch
*   Correspondence: emanuela.felley-bosco@usz.ch; Tel.: +41-44-255-2771

Received: 7 March 2018; Accepted: 10 April 2018; Published: 11 April 2018

Abstract: Mesothelioma is an aggressive, rapidly fatal cancer and a better understanding of its molecular heterogeneity may help with making more efficient therapeutic strategies. Non-coding RNAs represent a larger part of the transcriptome but their contribution to diseases is not fully understood yet. We used recently obtained RNA-seq data from asbestos-exposed mice and performed data mining of publicly available datasets in order to evaluate how non-coding RNA contribute to mesothelioma heterogeneity. Nine non-coding RNAs are specifically elevated in mesothelioma tumors and contribute to human mesothelioma heterogeneity. Because some of them have known oncogenic properties, this study supports the concept of non-coding RNAs as cancer progenitor genes.

Keywords: mesothelioma heterogeneity; non-coding RNA; long-non-coding RNA

1. Introduction

Protein coding genes make up only 2% of the human genome. In the remaining part of the genome, many transcriptionally active regions are found that give rise to non-coding RNA (ncRNA) [1]. Long non-coding RNAs (lncRNAs) are defined as longer than 200 nucleotides and represent the major class of ncRNAs since there are nearly three times as many lncRNA genes as protein-coding genes [2,3], and recently there has been a steep increase in research focusing on lncRNAs owing to their impact in several biological processes [4,5]. The class of non-coding RNAs that are smaller than 200 nucleotide includes the microRNA (miRNA, 19–25 nucleotides) that post-transcriptionally regulate gene expression via the suppression of specific target mRNAs [6].

LncRNA expression plays a crucial role in regulating the gene expression during differentiation and development [7,8]. For a few lncRNAs, functional characterization is available and indicates an association with transcriptional regulation and post-transcriptional processing of coding regions. Specifically, these lncRNAs affect miRNA expression, mRNA stability, and translation [9]. One of the first lncRNAs described to contribute to cancer was the HOX antisense intergenic RNA (*HOTAIR*)—this lncRNA interacts with chromatin and represses the transcription of human HOX genes, thus regulating development [10]. Several lncRNAs have been identified to be involved in the various hallmarks of cancer causing various tumor types including lung, liver, prostate, breast, and ovarian cancers [11–13].

Mesothelioma is a rare, aggressive cancer developing from the mesothelium and it is mostly associated with exposure to asbestos [14]. Recent molecular analyses have defined four different types of mesothelioma on the basis of gene expression [15], and two molecularly defined groups associated with different prognosis [16]. In this study, we explore the variation of non-coding RNA expression associated with this heterogeneity. In order to prioritize which ncRNA might be the most relevant in a given cancer type, it has been suggested that by using the The Cancer Genome Atlas (TCGA) ncRNAome information as a clinical filter, one would be able to generate a reduced and clinically relevant ncRNA list that could be used for a candidate-oriented functional screening. Here, we take the opportunity of our recent study in asbestos-exposed C57Bl/6J $Nf2^{+/-}$ mice [17], to identify lncRNAs and miRNAs associated with tumor development and scrutinize their expression and heterogeneity in human mesothelioma and human mesothelioma TCGA RNAome. $Nf2$ heterozygote background was chosen based on the fact that $NF2$ mutations are often observed in mesothelioma [18–21], and a previous study showing its contribution to tumor development [22].

2. Results

We analyzed the expression of non-protein-coding RNA in the RNA-seq data [17] obtained in tissue extracted from either C57Bl/6J $Nf2^{+/-}$ mice that were exposed eight times to crocidolite (blue asbestos) every three weeks, or sham-treated mice. Mice had been sacrificed 33 weeks after first crocidolite exposure in order to have the possibility of investigating pre-cancer and cancer stages. In order to identify gene expression changes during mesotheliomagenesis, we have analyzed three treatment groups by RNA-seq: sham, age-matched crocidolite-exposed, and age-matched crocidolite-exposed with observable tumors. We performed differential expression analysis between crocidolite-exposed and sham, and identified 108 non-protein-coding genes with more than 2-fold expression ($p < 0.01$, False Discovery Rate (FDR) < 0.017). Differential expression analysis between crocidolite-exposed with tumors and crocidolite-exposed, identified 366 non-protein-coding genes with more than 2-fold expression ($p < 0.01$, FDR < 0.024). 33 genes were found in both comparisons, as shown in the Venn diagram (Figure 1).

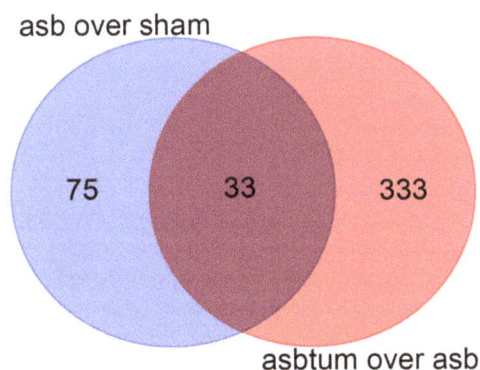

Figure 1. Overlap of the differentially expressed non-coding genes (more than 2-fold change, $p < 0.01$) in crocidolite-exposed vs. sham (asb over sham) and crocidolite-exposed with tumors vs. crocidolite-exposed (asbtum over asb) comparisons visualized as a Venn diagram.

We selected some of them based on (a) the significance of their differential expression in tumor vs. crocidolite-exposed inflamed mesothelium and (b) the availability of some functional knowledge about them (Table 1).

Table 1. Selected non-coding RNA more than 2-fold upregulated in murine mesothelioma compared to inflamed crocidolite-exposed mesothelium.

Gene Name	Type	p-Value	FDR	Chromosome Location (GRCm38.p5)	Human Ortholog	Upregulation in Crocidolite vs. Sham
Fendrr	Divergent lincRNA, nuclear	1.94×10^{-15}	1.4×10^{-14}	Chromosome 8: 121,054,882-121,083,110	yes	no
Gm26902	lincRNA	1.16×10^{-9}	4.91×10^{-9}	Chromosome 19: 34,474,808-34,481,546	no	no
Gm17501	lincRNA	3.84×10^{-5}	8.33×10^{-5}	Chromosome 3: 145,650,312-145,677,580	no	no
Meg3	lincRNA	7.97×10^{-5}	0.0001805	Chromosome 12: 109,541,001-109,571,726	yes	no
miR 17-92 cluster	lincRNA	7.02×10^{-10}	3.05×10^{-9}	Chromosome 14: 115,042,879-115,046,727	yes	no
Dio3os	antisense	0.003026	0.005339	Chromosome 12: 110,275,384-110,278,068	yes	yes
Dubr	linRNA, nuclear	9.36×10^{-7}	2.79×10^{-6}	Chromosome 16: 50,719,294-50,732,773	yes	yes
Malat1	antisense, nuclear	6.09×10^{-7}	1.86×10^{-6}	Chromosome 19: 5,795,690-5,802,672	yes	no
Dnm3os	antisense	2.26×10^{-16}	1.87×10^{-15}	Chromosome 1: 162,217,623-162,225,550	yes	no
Hoxaas2	antisense	5.73×10^{-7}	1.76×10^{-6}	Chromosome 6: 52,165,674-52,169,564	yes	no
Firre	lincRNA, nuclear	4.09×10^{-7}	1.28×10^{-6}	Chromosome X: 50,555,744-50,635,321	yes	no
Morrbid	nuclear	1.18×10^{-7}	3.92×10^{-7}	Chromosome 2: 128,178,319-128,502,765	yes	yes
miRlet7b	miRNA	0.000884	0.001697	Chromosome 15: 85,707,319-85,707,403	yes	no
Mir214	mirRNA	1.08×10^{-5}	2.78×10^{-5}	Chromosome 1: 162,223,368-162,223,477	yes	no

Then we compared the selected ncRNAs to differentially expressed genes with more than two-fold increased expression between inflamed tissue from crocidolite and sham (Table 1, last column). Of the 14 selected genes, three (*Dios3os*, *Dubr*, and *Morrbid*) were also overexpressed in inflamed crocidolite-exposed tissues compared to tissues from sham-treated mice.

The ncRNA gene with the highest upregulation in mesothelioma tumor in mice exposed to asbestos was *Fendrr* (*Fetal-lethal noncoding developmental regulatory RNA*) and we validated this finding by quantitative-PCR (Figure 2a). We then took the opportunity to investigate its expression in tumor tissue collected at different time (Figure 2b) during tumor progression in nine patients. We have recently deeply characterized genomic alterations in two out of these nine patients [23]. Interestingly, *FENDRR* expression was increased in the tissue of the patient, which had maintained epithelioid histology (P236A_tum and P236B_tum), compared to the patient that had initially been diagnosed as epitheloid mesothelioma (P95A_tum) but where we have observed epithelial to mesenchymal transition (EMT) during tumor progression. We could detect FENDRR expression in all first tumor samples from patients diagnosed with epithelioid mesothelioma but not in patient P399, who had been diagnosed with biphasic histology.

(a)

(b)

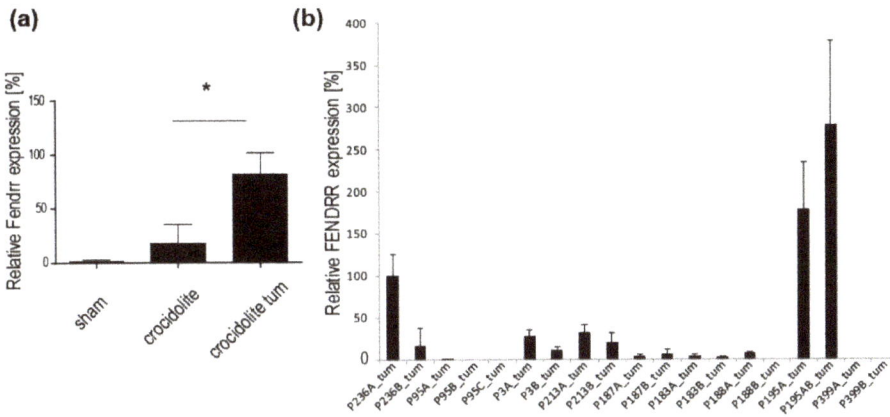

Figure 2. Fendrr is overexpressed in mice mesothelioma and associates with epithelial histotype commitment in human mesothelioma. (**a**) q-PCR of *Fendrr* expression was performed in sham, crocidolite-exposed mice without malignant tumors. Mean ± SE, N = 5–8 mice. * $p < 0.05$, Mann–Whitney test. (**b**) FENDRR gene expression analysis in tumor samples from nine patients for whom tissue is available at different time points during the progression of the disease. Mean ± SD, N = 3.

The existence of known orthologs in human for 13 of the selected ncRNAs allowed us to evaluate their contribution to tumor heterogeneity by interrogating publicly available TCGA data of 87 MPM samples (MESO) through the cBioPortal [24,25] together with five tumor suppressor genes frequently mutated in mesothelioma (Figure 3). For *HOXA-AS2* (*lncRNA–HOXA cluster antisense RNA 2*), *FIRRE* (*functional intergenic repeating RNA element*) and *MORBIDD* (*myeloid RNA regulator of Bim-induced death*), no differences were detected in TCGA data; therefore, they were not included in the figure. All other ncRNAs contribute to tumor heterogeneity.

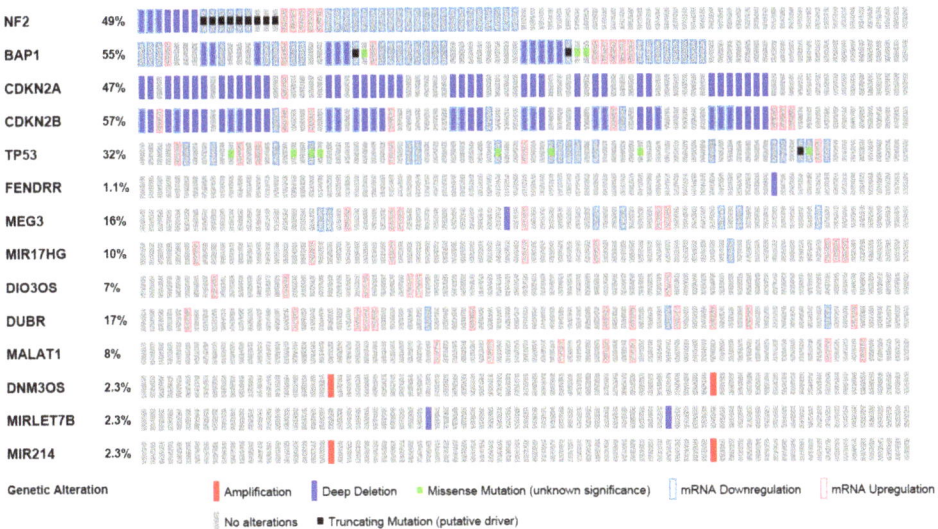

Figure 3. Non-coding RNAs contribute to mesothelioma heterogeneity. "Oncoprint" analysis performed using cBioportal of selected ncRNAs and five tumor suppressor genes frequently mutated in mesothelioma.

Interestingly *DNM3OS* (*dynamin 3 (Dnm3) gene antisense*) is amplified in two patients and consistent with *DNM3OS* being a precursor for *miR214*, the latter is amplified as well. Although DNM3OS overexpression is associated with enrichment is sarcomatoid histotype compared to epithelioid histotype in Bueno et al [15], in TCGA samples it is amplified in a patient bearing a biphasic and a patient bearing an epithelioid tumor.

In human mesothelioma miRlet7b was deleted in a patient bearing a biphasic and a patient bearing an epithelioid tumor, indicating that it possibly contributes to epithelial heterogeneity. *FENDRR* was deleted in a patient with biphasic histotype, which would fit with the observation that it is enriched in epithelioid mesothelioma, but this observation is based only on a single patient.

Interestingly there is a significant co-occurrence of alterations of *BAP1* and *DIO3OS* ($p = 0.024$), and of *DUBR* and *miRlet7b* ($p = 0.028$).

3. Discussion

In order to improve the treatment of mesothelioma, it is necessary to better understand how molecular heterogeneity contributes to tumor growth.

We report here the likely contribution of ncRNAs to the heterogeneity profile and suggest that oncogenic driver events in mesothelioma development are associated with lncRNA expression. This extends the current view that focuses on the loss of tumor suppressor functions as drivers.

Fendrr is transcribed divergently from the transcription factor-coding gene *Foxf1*. *Fendrr*-deficiency results in mice lethality due to lack of proper differentiation of mesenchymal derived tissue [26,27]. This lincRNA is predominantly nuclear and physically associates with the PRC2 Polycomb complex [28]). In humans the orthologous transcript is expressed from a syntenic region [29]. Silencing *FENDRR* increases FN1 expression in gastric cancer cells and increases their migration [30]. Interestingly, *FENDRR* is among the genes enriched in the epithelioid compared to sarcomatoid mesothelioma cluster based on gene-expression profile [15].

Not much is known about *gm26902*, except that its expression characterizes a subset of microglia CD11c+ population, which sustains brain development [31], while expression of *gm17501* has been associated with cardiac hypertrophy [32].

Meg3 (maternally expressed 3) binds to p53 and activates the transcription of a part of p53-regulated genes [33]. In gastric cancer, MEG3 increases Bcl-2 levels by sequestering miR-181-a [34]. In addition, MEG3 modulates the activity of TGF-β pathway genes by binding to distal regulatory elements, which have GA-rich sequences, allowing MEG3 specific binding to the chromatin through RNA–DNA triplex formation [35,36].

miR17-92 cluster (*miR-17-92a-1 cluster host gene*) binds HuR, a member of the ELAVL family, which has been reported to contribute to the stabilization of AU-rich elements (ARE)-containing mRNAs, possibly modulating HuR activity on target mRNA stability [37]. MiR 17-92 cluster is amplified in high-grade B-cell lymphoma with Burkitt lymphoma signature, resulting in higher expression of miR17-92 and lower expression of *BIM* and *PTEN* and increased BCR signaling [38]. It is noteworthy that miR17-92 expression is increased in mesothelioma [39].

Dio3os is transcribed in the antisense orientation to Dio3, which codes for the type 3 deiodinase, an enzyme-inactivating thyroid hormones that is highly expressed during pregnancy and development [40].

Dubr (also called Dum: developmental pluripotency-associated 2 (Dppa2) Upstream binding Muscle lncRNA) silences its neighboring gene, Dppa2, in *cis* through the recruitment of Dnmt1, Dnmt3a and Dnmt3b, thereby promoting myoblast differentiation and damage-induced muscle regeneration [41].

Malat1 (Metastasis-associated lung adenocarcinoma transcript 1) expression results in alternatively spliced transcripts [42]. It is for example necessary for correct splicing of B-Myb, a transcription factor involved in G2/M transition [43]. In patients with early-stage non-small cell lung cancer high levels of MALAT1 predict a high risk of metastatic progression [44]. *Malat1* loss of function in mouse revealed that it is a nonessential gene in development or for adult normal tissue homeostasis [45,46], but depletion of *MALAT1* in lung carcinoma cells impairs cellular motility in vitro and metastasis in mice [47]. Therefore, it has been suggested that *MALAT1* overexpression in cancer may drive gain-of-function phenotypes not observed during normal tissue development or homeostasis. Its action seems mediated not only by regulation of alternative splicing, as mentioned above, but also possibly through interaction with HuR [48] like for miR17-92 cluster.

Firre-encoded lncRNA serves as a platform for trans-chromosomal association by interacting with the nuclear matrix factor heterogeneous nuclear ribonucleoproteins U (hnRNPU) through a 156-bp repeating sequence and localizes across a ~5-Mb domain on the X chromosome [49]. It was suggested that it modulates nuclear architecture across chromosomes [49]. Transcription of *FIRRE* is regulated by NF-κB signaling in macrophages and intestinal epithelial cells [50]. Indeed, FIRRE positively regulates the expression of several inflammatory genes in macrophages or intestinal epithelial cells in response to lipopolysaccharide stimulation via posttranscriptional mechanisms including interaction with hnRNPU, which controls the stability of mRNAs of selected inflammatory genes through targeting the adenine-rich element of their mRNAs [50].

Dnm3os is essential for skeletal muscle formation and body growth during development and it serves as precursor of miR214 [51,52]. It is enriched in the sarcomatoid mesothelioma subtype cluster compared to epithelioid [15].

Hoxaas2 directly interacts with enhancer of zeste homolog 2 (EZH2) and lysine-specific demethylase 1 (LSD1), promoting pancreatic cell growth [53].

Morrbid is highly and distinctively expressed by mature eosinophils, neutrophils, and classical monocytes in both mice and humans [54]. Interestingly it could be a marker of exposure to carcinogenic fibers since it is overexpressed in tissues of mice exposed to long carcinogenic compared to short non-carcinogenic asbestos and also long compared to short nanotubes [55].

miRlet7 downregulates interferon β (IFNβ) and is upregulated in macrophages upon IFNβ treatment [56].

Although the method that we have used to extract RNA was not optimal for miRNA analysis we detected the overexpression of miR214, likely because of Dnm3os overexpression. MiR214 downregulates PTEN [57] and Sufu [58].

Although only Malat1, from the ncRNAs mentioned, is an lncRNA for which a clear genetic link with tumorigenesis has been established [59,60], it is likely that ncRNAs function in mesothelioma as "cancer progenitors genes" [61]. In addition to *MALAT1*, overexpression of *miR17-92 cluster* is likely oncogenic and of potential therapeutic interest because it activates druggable pathways. Similarly, overexpression of miR214 possibly indicates activation of Hedgehog and PI3K signaling.

Although for FENDRR the contribution to heterogeneity is based on the observation that it is enriched in epitheloid histotype and that one patient has a deletion in this gene, the fact that it is overexpressed in tumors and associates with epithelioid commitment in the patients analyzed indicate that further studies should explore the role of this lncRNA in mesothelioma.

Because MEG3 has been found to modulate TGF-β activity and it has an heterogeneous expression, it would be interesting to investigate whether its expression plays a role in the EMT signature that we observed in the mesothelioma tumors developing in asbestos-exposed mice [17] and also if it contributes to mesothelioma's so-called transitional state [62].

In an era where immunotherapy is also being intensively explored in mesothelioma treatment [63], it might be wise to consider the deletion of miRlet7b as a possible biomarker for response.

In summary, we were able to identify lncRNAs that are overexpressed in mesothelioma and we found that they contribute to human mesothelioma heterogeneity. We suggest that they may indicate pathways for precision medicine. One limitation of our approach might be the fact that in our experimental model we observed only spindeloid tumors, which is the opposite of what is observed in human mesothelioma, where epithelioid histotypes are the most frequent.

Appropriate functional experiments need to be carried out and it would make sense to establish consortia to validate our hypotheses. There is a plethora of ncRNA genes whose functions we need to

Int. J. Mol. Sci. **2018**, *19*, 1163

understand better. In addition, very instructive functional studies rely on animal models but modeling lncRNA function in mice might be difficult because lncRNAs are conserved at much lower rates compared to protein-coding genes, and therefore orthologs are more difficult to identify.

4. Materials and Methods

4.1. Analysis of RNA-Seq Data from Tissue Samples from Asbestos-Exposed Mice

RNA was extracted and analyzed as described in our previous paper, where we characterized the overall transcriptome profile of the same samples [17]. Assessment of miR expression was not optimal because the Qiagen RNeasy kit was used to extract RNA, which does not preserve very short RNAs.

4.2. Relative Gene Expression

Fendrr gene expression was conducted as previously described [17] [64–66] using the following primers (5′–3′): human: AGTGCACTGTGTGCTCTTAG and GAGGATCTGTGGTTGGGTATTT mouse GAAACCAGAGAGCTCCGAATAG and CTTCTGGTGGAGTCAGATCAAA. As in previous studies, histone 3 and β-actin were used as normalizer genes for human and murine gene expression, respectively. RNA was extracted from human mesothelioma tumors and cDNA was prepared as we recently described [23].

4.3. Analysis of Publicly Available Datasets

To analyze the expression and genetic alterations of selected non-coding RNA together with five tumor suppressor genes frequently mutated in mesothelioma, we obtained the data from TCGA, using www.cbioportal.org. For mRNA differential expression we used a z score of 1.2, where the z-score is the standard deviation of static levels of transcript expression in a given case compared to the mean transcript expression in diploid tumors.

Acknowledgments: We thank Manuel Ronner for performing some q-PCR analysis. Emanuela Felley-Bosco's research was supported by the Stiftung für Angewandte Krebsforschung, the Krebsliga Zürich, and the Swiss National Science Foundation (CRSII3_147697).

Author Contributions: All authors contributed to this manuscript and approved the final version.

Conflicts of Interest: The authors declare no conflict of interest.

References

1. Djebali, S.; Davis, C.A.; Merkel, A.; Dobin, A.; Lassmann, T.; Mortazavi, A.; Tanzer, A.; Lagarde, J.; Lin, W.; Schlesinger, F.; et al. Landscape of transcription in human cells. *Nature* **2012**, *489*, 101–108. [CrossRef] [PubMed]
2. Carninci, P.; Hayashizaki, Y. Noncoding RNA transcription beyond annotated genes. *Curr. Opin. Genet. Dev.* **2007**, *17*, 139–144. [CrossRef] [PubMed]
3. Iyer, M.K.; Niknafs, Y.S.; Malik, R.; Singhal, U.; Sahu, A.; Hosono, Y.; Barrette, T.R.; Prensner, J.R.; Evans, J.R.; Zhao, S.; et al. The landscape of long noncoding RNAs in the human transcriptome. *Nat. Genet.* **2015**, *47*, 199–208. [CrossRef] [PubMed]
4. Ulitsky, I.; Bartel, D.P. LincRNAs: Genomics, evolution, and mechanisms. *Cell* **2013**, *154*, 26–46. [CrossRef] [PubMed]
5. Kapranov, P.; Cheng, J.; Dike, S.; Nix, D.A.; Duttagupta, R.; Willingham, A.T.; Stadler, P.F.; Hertel, J.; Hackermuller, J.; Hofacker, I.L.; et al. RNA maps reveal new RNA classes and a possible function for pervasive transcription. *Science* **2007**, *316*, 1484–1488. [CrossRef] [PubMed]
6. Schwarzenbach, H.; Nishida, N.; Calin, G.A.; Pantel, K. Clinical relevance of circulating cell-free microRNAs in cancer. *Nat. Rev. Clin. Oncol.* **2014**, *11*, 145–156. [CrossRef] [PubMed]
7. Taft, R.J.; Pang, K.C.; Mercer, T.R.; Dinger, M.; Mattick, J.S. Non-coding RNAs: Regulators of disease. *J. Pathol.* **2010**, *220*, 126–139. [CrossRef] [PubMed]

8. Huarte, M.; Rinn, J.L. Large non-coding RNAs: Missing links in cancer? *Hum. Mol. Genet.* **2010**, *19*, R152–R161. [CrossRef] [PubMed]

9. Kornienko, A.E.; Guenzl, P.M.; Barlow, D.P.; Pauler, F.M. Gene regulation by the act of long non-coding RNA transcription. *BMC Biol.* **2013**, *11*, 1–14. [CrossRef] [PubMed]

10. Rinn, J.L.; Kertesz, M.; Wang, J.K.; Squazzo, S.L.; Xu, X.; Brugmann, S.A.; Goodnough, L.H.; Helms, J.A.; Farnham, P.J.; Segal, E.; et al. Functional demarcation of active and silent chromatin domains in human HOX loci by noncoding RNAs. *Cell* **2007**, *129*, 1311–1323. [CrossRef] [PubMed]

11. Lin, C.; Yang, L. Long noncoding RNA in cancer: Wiring signaling circuitry. *Trends Cell Boil.* **2017**, *28*, 287–301. [CrossRef] [PubMed]

12. Bhan, A.; Soleimani, M.; Mandal, S.S. Long noncoding RNA and cancer: A new paradigm. *Cancer Res.* **2017**, *77*, 3965–3981. [CrossRef] [PubMed]

13. Renganathan, A.; Felley-Bosco, E. Long noncoding RNAs in cancer and therapeutic potential. *Adv. Exp. Med. Biol.* **2017**, *1008*, 199–222. [PubMed]

14. Delgermaa, V.; Takahashi, K.; Park, E.K.; Le, G.V.; Hara, T.; Sorahan, T. Global mesothelioma deaths reported to the world health organization between 1994 and 2008. *Bull. World Health Organ.* **2011**, *89*, 716–724, 724A–724C. [CrossRef] [PubMed]

15. Bueno, R.; Stawiski, E.W.; Goldstein, L.D.; Durinck, S.; De Rienzo, A.; Modrusan, Z.; Gnad, F.; Nguyen, T.T.; Jaiswal, B.S.; Chirieac, L.R.; et al. Comprehensive genomic analysis of malignant pleural mesothelioma identifies recurrent mutations, gene fusions and splicing alterations. *Nat. Genet.* **2016**, *48*, 407–416. [CrossRef] [PubMed]

16. de Reynies, A.; Jaurand, M.C.; Renier, A.; Couchy, G.; Hysi, I.; Elarouci, N.; Galateau-Salle, F.; Copin, M.C.; Hofman, P.; Cazes, A.; et al. Molecular classification of malignant pleural mesothelioma: Identification of a poor prognosis subgroup linked to the epithelial-to-mesenchymal transition. *Clin. Cancer Res.* **2014**, *20*, 1323–1334. [CrossRef] [PubMed]

17. Rehrauer, H.; Wu, L.; Blum, W.; Pecze, L.; Henzi, T.; Serre-Beinier, V.; Aquino, C.; Vrugt, B.; de Perrot, M.; Schwaller, B.; et al. How asbestos drives the tissue towards tumors: Yap activation, macrophage and mesothelial precursor recruitment, RNA editing, and somatic mutations. *Oncogene* 2018. [CrossRef] [PubMed]

18. Bianchi, A.B.; Mitsunaga, S.I.; Cheng, J.Q.; Klein, W.M.; Jhanwar, S.C.; Seizinger, B.; Kley, N.; Klein-Szanto, A.J.; Testa, J.R. High frequency of inactivating mutations in the neurofibromatosis type 2 gene (NF2) in primary malignant mesotheliomas. *Proc. Natl. Acad. Sci. USA* **1995**, *92*, 10854–10858. [CrossRef] [PubMed]

19. Sekido, Y.; Pass, H.I.; Bader, S.; Mew, D.J.; Christman, M.F.; Gazdar, A.F.; Minna, J.D. Neurofibromatosis type 2 (NF2) gene is somatically mutated in mesothelioma but not in lung cancer. *Cancer Res.* **1995**, *55*, 1227–1231. [PubMed]

20. Deguen, B.; Goutebroze, L.; Giovannini, M.; Boisson, C.; van der Neut, R.; Jaurand, M.C.; Thomas, G. Heterogeneity of mesothelioma cell lines as defined by altered genomic structure and expression of the nf2 gene. *Int. J. Cancer* **1998**, *77*, 554–560. [CrossRef]

21. Thurneysen, C.; Opitz, I.; Kurtz, S.; Weder, W.; Stahel, R.A.; Felley-Bosco, E. Functional inactivation of nf2/merlin in human mesothelioma. *Lung Cancer* **2009**, *64*, 140–147. [CrossRef] [PubMed]

22. Jongsma, J.; van Montfort, E.; Vooijs, M.; Zevenhoven, J.; Krimpenfort, P.; van der Valk, M.; van de Vijver, M.; Berns, A. A conditional mouse model for malignant mesothelioma. *Cancer Cell* **2008**, *13*, 261–271. [CrossRef] [PubMed]

23. Oehl, K.; Kresoja-Rakic, J.; Opitz, I.; Vrugt, B.; Weder, W.; Stahel, R.; Wild, P.; Felley-Bosco, E. Live-cell mesothelioma biobank to explore mechanisms of tumor progression. *Front. Oncol.* **2018**, *8*, 40. [CrossRef] [PubMed]

24. Gao, J.; Aksoy, B.A.; Dogrusoz, U.; Dresdner, G.; Gross, B.; Sumer, S.O.; Sun, Y.; Jacobsen, A.; Sinha, R.; Larsson, E.; et al. Integrative analysis of complex cancer genomics and clinical profiles using the cBioPortal. *Sci. Signal* **2013**, *6*, pl1. [CrossRef] [PubMed]

25. Cerami, E.; Gao, J.; Dogrusoz, U.; Gross, B.E.; Sumer, S.O.; Aksoy, B.A.; Jacobsen, A.; Byrne, C.J.; Heuer, M.L.; Larsson, E.; et al. The cbio cancer genomics portal: An open platform for exploring multidimensional cancer genomics data. *Cancer Discov.* **2012**, *2*, 401–404. [CrossRef] [PubMed]

26. Grote, P.; Wittler, L.; Hendrix, D.; Koch, F.; Wahrisch, S.; Beisaw, A.; Macura, K.; Blass, G.; Kellis, M.; Werber, M.; et al. The tissue-specific lncRNA fendrr is an essential regulator of heart and body wall development in the mouse. *Dev. Cell* **2013**, *24*, 206–214. [CrossRef] [PubMed]

27. Sauvageau, M.; Goff, L.A.; Lodato, S.; Bonev, B.; Groff, A.F.; Gerhardinger, C.; Sanchez-Gomez, D.B.; Hacisuleyman, E.; Li, E.; Spence, M.; et al. Multiple knockout mouse models reveal lincRNAs are required for life and brain development. *Elife* **2013**, *2*, e01749. [CrossRef] [PubMed]

28. Khalil, A.M.; Guttman, M.; Huarte, M.; Garber, M.; Raj, A.; Rivea Morales, D.; Thomas, K.; Presser, A.; Bernstein, B.E.; van Oudenaarden, A.; et al. Many human large intergenic noncoding RNAs associate with chromatin-modifying complexes and affect gene expression. *Proc. Natl. Acad. Sci. USA* **2009**, *106*, 11667–11672. [CrossRef] [PubMed]

29. Cabili, M.N.; Trapnell, C.; Goff, L.; Koziol, M.; Tazon-Vega, B.; Regev, A.; Rinn, J.L. Integrative annotation of human large intergenic noncoding RNAs reveals global properties and specific subclasses. *Genes Dev.* **2011**, *25*, 1915–1927. [CrossRef] [PubMed]

30. Xu, T.P.; Huang, M.D.; Xia, R.; Liu, X.X.; Sun, M.; Yin, L.; Chen, W.M.; Han, L.; Zhang, E.B.; Kong, R.; et al. Decreased expression of the long non-coding RNA fendrr is associated with poor prognosis in gastric cancer and fendrr regulates gastric cancer cell metastasis by affecting fibronectin1 expression. *J. Hematol. Oncol.* **2014**, *7*, 63. [CrossRef] [PubMed]

31. Wlodarczyk, A.; Holtman, I.R.; Krueger, M.; Yogev, N.; Bruttger, J.; Khorooshi, R.; Benmamar-Badel, A.; de Boer-Bergsma, J.J.; Martin, N.A.; Karram, K.; et al. A novel microglial subset plays a key role in myelinogenesis in developing brain. *EMBO J.* **2017**, *36*, 3292–3308. [CrossRef] [PubMed]

32. Zhang, J.; Feng, C.; Song, C.; Ai, B.; Bai, X.; Liu, Y.; Li, X.; Zhao, J.; Shi, S.; Chen, X.; et al. Identification and analysis of a key long non-coding RNAs (lncRNAs)-associated module reveal functional lncRNAs in cardiac hypertrophy. *J. Cell Mol. Med.* **2018**, *22*, 892–903. [CrossRef] [PubMed]

33. Zhou, Y.; Zhong, Y.; Wang, Y.; Zhang, X.; Batista, D.L.; Gejman, R.; Ansell, P.J.; Zhao, J.; Weng, C.; Klibanski, A. Activation of p53 by meg3 non-coding RNA. *J. Biol. Chem.* **2007**, *282*, 24731–24742. [CrossRef] [PubMed]

34. Peng, W.; Si, S.; Zhang, Q.; Li, C.; Zhao, F.; Wang, F.; Yu, J.; Ma, R. Long non-coding RNA meg3 functions as a competing endogenous RNA to regulate gastric cancer progression. *J. Exp. Clin. Cancer Res.* **2015**, *34*, 79. [CrossRef] [PubMed]

35. Mondal, T.; Subhash, S.; Vaid, R.; Enroth, S.; Uday, S.; Reinius, B.; Mitra, S.; Mohammed, A.; James, A.R.; Hoberg, E.; et al. Meg3 long noncoding RNA regulates the TGF-β pathway genes through formation of RNA-DNA triplex structures. *Nat. Commun.* **2015**, *6*, 7743. [CrossRef] [PubMed]

36. Iyer, S.; Modali, S.D.; Agarwal, S.K. Long noncoding RNA meg3 is an epigenetic determinant of oncogenic signaling in functional pancreatic neuroendocrine tumor cells. *Mol. Cell Biol.* **2017**, *37*. [CrossRef] [PubMed]

37. Kim, J.; Abdelmohsen, K.; Yang, X.; De, S.; Grammatikakis, I.; Noh, J.H.; Gorospe, M. LncRNA oip5-as1/cyrano sponges RNA-binding protein HuR. *Nucleic Acids Res.* **2016**, *44*, 2378–2392. [CrossRef] [PubMed]

38. Bouska, A.; Bi, C.; Lone, W.; Zhang, W.; Kedwaii, A.; Heavican, T.; Lachel, C.M.; Yu, J.; Ferro, R.; Eldorghamy, N.; et al. Adult high-grade b-cell lymphoma with burkitt lymphoma signature: Genomic features and potential therapeutic targets. *Blood* **2017**, *130*, 1819–1831. [CrossRef] [PubMed]

39. Balatti, V.; Maniero, S.; Ferracin, M.; Veronese, A.; Negrini, M.; Ferrocci, G.; Martini, F.; Tognon, M.G. MicroRNAs dysregulation in human malignant pleural mesothelioma. *J. Thorac. Oncol.* **2011**, *6*, 844–851. [CrossRef] [PubMed]

40. Hernandez, A.; Martinez, M.E.; Croteau, W.; St. Germain, D.L. Complex organization and structure of sense and antisense transcripts expressed from the dio3 gene imprinted locus. *Genomics* **2004**, *83*, 413–424. [CrossRef] [PubMed]

41. Wang, L.; Zhao, Y.; Bao, X.; Zhu, X.; Kwok, Y.K.; Sun, K.; Chen, X.; Huang, Y.; Jauch, R.; Esteban, M.A.; et al. LncRNA dum interacts with dnmts to regulate dppa2 expression during myogenic differentiation and muscle regeneration. *Cell Res.* **2015**, *25*, 335–350. [CrossRef] [PubMed]

42. Tripathi, V.; Ellis, J.D.; Shen, Z.; Song, D.Y.; Pan, Q.; Watt, A.T.; Freier, S.M.; Bennett, C.F.; Sharma, A.; Bubulya, P.A.; et al. The nuclear-retained noncoding RNA malat1 regulates alternative splicing by modulating sr splicing factor phosphorylation. *Mol. Cell* **2010**, *39*, 925–938. [CrossRef] [PubMed]

43. Tripathi, V.; Shen, Z.; Chakraborty, A.; Giri, S.; Freier, S.M.; Wu, X.; Zhang, Y.; Gorospe, M.; Prasanth, S.G.; Lal, A.; et al. Long noncoding RNA malat1 controls cell cycle progression by regulating the expression of oncogenic transcription factor b-myb. *PLoS Genet* **2013**, *9*, e1003368. [CrossRef] [PubMed]

44. Ji, P.; Diederichs, S.; Wang, W.; Boing, S.; Metzger, R.; Schneider, P.M.; Tidow, N.; Brandt, B.; Buerger, H.; Bulk, E.; et al. Malat-1, a novel noncoding RNA, and thymosin β4 predict metastasis and survival in early-stage non-small cell lung cancer. *Oncogene* **2003**, *22*, 8031–8041. [CrossRef] [PubMed]

45. Zhang, B.; Arun, G.; Mao, Y.S.; Lazar, Z.; Hung, G.; Bhattacharjee, G.; Xiao, X.; Booth, C.J.; Wu, J.; Zhang, C.; et al. The lncRNA malat1 is dispensable for mouse development but its transcription plays a cis-regulatory role in the adult. *Cell Rep.* **2012**, *2*, 111–123. [CrossRef] [PubMed]

46. Nakagawa, S.; Ip, J.Y.; Shioi, G.; Tripathi, V.; Zong, X.; Hirose, T.; Prasanth, K.V. Malat1 is not an essential component of nuclear speckles in mice. *RNA* **2012**, *18*, 1487–1499. [CrossRef] [PubMed]

47. Gutschner, T.; Hammerle, M.; Eissmann, M.; Hsu, J.; Kim, Y.; Hung, G.; Revenko, A.; Arun, G.; Stentrup, M.; Gross, M.; et al. The noncoding RNA malat1 is a critical regulator of the metastasis phenotype of lung cancer cells. *Cancer Res.* **2013**, *73*, 1180–1189. [CrossRef] [PubMed]

48. Li, L.; Chen, H.; Gao, Y.; Wang, Y.W.; Zhang, G.Q.; Pan, S.H.; Ji, L.; Kong, R.; Wang, G.; Jia, Y.H.; et al. Long noncoding RNA malat1 promotes aggressive pancreatic cancer proliferation and metastasis via the stimulation of autophagy. *Mol. Cancer Ther.* **2016**, *15*, 2232–2243. [CrossRef] [PubMed]

49. Hacisuleyman, E.; Goff, L.A.; Trapnell, C.; Williams, A.; Henao-Mejia, J.; Sun, L.; McClanahan, P.; Hendrickson, D.G.; Sauvageau, M.; Kelley, D.R.; et al. Topological organization of multichromosomal regions by the long intergenic noncoding RNA firre. *Nat. Struct. Mol. Biol.* **2014**, *21*, 198–206. [CrossRef] [PubMed]

50. Lu, Y.; Liu, X.; Xie, M.; Liu, M.; Ye, M.; Li, M.; Chen, X.M.; Li, X.; Zhou, R. The nf-kappab-responsive long noncoding RNA firre regulates posttranscriptional regulation of inflammatory gene expression through interacting with hnrnpu. *J. Immunol.* **2017**, *199*, 3571–3582. [CrossRef] [PubMed]

51. Watanabe, T.; Sato, T.; Amano, T.; Kawamura, Y.; Kawamura, N.; Kawaguchi, H.; Yamashita, N.; Kurihara, H.; Nakaoka, T. Dnm3os, a non-coding RNA, is required for normal growth and skeletal development in mice. *Dev. Dyn.* **2008**, *237*, 3738–3748. [CrossRef] [PubMed]

52. Juan, A.H.; Kumar, R.M.; Marx, J.G.; Young, R.A.; Sartorelli, V. Mir-214-dependent regulation of the polycomb protein ezh2 in skeletal muscle and embryonic stem cells. *Mol. Cell* **2009**, *36*, 61–74. [CrossRef] [PubMed]

53. Lian, Y.; Li, Z.; Fan, Y.; Huang, Q.; Chen, J.; Liu, W.; Xiao, C.; Xu, H. The lncRNA-hoxa-as2/ezh2/lsd1 oncogene complex promotes cell proliferation in pancreatic cancer. *Am. J. Transl. Res.* **2017**, *9*, 5496–5506. [PubMed]

54. Kotzin, J.J.; Spencer, S.P.; McCright, S.J.; Kumar, D.B.U.; Collet, M.A.; Mowel, W.K.; Elliott, E.N.; Uyar, A.; Makiya, M.A.; Dunagin, M.C.; et al. The long non-coding RNA morrbid regulates bim and short-lived myeloid cell lifespan. *Nature* **2016**, *537*, 239–243. [CrossRef] [PubMed]

55. Chernova, T.; Murphy, F.A.; Galavotti, S.; Sun, X.M.; Powley, I.R.; Grosso, S.; Schinwald, A.; Zacarias-Cabeza, J.; Dudek, K.M.; Dinsdale, D.; et al. Long-fiber carbon nanotubes replicate asbestos-induced mesothelioma with disruption of the tumor suppressor gene *cdkn2a* (*ink4a/arf*). *Curr. Biol.* **2017**, *27*, 3302–3314e6. [CrossRef] [PubMed]

56. Witwer, K.W.; Sisk, J.M.; Gama, L.; Clements, J.E. MicroRNA regulation of ifn-β protein expression: Rapid and sensitive modulation of the innate immune response. *J. Immunol.* **2010**, *184*, 2369–2376. [CrossRef] [PubMed]

57. Yang, H.; Kong, W.; He, L.; Zhao, J.J.; O'Donnell, J.D.; Wang, J.; Wenham, R.M.; Coppola, D.; Kruk, P.A.; Nicosia, S.V.; et al. MicroRNA expression profiling in human ovarian cancer: Mir-214 induces cell survival and cisplatin resistance by targeting pten. *Cancer Res.* **2008**, *68*, 425–433. [CrossRef] [PubMed]

58. Alimirah, F.; Peng, X.; Gupta, A.; Yuan, L.; Welsh, J.; Cleary, M.; Mehta, R.G. Crosstalk between the vitamin d receptor (vdr) and mir-214 in regulating sufu, a hedgehog pathway inhibitor in breast cancer cells. *Exp. Cell Res.* **2016**, *349*, 15–22. [CrossRef] [PubMed]

59. Yildirim, E.; Kirby, J.E.; Brown, D.E.; Mercier, F.E.; Sadreyev, R.I.; Scadden, D.T.; Lee, J.T. Xist RNA is a potent suppressor of hematologic cancer in mice. *Cell* **2013**, *152*, 727–742. [CrossRef] [PubMed]

60. Arun, G.; Diermeier, S.; Akerman, M.; Chang, K.C.; Wilkinson, J.E.; Hearn, S.; Kim, Y.; MacLeod, A.R.; Krainer, A.R.; Norton, L.; et al. Differentiation of mammary tumors and reduction in metastasis upon malat1 lncRNA loss. *Genes Dev.* **2016**, *30*, 34–51. [CrossRef] [PubMed]

61. Feinberg, A.P.; Koldobskiy, M.A.; Gondor, A. Epigenetic modulators, modifiers and mediators in cancer aetiology and progression. *Nat. Rev. Genet.* **2016**, *17*, 284–299. [CrossRef] [PubMed]

62. Husain, A.N.; Krausz, T. Morphological alterations of serous membranes of the mediastinum in reactive and neoplastic settings. In *Pathology of the Mediastinum*; Marchevsky, A.M., Wick, M., Eds.; Cambridge University Press: Cambridge, UK, 2014.

63. Alley, E.W.; Lopez, J.; Santoro, A.; Morosky, A.; Saraf, S.; Piperdi, B.; van Brummelen, E. Clinical safety and activity of pembrolizumab in patients with malignant pleural mesothelioma (keynote-028): Preliminary results from a non-randomised, open-label, phase 1b trial. *Lancet Oncol.* **2017**, *18*, 623–630. [CrossRef]

64. Andre, M.; Felley-Bosco, E. Heme oxygenase-1 induction by endogenous nitric oxide: Influence of intracellular glutathione. *FEBS Lett.* **2003**, *546*, 223–227. [CrossRef]

65. Sidi, R.; Pasello, G.; Opitz, I.; Soltermann, A.; Tutic, M.; Rehrauer, H.; Weder, W.; Stahel, R.A.; Felley-Bosco, E. Induction of senescence markers after neo-adjuvant chemotherapy of malignant pleural mesothelioma and association with clinical outcome: An exploratory analysis. *Eur. J. Cancer* **2011**, *47*, 326–332. [CrossRef] [PubMed]

66. Shi, Y.; Moura, U.; Opitz, I.; Soltermann, A.; Rehrauer, H.; Thies, S.; Weder, W.; Stahel, R.A.; Felley-Bosco, E. Role of hedgehog signaling in malignant pleural mesothelioma. *Clin. Cancer Res.* **2012**, *18*, 4646–4656. [CrossRef] [PubMed]

MDPI

St. Alban-Anlage 66

4052 Basel

Switzerland

Tel. +41 61 683 77 34

Fax +41 61 302 89 18

www.mdpi.com

International Journal of Molecular Sciences Editorial Office

E-mail: ijms@mdpi.com

www.mdpi.com/journal/ijms

www.ingramcontent.com/pod-product-compliance
Lightning Source LLC
Chambersburg PA
CBHW051850210326
41597CB00033B/5840